APPLIED NAVAL ARCHITECTURE

APPLIED NAVAL ARCHITECTURE

Robert B. Zubaly

CORNELL MARITIME PRESS
CENTREVILLE, MARYLAND

Copyright © 1996 by Cornell Maritime Press, Inc.

All rights reserved. No part of this book may be used or reproduced in any manner whatsoever without written permission except in the case of brief quotations embodied in critical articles or reviews. For information, address Cornell Maritime Press, Inc., Centreville, Maryland 21617 or go to our website at www.cmptp.com.

Library of Congress Cataloging-in-Publication Data

Zubaly, R. B.
    Applied naval architecture / Robert B. Zubaly.
      p. cm.
    Includes bibliographical references and index.
    ISBN 978-0-87033-475-7 (hardcover)
    1. Naval architecture.   I. Title.
VM156.Z83 1996
623.8¢1—dc20   95-48004
                                                                             CIP

Manufactured in the United States of America
First edition, 1996; fourth printing, 2009

To Marylee

CONTENTS

Preface, ix

Chapter 1. Cargo Ships, 3

Chapter 2. Hull Form, 22

Chapter 3. Static Equilibrium and Stability, 48

Chapter 4. Stability at Large Angles, 100

Chapter 5. Trim and Longitudinal Stability, 133

Chapter 6. Flooding and Subdivision, 169

Chapter 7. Ship Strength, 195

Chapter 8. Ship Resistance, 238

Chapter 9. Ship Propulsion, 271

Chapter 10. Ship Dynamics, 299

Appendix A. Excerpts from Trim and Stability Booklet for Single Screw Cargo Vessel, Mariner Class C4–S–1a, 330

Appendix B. Properties of Salt Water and Fresh Water, and Friction Formulations, 335

Answers to Selected Problems, 340

Bibliography, 341

Index, 343

About the Author, 350

PREFACE

This book is an introduction to those practical elements of the theories of naval architecture that are most helpful and essential to ship's officers (deck and engine), marine engineers, shipyard engineers and technicians, fishermen, and others who work on seagoing vessels or in the professions that service them. The inspiration for this work came from untold numbers of students whose inquisitive faces in countless classrooms have challenged me to explain these concepts in simple language that they could understand.

Textbooks that are intended to cover naval architecture subjects strictly for ship deck officers are often focused so specifically on the duties of the mate who is in charge of assessing the ship's stability and on the various calculation aids prepared especially for that mate that they neglect to develop the underlying physical principles to the extent that the student could do the calculations without the use of the prepared aids. On the other extreme are the textbooks written with the professional naval architect in mind, which contain so much detailed material needed for ship design or for research and development that they are not suitable for guidance in ship operation.

In this book I have tried to steer the middle course between the two extremes. Derivations of the equations that express the principles of flotation, equilibrium, stability, ship strength, ship resistance and propulsion, and ship motions at sea are emphasized to a greater extent than is typical of the books written for mates, and problems are solved by referring to the basic principles rather than by using calculation aids, standard forms, and shipboard computers. An understanding of the technical basis of ship-loading calculations will give the student the confidence to make rational decisions that will improve the ship's stability, as opposed to using trial-and-error procedures at a computer keyboard. Ideally, the approach used in this book should be mastered first, to be followed by instruction in the use of the standard forms and shipboard computers. The student will then know why the computations are done and what their limitations are, as well as how to do them in a rote manner.

The topics of ship resistance and propulsion and ship motions at sea, which are often omitted from books intended solely for mates, are included to emphasize their importance to the safe and economical operation of modern cargo ships in these days of high-cost fuel and high-tech ship design.

To benefit from this book and to make practical use of the material in it, the reader need only have an interest in ships and their safe operation and have a general knowledge of physics and mathematics at the level of a high school graduate. Calculus at the level of a sophomore in a technical engineering program and engineering physics are helpful, but not essential. I have tried to describe the application of integration to ship form calculations so that it can be understood by students who have not had a formal course in integral calculus.

The problems have been selected to be representative of practical real-life problems. Answers to half of the problems are furnished, a real help to those who might use the book for a self-study program. Most of the problems have been put to the test (literally) of workability—they are drawn from the files of examination problems that I have been writing for the past forty years.

The photographs of cargo ship types in the first chapter are reproduced here with the kind permission of maritime photographer and lecturer Francis J. Duffy of Granard Associates.

APPLIED NAVAL ARCHITECTURE

CHAPTER ONE

CARGO SHIPS

Since before the beginning of recorded history, man has been inventing ways to travel by sea. And for at least as long a time, he has had the urge to engage in trading goods with his fellow men in remote places. Thus the activity of moving goods by sea for the purpose of commerce has a very long history. For millennia men have been building and improving on the design of ships of commerce. Modern merchant ships, with their high level of technological development and their remarkable diversity of function and type, are the result of that long evolutionary process. Naval architects and marine engineers continue to develop new ways to design ships and their engines to move cargo more efficiently while maintaining the high standards of safety of cargo and crew demanded by national and international regulatory bodies.

When mechanical propulsion replaced sails and steel replaced wood as a material for hull construction in the nineteenth century, merchant ships as we know them today had their modest beginnings. Those two technological developments made it possible to achieve vast increases in the size and speed of ships, and since the ships were no longer subject to the unpredictability of the winds for their propulsion, scheduled sailings and dependable arrivals were made possible. The establishment of regularly scheduled trade routes had a significant effect on ship design. A modern ship's cargo capacity, speed, power, and bunker capacity (space allocated to carrying fuel) can be chosen to suit the annual amount of cargo a shipowner wants each ship or a fleet of sister ships to carry on a particular trade route, with ships calling at each port on the route on a regular published schedule. Ships operating in this kind of service are known as cargo liners. Cargo ships that do not operate on regular schedules and designated trade routes, but that are prepared to call at any port as needed and to transport any type of cargo to any other port within their operating range, are called tramp ships. Tramp ships tend to be somewhat smaller and slower than cargo liners.

In the early days of self-propelled steel ships, virtually all cargo ships were general cargo ships or breakbulk ships, as they are also called. They were designed to carry any kind of dry cargo, lifting it aboard with their own cargo-handling gear. The modern trend in cargo ship design is toward specialization, either in regard to how the cargo is carried (in bulk, in containers, in barges, in vehicles) or to the type of cargo (oil, chemicals, ore, grain, lumber, liquefied gases). These special types have not supplanted the general cargo ship entirely, but for many trades they predominate because they are more efficient as part of an overall transportation system.

In the ship descriptions that follow, the features of each special type that distinguish it from other types are emphasized. Differences will be found mostly in the size and arrangement of the cargo spaces which make them suitable for special kinds of cargo, in the type of cargo-handling gear installed, and in the nature of the provisions made for access to the cargo spaces for efficient loading and discharging of the cargo.

THE GENERAL CARGO SHIP

General cargo ships transport all kinds of packaged goods. The cargo spaces, or holds, are separated by transverse bulkheads to limit flooding in case an accident should occur. A double bottom for protection against flooding caused by grounding also forms the main fuel oil tanks, called double bottom tanks. General cargo ships range from about 430 to 560 feet in length (130–170 meters), and their deadweight (total weight capacity of cargo, fuel, water, and stores) might be between 12,000 and 17,000 tons. Sea speeds of 14 to 25 knots are common, the lower speeds pertaining to tramps and the higher speeds to liner services. Commonly, the main spaces on the ship are arranged with from three to five cargo holds forward of the deckhouse and machinery spaces, and one or two holds aft. In some cases the machinery space is aft and all cargo spaces are forward of it.

Each main cargo hold space consists of a deep lower hold just above the double bottom space, plus one or two 'tween deck spaces above the hold, depending on the size of the ship and the number of decks. Access to the holds is through weathertight hatches in the decks. Hatches are made as large as possible without weakening the deck structure, so that the need for moving cargo longitudinally and laterally after it has been lowered through the hatch is minimized, thus making cargo handling as efficient as possible. The traditional arrangement of a single line of hatches, one centered on each hold, has been largely replaced by twin and three-across hatch arrangements to open more of the cargo space below to direct access by the cargo-handling gear. Even with the largest possible hatches, however, some of the cargo loaded into a general cargo ship has to be handled individually inside the hold space by manhandling or by forklift trucks, and all of it has to be secured to or wedged against the internal ship structure or to other

General cargo ship. Photography by Granard Associates, F.J. Duffy

cargo to prevent it from shifting and sustaining damage after the ship puts to sea. Cargo stowage of breakbulk cargo, which arrives in individual lots as opposed to being consolidated into larger units on pallets or in containers, is thus a process that requires considerable time, man-hours, and the skills of trained stevedores.

General cargo ships are outfitted with shipboard cargo-handling gear capable of lifting aboard and lowering into the hatches a great variety of dry cargoes. The traditional gear consists of cargo derricks mounted on deck between the hatches. The derricks have stationary vertical posts (called masts if they are single and on centerline, kingposts or samson posts if they are in pairs) and movable booms pivoted to the masts or kingposts near their bottoms. Cargo booms rigged with wire ropes and hooks are positioned over the pier and over the hatch during cargo loading so that the cargo can be lifted aboard. Numerous patented pivoting cranes and gantry cranes that run on tracks along the sides of the deck are also employed on modern cargo ships instead of derricks. Compared to derricks, cranes require minimal rigging, and many designs have been shown to be superior to the traditional gear in cargo-handling efficiency. Cargo-loading efficiency is also improved on some ships by having cargo side ports in addition to the hatches in the deck.

Flexibility and adaptability to all kinds of cargo are the hallmarks of the general cargo ship, so they usually have some provisions for cargoes other than packaged dry goods. For example, they are often equipped to carry liquid cargoes in cargo tanks, and they may have one or more insulated refrigerated holds or 'tween decks for foodstuffs requiring refrigeration.

6 APPLIED NAVAL ARCHITECTURE

Refrigerated cargo ship. Photography by Granard Associates, F.J. Duffy

UNITIZED CARGO SHIPS

If a group of small packages of general cargo is combined into a larger unit before being placed aboard ship, the cargo is said to be unitized. Various methods of unitizing cargo have been developed and special ships have evolved to take advantage of the increase in efficiency of cargo handling that unitization can achieve.

Containerships

The most ubiquitous of the unitized cargo ships are containerships, which are specially designed and outfitted to carry cargo that has been unitized by packing it into standard size containers. The advantages to be gained by containerization of cargo are numerous, for example:

- Cargo-handling time and the manpower required for cargo handling are reduced significantly, especially in the vertical cellular type of containership, in which containers never have to be shifted athwartships or fore and aft after they are lowered into the ship, and lashing, packing, and tying down of cargo inside the holds are eliminated.
- Containerized cargoes allow for the intermodal (road, rail, sea) transport of goods with minimum time spent in transferring cargo between modes.
- Containerized cargo, if it is properly stowed in containers at the point of origin, is less prone to damage of goods during transit than breakbulk cargo is.

CARGO SHIPS 7

Containership. Photography by Granard Associates, F.J. Duffy

- Since the containers can be locked and sealed at the point of origin and not opened except at the final destination point, containerized cargo is less subject to pilferage than breakbulk cargo is.
- Containers keep the cargo inside protected against the weather, so a ship's cargo capacity can be extended when carrying containers by stowing significant numbers of containers on the weather deck.

The concept of the containership arose in the 1950s out of the desire to reduce the time that a cargo ship spent in port, which was about 50 percent of its time for a typical breakbulk operation. Early containerships, which were converted from other types, soon gave way to the special designs of containerships that are so common in the major finished goods trade routes of the world today. In the pure containership, all cargo is unitized, and the containers are built to standard dimensions developed by international agreement so that they are compatible with the cell guides inside any containership and with the special fittings that are installed on the decks of the ships and on the flatbeds of trucks and railcars for transport over highways and railroads. The universally adopted container width is 8 feet (2.435 meters). Several modular lengths are provided for in the standards, the most common lengths in use being 20 feet (6.055 m) and 40 feet (12.190 m). Longer containers in the 45-foot range are coming into use in the 1990s. When container sizes were first standardized, a height of 8 feet (2.435 m) was adopted, but in fact, it is not necessary to standardize the height since it does not affect the design of any hardware or fittings, but only the height of a stack of containers, which may be variable. Dry cargo boxes of 8.5 feet height are common, and special-purpose boxes for liquid tanks, granular materials, refrigerated goods, and open-top boxes for odd-sized loads may have heights that vary considerably.

The typical large containership has most or all of the container holds forward of the deckhouse and engine spaces, so that the engine room, deckhouse, and navigating bridge are in either the aft or nearly aft position. There are exceptions: some of the larger ships have been provided with a small house and navigating bridge forward to improve the forward visibility, which becomes a problem when many tiers of containers are stacked on the deck forward of the wheelhouse. Most of the pure containerships have no shipboard container-handling gear, since they depend on the giant container-handling cranes located ashore at the container ports to which they trade. Many kinds of special vehicles and lifting devices have been developed to provide for efficient handling of containers within the terminal where the boxes are taken off trucks or railcars, stacked temporarily, and ultimately moved to and placed aboard the containership. The operation has become so efficient compared with the old breakbulk methods that containerships have supplanted almost all breakbulk cargo liners in the major trade routes for manufactured goods. Breakbulk ships remain essential, of course, to serve ports where container-handling equipment is not available, and to carry cargoes that are not suitable to be stowed in containers (steel rails, plates and shapes, and timber, for example).

The "cellular" type of containership, which is the most common of the pure containership types, is characterized by the fact, mentioned among the advantages above, that no containers have to be moved or handled after they are lowered into the hold of the ship. To accomplish this, the holds are fitted with container cell guides made of steel angle bars standing vertically and positioned so that a standard size container fits into a cell without jamming, but securely enough against shifting that it needs no provisions for being tied down or otherwise secured. Each container in a given cell stacks on top of the one below until the cell is filled to the deck. Since the deck must be open above all cells, hatches of containerships are very large. They are made as wide as possible, but some space at the ship sides outboard of the container cells is necessary in order that there be adequate deck structure to provide for the required longitudinal structural strength of the ship. There are no lower decks, only the tank top above the double bottom tanks and the main or weather deck. After the holds or cells are filled, hatch covers are fitted, and special fixtures on the deck and on the hatch covers provide for stacking from two to four tiers of containers on the deck. The deck-stowed containers are secured by lashing or by specially designed buttress structures. The cargo capacity gained by stowing containers on the deck makes up for the inefficient utilization of the cubic space inside the ship that occurs because the box-shaped containers cannot be fitted neatly into curved, ship-shaped holds.

Barge Carriers

Barge-carrying cargo ships take the concept of unitization to the limit in terms of the size of the unit load placed aboard the ship. A fully loaded 40-foot container may weigh as much as 30 tons, but the largest of the barges carried by some barge carriers can contain as much as 834 tons of cargo and weigh as much as 1,000 tons when loaded to capacity. Thus the barge carriers have reduced port time to the minimum possible so far. What the barge carrier achieves is a sort of physical separation of the self-propelled ship from its cargo holds, each of which can float and be handled by a tugboat or pushboat in the port area and on inland navigable river routes as well. The barges are so large that they can be stowed with virtually any kind of cargo, even including loaded cargo containers.

There are two methods of loading large barges onto the barge ships, and special ship designs have been created for each type. In one system, the loaded barges are lifted aboard the barge ship over the stern. This system is popularly known as the LASH system, which stands for "Lighter Aboard Ship." Each barge, or lighter, can carry up to 370 tons of cargo. A shipboard gantry crane of 500-ton capacity rides on tracks that run along the edges of the deck and along cantilevered extensions of the deck that protrude from the stern of the ship. The barge to be loaded is floated up to the stern of the ship and lifted by the gantry crane; when it is clear of the deck, the crane transports it forward and lowers it

LASH barge carrier. Photography by Granard Associates, F.J. Duffy

into one of the holds of the ship. Additional barges are stowed on deck on top of the hatch covers. Several sizes of LASH ships have been built, the largest of which can carry 89 barges.

The second principal barge-carrying ship system, called the Seabee (for sea barge), is designed to carry even larger barges—the 834-ton-capacity barges mentioned above. The loading system for these extremely heavy barges consists of a large elevator, the platform of which is large enough to carry two of the barges, each of which is 97.5 feet (29.7 m) long by 35 feet (10.7 m) wide. The elevator spans the stern of the ship and it can be submerged during loading so that the barge can be floated in place over the elevator platform. The Seabee ship has three decks with no hatches, since access to the decks is through the stern. After a pair of barges is lifted by the elevator to the level of the deck that is being loaded, rolling transporters engage each of the barges and move them forward to a stowed position on the deck.

The rapid turnaround of barge carriers in port is attributed not only to the fact that relatively few units each of very large size need to be handled, but also because these ships do not require dock or pier space or any special handling equipment furnished by the port. Thus they are not subject to delays caused by port congestion that require other types of ships to line up and wait their turn at a loading facility. Barge carriers have been found to be most useful in trades that involve ports that give access to substantial navigable river networks. For example, the port of New Orleans in the United States is the principal barge ship port because it is the entrance to the vast Mississippi River system.

Roll-on/Roll-off Ships

Ships on which wheeled vehicles are loaded by driving them aboard using special ramps and doors for the purpose are called "RO/RO" ships, for roll-on/roll-off. Prior to the development of the containership, RO/RO ships (then usually called "trailerships") were used to carry truck trailers loaded with high-value cargo. When cellular containerships came along, however, trailerships were not competitive with them, because the large amount of wasted space taken up by the wheels and chassis of the trailers made them inefficient as compared to the lift-on/lift-off arrangement of the containerships. In contrast to the inefficient use of cargo volume, however, RO/RO ships are extremely efficient in cargo-handling. They have faster cargo-handling rates and quicker turnaround times (shorter port time) than most other types of ships.

To reduce wasted cargo space volume, many modern RO/RO ships employ special wheeled dollies carrying cargo-filled containers to move onto and about the ship, rather than the regular over-the-highway truck bodies carried by the original trailerships. Also, as the RO/RO ships evolved and were custom designed for particular trade routes, many of them have become combination carriers, perhaps with containers stacked on deck and RO/RO decks below. Today,

CARGO SHIPS 11

A special type of RO/RO ship—the pure car carrier.
Photography by Granard Associates, F.J. Duffy

the RO/RO feature is also used when the payload itself consists of vehicles—that is, for the transportation of automobiles, trucks, and military vehicles.

The particular configurations of RO/RO vessels are endless in the variety of types and locations of ramps and loading ports that they employ, but all of them must have some ramps or elevators that enable the cargo to be loaded by driving vehicles about the ship. Unlike a barge ship that can load and discharge barges far from a pier or roadstead, the RO/RO ship must be able to berth right next to a suitable shoreside facility. Those that employ only stern ramps, however, need only a minimal amount of berth space, if their anchoring gear is sufficient to keep them in position while loading and discharging cargo. Ramps carried aboard ship are unfolded and extended to the shore to make roadways from ship to shore on which the vehicles can be driven. Ramps and the doors into the ship may be located at the stern, on the side, and sometimes even through the bow. The ramps must be adjustable so that the ship can load at varying stages of the tide. Traffic lanes within the ship and ramps are usually sufficiently wide that loading and discharging of the vehicles can take place simultaneously. This capability contributes to the high cargo-handling efficiency of these ships.

RO/RO ships usually have many decks, since only one layer of cargo can be placed on each deck. If some of the cargo spaces are dedicated exclusively to the carriage of ordinary passenger cars, they can be built with very low overhead clearance, or removable platform decks may be used to convert one deep hold space into two or more car decks. There are other special features of RO/RO ships that do not have to be contended with in other types of ships. Among them are the need for very large openings in the transverse bulkheads (the bulkheads are needed to limit flooding in the event of a collision) for the vehicles to pass through. The openings must be fitted with heavy watertight doors that must be gasketed and secured before the ship puts to sea. A substantial and reliable ventilation system must also be installed throughout the cargo holds to clear out the noxious exhaust fumes produced by the vehicles during loading and discharging operations. The deck structure requires special attention as well, since the decks must support very heavy vehicle loads.

LIQUID-CARGO CARRIERS

Crude Oil Tankers

During the post–World War II time period when dry cargo ships were evolving in complexity, degree of specialization, and speed, a different sort of evolution was taking place in the design of oil tankers—an evolution in size. The growth of the crude oil tanker from the typical 20,000-deadweight-ton size of the 1940s was at first gradual, to about 100,000 tons by 1960 and 150,000 tons by 1965, but it became explosive during the time the Suez Canal was closed (for seven years, from 1967 to 1975), because the long route from the oilfields in the vicinity of the Persian Gulf around the Cape of Good Hope to Europe and North America made the smaller ships uneconomical. Since it was not necessary to limit their size to navigate the canal, economies of scale took over and the deadweight of the crude oil tankers increased to about 350,000 tons by 1970, and ulti-

Crude oil tanker. Photography by Granard Associates, F.J. Duffy

mately to 560,000 tons in 1981. Inventing new superlative terms to describe these giants became quite a challenge, as "jumbo" tankers were supplanted by "super," "mammoth," VLCC (very large crude carrier), and ULCC (ultra large crude carrier). While such enormous ships can transport massive quantities of oil economically, they are not without their problems. Since loaded ULCCs have drafts that exceed 90 feet, very few ports have water depths sufficient to accommodate them, so special offshore mooring stations connected to the mainland by pipeline have had to be built to off-load them. Furthermore, accidental spills of cargo can pollute large areas of the sea and shorelines because of the sheer quantity of oil that may be released. Another problem is that very few dry-docking facilities are available to handle the largest of the ULCCs.

All tankers, whatever their size, have some characteristics in common. The standard arrangement has the engine room, deckhouse, and navigating bridge aft, even on the largest tankers. The cargo spaces are divided into three tanks athwartships by a pair of longitudinal oil-tight bulkheads. The number of such sets of three tanks within the cargo section of a tanker depends on the ship's length, the structural need for transverse bulkheads, and tank size requirements to limit the amount of pollution that might result if the tank is damaged. Cargo tanks in most tankers extend from the bottom plating to the weather deck, there being no need for intermediate decks as in dry cargo ships. The traditional tanker has no double bottom tanks within the cargo tank section of the ship, but antipollution regulations in the 1990s are becoming ever more strict, and tankers with double bottoms and even with double side skins are being built and proposed to minimize the possible extent of pollution of the oceans following marine accidents. Antipollution regulations also require that tankers have segregated ballast arrangements—cargo tanks may not be used for seawater ballast during the ballasted voyages when no cargo is aboard. This procedure reduces the hazard of pollution caused by discharging oily ballast into the sea. It has the added advantage of reducing corrosion of the steel tank structures.

Liquid cargo is loaded and discharged from tankers by high-capacity cargo pumps located in special pump rooms on board ship and connected to the tanks by a piping system. Special provision must be made to prevent explosive mixtures of air and oil vapor from developing during cargo-discharging and tank-cleaning operations. Inert gas systems, which replace discharged oil from a tank with a gas containing no oxygen, rather than with air, are typical of such special provisions.

Although the extraordinarily large crude oil tankers mentioned above may be the most spectacular development in tanker design and construction in recent decades, they are not the only types worthy of mention. Crude oil tankers of about 80,000 tons deadweight, built to comply with strict antipollution regulations and with drafts and lengths restricted so that they can serve U.S. East Coast and Gulf Coast ports, are numerous, as are "Suezmax" tankers, the largest

Parcel tanker. Photography by Granard Associates, F.J. Duffy

tankers that can transit the Suez Canal, which are up to about 160,000 tons deadweight.

Product Carriers and Parcel Tankers
Smaller tankers (typically 15,000 to 40,000 tons deadweight) are also the norm for transporting refined petroleum products. Product carriers are tankers that are outfitted to carry several different grades of refined products simultaneously without ever having different products share the same pipelines or pumps, so that contamination of products cannot take place. If pump rooms and piping systems similar to single product tankers are fitted, the maximum number of different products is usually four, limited by the number of separate cargo pumps that can practically be installed. In some such tankers, often called parcel tankers, each tank is fitted with its own submerged cargo pump, and more products can be carried without fear of contamination. Special materials and tank coatings can be installed so that nonpetroleum products and edible oils and liquids can also be handled.

Chemical Tankers
The most complex pumping and piping systems are those on the chemical tankers. The deck of a chemical tanker is covered with an elaborate maze of pipes, fittings, and cylindrical tanks made of special materials to carry small quantities of hazardous liquid cargoes. Many different chemicals, including those that are noxious, poisonous, highly corrosive, caustic, or otherwise very hazardous, are carried by such ships. Needless to say, their tank materials and coatings must

Chemical tanker. Photography by Granard Associates, F.J. Duffy

often be quite specialized, and the measures taken to prevent contamination and accidental discharge of such liquids are highly sophisticated.

LIQUEFIED GAS CARRIERS

Fuel gases that are transported by sea in bulk form as liquids are classed either as LPG (liquid petroleum gas) or LNG (liquid natural gas). LPG cargoes are principally either propane (C_3H_8) or butane (C_4H_{10}), two fuel gases that can be liquefied at ambient temperature by pressurizing them, or at atmospheric pressure by refrigeration to about −50°C (−58°F). When liquefied, LPG is a little heavier than water. LNG is a natural mixture of gases, its principal component being methane (CH_4), which cannot be liquefied at normal temperatures by pressurizing. It requires the extremely low temperature of −162°C (−260°F) to liquefy it. As a liquid, it is quite light, weighing about half the weight of water. Although some of the characteristics of LNG and LPG are similar, because of the very different boiling temperatures and densities of the two materials, the ships designed to carry them are quite different from one another.

LPG Ships
Before the advent of the LPG tanker, pressurized tanks of LPG at ambient temperatures were transported by ships. Beginning in the late 1950s the demand for these fuels justified special ships that could carry them in bulk. Tank systems that controlled either the pressure or the temperature, or both, to maintain the cargo in the liquid state were developed. The majority of modern LPG ships employ fully refrigerated tanks, in which refrigeration alone is sufficient to liquefy the cargo. Special steel alloys must be used for the tanks, because ordinary structural steel would become brittle and crack at the −50°C temperature of the liquefied gas. The tanks are insulated, typically with polyurethane foam, to minimize boil-off. The gases that do boil off are usually reliquefied and returned to the tanks. A double skin or some other form of secondary barrier is installed to contain spilled cargo without allowing it to reach and fracture the ship hull in case a tank should fail.

LNG Ships
The design of LNG ships is vastly more difficult and the ships are far more costly than LPG ships because of the extremely low temperature of the liquefied cargo. Problems involving costly materials, differential expansion of cold and warmer structures, the extent of insulation required, secondary barriers, and the control of boil-off took many years of research before specially designed ships that could transport LNG safely became a reality in 1964.

Because LNG is so light, having a density about half that of water, ship carrying capacity is designated in cubic meters rather than tons. The typical LNG ship has a cargo capacity of about 125,000 cubic meters. They are large ships, about the size of a 100,000-deadweight-ton conventional tanker—about 900 feet long (274 m). They operate at relatively light draft because of the low-density cargo, which makes them particularly vulnerable to problems of wind-induced heeling when beam winds blow against the large exposed area of the ship sides and tanks.

A number of different kinds of tank systems have been developed. Self-supporting tanks that are spherical, cylindrical, or prismatic in shape have been tried, made of aluminum or very special nickel steel alloys. The tanks are very large—

the spherical tanks in a 125,000 m^3 ship are more than 120 feet (36.5 m) in diameter. Thick insulation layers of balsa wood or cork backed with or sandwiched between layers of plywood are used around the tanks, the plywood forming the secondary barrier. Elaborate systems of supporting the tanks in the ship hull while allowing for expansion and contraction of the tanks are essential. Boil-off of the cargo cannot be reliquefied, because the ships do not carry refrigerating equipment that can achieve the extremely low temperatures required. Instead, the boil-off, which can amount to up to 10 percent of the cargo on a long voyage, is usually piped to the boilers of the steam turbine main propulsion plant where it is useful as a very efficient and clean fuel.

Another type of tank system that has been successful is the membrane tank, in which the containment tank is not structurally self-supporting. It consists of thin sheets of stainless steel, nickel, or aluminum deformed into wafflelike ridges that allow for expansion, backed and supported by insulation of balsa or perlite with plywood. Secondary barriers might be of plywood or a nickel steel alloy. Other containment systems that are variants of the two described above have also been developed.

DRY BULK CARRIERS

Second to tankers in the worldwide tonnage of cargoes carried each year are the dry bulk carriers. Their cargoes are contained in their holds without packaging, and they are usually loaded aboard and discharged by shoreside cargo-handling gear. Typical cargoes carried by dry bulk ships are iron ore, coal, bauxite (aluminum ore), phosphate (a rock used to make fertilizers), grains, and raw sugar. Forest products, steel products, and cement are also transported as bulk cargoes. Although shipowners try to engage their bulk carriers in trades in which some kind of bulk cargo is moving in both directions, this is not always possible, and many trades are one way, the return trip being made in ballast, like crude oil tankers.

Ship characteristics typical of all bulk ships are that they tend to be large with a single deck, have machinery and deckhouse aft, large hatches, and no cargo gear. Although the densities of the various cargoes carried vary considerably, most bulk materials are more dense when stowed than typical packaged or general dry cargo. Therefore the cargo spaces need not be so voluminous. The result is that bulk ships, like tankers, have a considerable amount of void space inside the hull when they are fully loaded, and the actual volume of cargo occupies only a small amount of the available internal space. The cargo spaces are concentrated about the ship centerline, with wing tanks on both sides. The lower parts of the holds are often hopper-shaped with sloping sides, so that the cargo settles to a central location as it is discharged, making the discharging operation efficient. All structural stiffeners that strengthen the hold sides and bottom are welded on

18   APPLIED NAVAL ARCHITECTURE

Dry bulk carrier. Photography by
Granard Associates, F.J. Duffy

the outside of the plating, making each hold a "smooth side inside" space which facilitates the complete removal of all cargo.

Ore Carrier
Iron ore is the most dense of the bulk cargoes carried in bulkers, and the single-purpose ore carrier has special features as a result. Ocean ore carriers in the size range from 25,000 to 100,000 deadweight tons are commonplace. The cargo is so dense and the space needed to contain it is so small that wing tanks are quite wide and the inner bottom is very deep. The deep double bottom is provided to keep the heavy cargo centered about at mid-depth of the ship hull, so that satisfactory stability is achieved. The bottoms of the holds require extra structural support underneath because of the large impact loads that occur during cargo

Self-unloading bulk carrier. Photography by
Granard Associates, F.J. Duffy

loading. Many ore ships have no cargo-handling gear and depend on shore-based equipment to load and discharge cargo. Some, however, are self-unloaders, which makes the unloading operation highly efficient. In a self-unloader, each hold has hopper doors in its bottom, which are opened to deposit the ore on a moving belt installed within the double bottom space. The ore is then transported to other belts or buckets at one end of the cargo section that lift it to deck level and deposit it on a long boom containing another belt that is swung over the side of the ship to deposit the ore on the shore. The modern trend seems to be more toward self-unloaders, so there are numerous versions of specialized unloading equipment.

A unique ore carrier design is the Great Lakes ore carrier. Because of special draft and other size restrictions in the ore route from the western reaches of Lake Superior through Lakes Huron and Erie or Michigan to their unloading ports, these ships are of very different proportions than their oceangoing counterparts. They are extremely long, narrow, and shallow, and very boxy in their hull shape.

The largest of these "boats" (a traditional Great Lakes appellation for these vessels, no matter how large they are) are sized solely by the dimensions of the Poe Lock at Saulte Ste. Marie (the "Soo") between Lakes Superior and Huron. They are 1,000 feet (305 m) long, 105 feet (32 m) wide, and may have a draft of no more than 32 feet (9.8 m).

Combination Carriers

Many types of combination bulk carriers have been developed in order to be able to carry cargoes in both directions on a trade route. The principal types are ore/bulk/oil carriers, or "OBO" ships. By adding pumping, piping, and cargo-oil heating systems to a bulker, and by careful design so that the hold spaces and wing tanks are sized, configured, and outfitted so that they can be utilized for a variety of cargoes on different legs of a voyage, the OBO can profitably carry cargoes of oil, grain, coal, or ore as the trade requires. Typical OBO ships range in size from 70,000 to 250,000 tons.

The largest bulkers are ore/oil ships. They have been built as large as 350,000 tons. Like the OBO, these ships must have oil-tight hatches in the spaces used for oil, and pumps and piping systems similar to those of a tanker.

Grain Ships

Grains were among the earliest bulk cargoes to be carried by ship, and they compose an extremely active trade today. Grain ships are configured much like other bulkers, but they are not so large, the typical size range being from 40,000 to 60,000 tons. Because grains are not as dense as most bulk cargoes, the hold size of a grain ship takes up most of the ship's cross section, and the double bottom is not so deep as that of an ore carrier. The grain cargoes do present a particular problem. As the ship rolls, the grain near the surface of the load can shift sideways and not return to its original position. The ship will then take a list (tilt) to one side, and later severe rolling motions may cause further shifting, ultimately resulting in the capsizing of the ship. Because of this unique hazard, very strict regulations are in place for the proper stowage of grain cargoes. Its influence on ship design is that the grain hold has triangular tank spaces at both sides of the upper corners of the hold which slope inward to form a narrow vertical trunk at the top of the much wider hold space. When loading the grain, care must be taken to be sure that the grain level stacks up well into the trunk. The width of the grain surface is thus much reduced, and very little grain can shift as the ship rolls.

OTHER SHIP TYPES

The ship types described above make up the great majority of the cargo carriers in service in the world's oceangoing commerce today. There are, of course, ex-

General cargo ship with heavy lift capability.
Photography by Granard Associates, F.J. Duffy

amples of special types of cargo ship that have not been mentioned. Inevitably, the needs of a special trade or product will justify the design of a different kind of ship. There have been ships designed, built, and operated quite successfully for special trades that have not been mentioned here, not because they are without merit, but only because it would be quite impossible to list, much less describe, every special ship of commerce ever put into service. There are or have been ships specially designed to carry lumber, wine, newsprint, and orange juice, to name a few. There are many types of cargo barges, which are transported by tugboats or towboats. Heavy lift ships capable of handling extremely large and heavy single items are in service throughout the world. And, of course, everyone is familiar with the passenger ships, from the super liners that are a mode of transportation that is a thing of the past, to the cruise ships, to the ferryboats that are so active in ports, harbors, and short sea runs worldwide. Nor are cargo-carrying vessels the only important commercial vessel types. Myriad service vessels (harbor tugs, fireboats) and industrial vessels (dredges, fishing boats, research ships) are in operation. When one considers all of these ships and adds to the list the noncommercial ships and boats, the combatant ships of the world's navies, and the enormous variety of recreational watercraft, one can appreciate the important contributions that waterborne vessels make to the benefit of mankind.

CHAPTER TWO

HULL FORM

Ship hull form refers to the shape of the hull, especially that part of the hull that is under water in normal operating conditions. Many of the calculations that a naval architect must make in order to design a ship are influenced by hull form. A great variety of hull forms have been successfully adapted to ships, each variation having evolved from experience as being particularly suited to the special mission of a particular ship. Even those who do not study naval architecture are aware of the differences in hull form among various ship types, to the extent that an untrained layman can often pick from a group of resistance models (showing hull form only) a cargo ship, a tugboat, a planing craft, a sailing yacht, and a destroyer, for example. The reason that their forms are so different from one another is because the requirements of their separate missions, especially as regards their required speeds and capacities, dictate that they should be different for each to operate efficiently.

Throughout this book and throughout the study of naval architecture, terms relating to ship hull form will be used frequently. An understanding of the language of ship form is a prerequisite to the mastery of naval architecture. Much of the material in this chapter is devoted to defining words that describe ship form. General terms that convey overall impressions like "full" and "fine" are defined in words; specific terms like "block coefficient" require mathematical definition. But neither words nor mathematical formulas are sufficient to describe completely and precisely the shape of a ship's hull in enough detail to actually build the ship or to assess its stability or powering requirements, to mention a few of the problems faced by the naval architect. For such precision and detail we must rely on another language—the language of graphics.

DELINEATING HULL FORM: SHIP'S LINES

The graphical display of the hull form of a ship is called a *lines drawing,* or *the lines*. A small scale sample of a lines drawing is shown in Figure 2-1. Consistent with the practice of standard engineering graphics, a lines drawing is orthographic; that is, it consists of three views of the hull as if seen from three orthogonal (mutually perpendicular) directions, not unlike the x, y, and z axes of classical analytical geometry. Since its purpose is only to define hull form, the lines drawing depicts the hull only up to the deck or decks to which the ship's side shell plating extends. Deckhouses are not included.

Geometrically speaking, the hull of a ship consists by design of curved surfaces with shapes that change gradually and smoothly from one region to another—it is said to be "fair"—with no sharp edges or discontinuities that would cause undesirable resistance to the flow of water around it as the ship moves through the sea. Because of this characteristic smooth shape, a drawing of its outline is not sufficient to define it with precision. The lines drawing therefore shows the shapes of the intersections of three orthogonal sets of planes with the hull, each set being shown in one of the views of the drawing. Each view thus depicts one set of intersections in true shape, that is, as if the observer's line of sight is normal or perpendicular to the cutting planes that produce the intersections. Measurements needed to make hull form calculations and to shape structural elements are taken from the principal reference planes defined below.

Views and Reference Planes

As shown in Figure 2-2, a ship hull is imagined to be resting on a horizontal plane called the *baseline plane,* which is the reference plane from which vertical measurements, or *heights above baseline,* are made to any point on the hull. The two symmetrical halves of the hull, starboard and port, are separated by the *centerline plane,* a vertical plane running longitudinally from bow to stern. Transverse or athwartships dimensions called *half-breadths* are measured from the centerline plane to the hull. The third reference plane, the *midship section plane,* is vertical and transverse, thus it is orthogonal to both the baseline and centerline planes. *Amidships,* denoted by the symbol ⊗, refers to the location of the midship section plane. Longitudinal measurements to points on the hull may be made from the midship section. Since longitudinal distances may be large, they may also be recorded as measurements from alternate cross-sectional planes at the bow (the *forward perpendicular*) or stern (the *after perpendicular*). The specific locations of the perpendiculars will be described later, but once they are located they define the location of the midship section, which is at the midpoint between the perpendiculars.

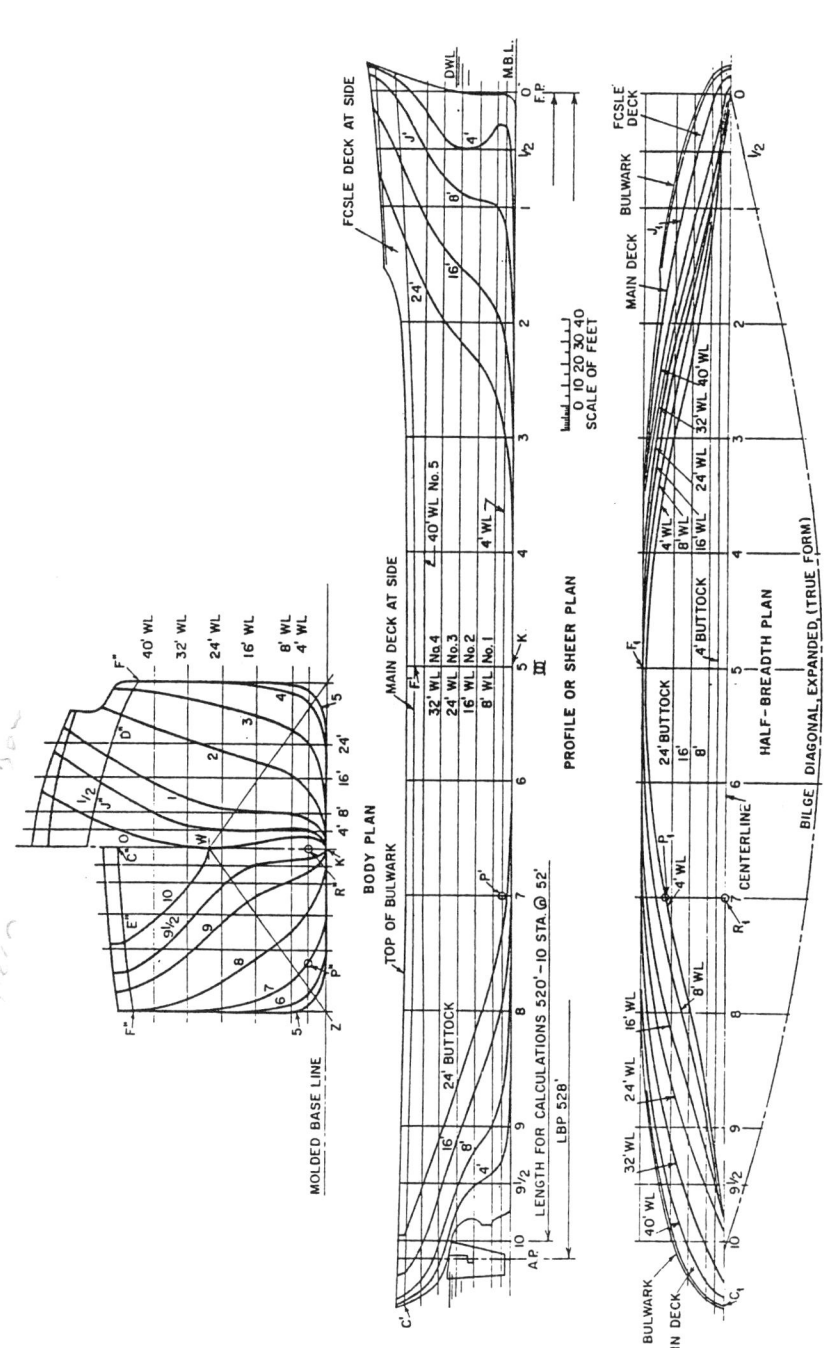

Figure 2-1. Lines drawing, Mariner class.

HULL FORM   25

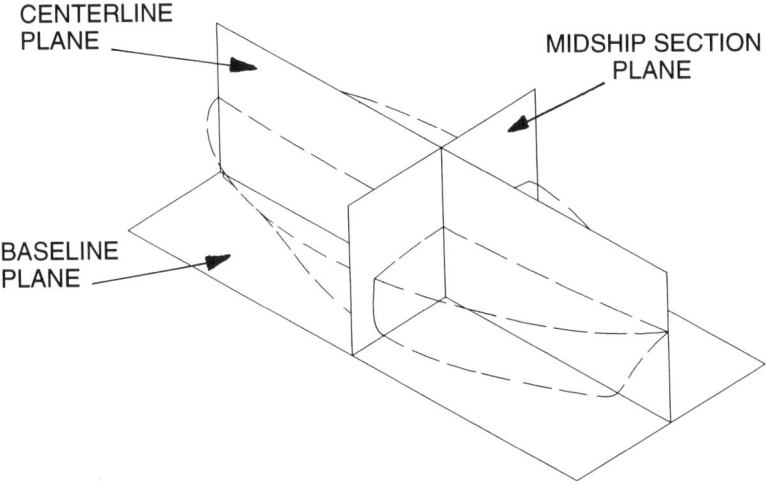

Figure 2-2. Reference planes for lines drawing.

The three sets of imaginary cutting planes that are established to define the hull shape are parallel respectively to the mutually perpendicular reference planes just described. Horizontal planes parallel to the baseline plane and at intervals of a few feet or meters above it are called *waterplanes,* and their intersections with the hull as shown in the lines drawing are *waterlines.* Planes parallel to the midship section plane, shown usually at either ten or twenty equal intervals along the ship's length, are called *station planes,* and the true shapes of their intersections with the hull are referred to as *stations.* Stations are identified by numbers, starting with zero at the bow, increasing aft (the convention in the United States and Great Britain), or with zero at the stern, increasing forward (the convention in Europe and Asia). The set of planes parallel to the centerline plane, at intervals defined by their distances off centerline, are the *buttock planes,* which intersect the hull in curves called *buttocks.*

The three views in a lines drawing have the same relationship to one another as the front, side, and top views in a typical orthographic engineering drawing, but their names are special to ship's lines. The view showing stations in true shape is called the *body plan.* Only half of each station is drawn, since the other half is symmetrical. By convention, stations from the bow to the midship section are drawn to the right of centerline, and those from amidships aft are drawn on the left side. Half-spaced stations are often added at the forward and aft ends where the waterline shape changes most rapidly. Waterlines are also drawn on one side of the centerline only, and they are all superimposed in the *half-breadth* plan. The other side of centerline in the half-breadth plan is often reserved for drawing the true shapes of intersections produced by *diagonal planes,* which are auxiliary planes, perpendicular only to station planes, but at arbitrary angles to the waterplanes and buttock planes. Diagonal planes are chosen so as to intersect sharply curved portions of the stations at nearly right angles in the body plan. Offsets measured along the edge view of the diagonal plane (shown as a straight line in the body plan) from centerline to the intersections with the stations are plotted opposite the waterlines in the half-breadth plan, and the points are adjusted as necessary until smooth curves can be drawn through them. This procedure helps the designer shape the stations, especially in the vicinity of the turn of the bilge. On the third view of the lines drawing, called the *profile plan* or the *sheer plan,* the true shapes of the buttocks are drawn. The outline of the ship's centerline profile (as seen from the side), showing bow and stern profile shapes, is also drawn in this view. The centerline plane is, in fact, the "zero foot buttock plane," so its intersection with the hull is properly one of the buttock curves.

The logic and clarity of presentation that results from this scheme of orthogonal sets of cutting planes is seen in the lines drawing of Figure 2-1. Here we observe that in each of the three views, one of the three sets of intersections is shown in true shape, the other two sets of cutting planes being shown as "edge views" or straight lines. Furthermore, every measurement or offset from a reference plane to the hull lines appears in two of the views, a fact that enables the design naval architect or draftsman to verify the accuracy of the drawing and to adjust the shapes until the lines are fair, that is, all curves are smooth and continuous, no local humps or hollows occur, and there is exact correspondence of each offset measurement in the two views in which it appears.

Molded Form, Dimensions, and Nomenclature

The shapes shown in a lines plan delineate what is called the *molded form* of the vessel. The term derives from the fact that before computer-controlled plate-cutting and frame-bending machines were developed, workers called loftsmen made wooden full-scale templates or *molds* to conform to a full-scale lines drawing laid out on the floor of the mold loft. Each mold defined the shape of a particular part of the hull structure. The molds were then used by the shipbuilders to bend, shape, and cut structural steel frames that formed a structural skeleton of the ship that was true to the shapes laid out in the lines drawing. Since the shell, or hull plating, of a ship is attached to the outside of the frames, the molded lines and molded dimensions define a hull as if it had no shell. *Displacement lines* and dimensions include the thickness of plating, which must be added to the molded dimensions to reflect the actual measurements that would be taken from the completed ship.

The sketches in Figure 2-3 illustrate pictorially the nomenclature of ship form. The *principal dimensions* of a ship are those important dimensions that define its basic size. Early in the design process the waterline at which the designer has es-

HULL FORM 27

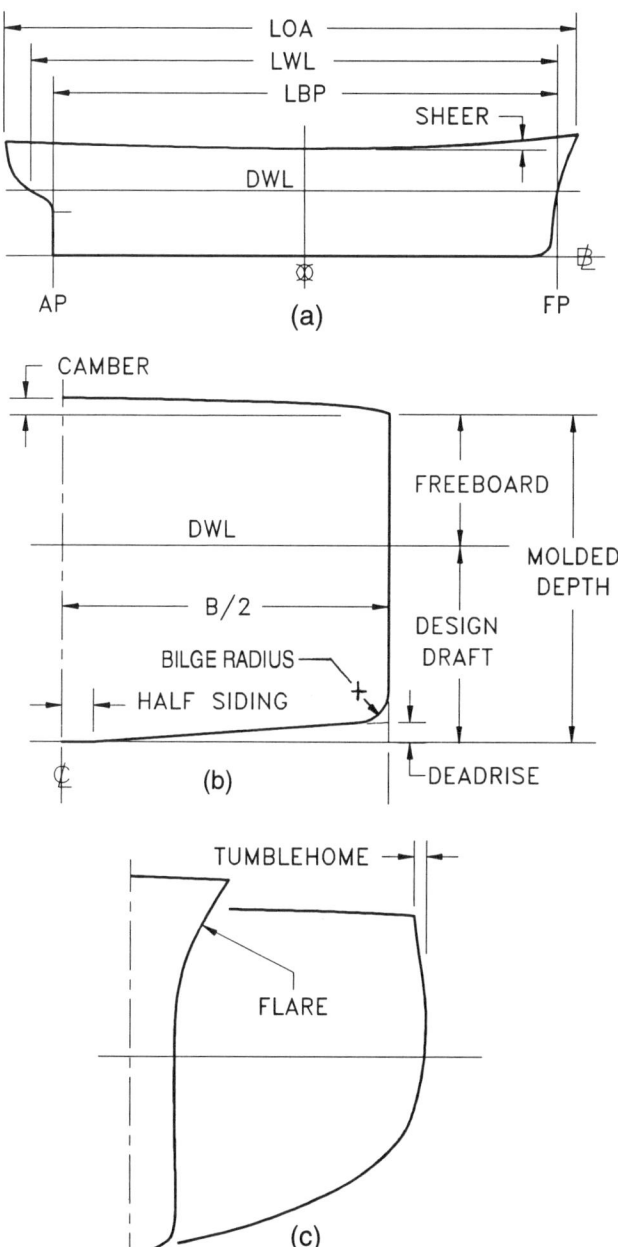

Figure 2-3. Molded dimensions.

timated the ship will float when fully loaded is determined. That waterline is called the *design waterline (DWL)* or the *load waterline*. At the point of intersection of the DWL and the forward extremity of the ship (called the *stem*), a vertical line called the *forward perpendicular (FP)* is drawn. The forward perpendicular thus defines the forward end of the immersed portion of the ship's hull. This definition of the location of the FP is universally applied to all ships. An *after perpendicular (AP)* is also defined for each ship, but its location is not specified by a unique definition for all vessels. It is intended to be representative of the after end of the ship's immersed body. Common choices for the AP are the centerline of the rudder stock, the after extremity of the design waterline, or the after edge of the sternpost (if it exists and is vertical). The length measured from the FP to the AP is designated the *length between perpendiculars (LBP)*, a principal dimension that is used to determine the ship's coefficients of form (defined later) and for structural calculations. For navigational and docking purposes, the extreme length of the ship, or *length overall (LOA)* is important. In certain hydrodynamic analyses, such as ship resistance calculations, the most characteristic length is the *length on waterline (LWL)*. All three lengths are illustrated in the profile view of Figure 2-3.

The transverse, or athwartships, dimension of a ship is called the *beam (B)* or *breadth*. Since ships are not box-shaped, the beam varies with position along the ship's length, but in a list of principal dimensions, the molded beam refers to the molded measurement at the ship's widest point. *Maximum beam* refers to the extreme breadth, from outside to outside of the shell plating, hence it is equal to the molded beam plus twice the thickness of the side shell plating. The maximum breadth dimension of large ships may extend for a considerable part of the ship's length forward and aft of amidships. On smaller vessels the maximum breadth may be reached at one point only, at or slightly aft of amidships.

In the half cross section shown in Figure 2-3(b), assumed to be at amidships, the *molded depth (D)*, or *depth at side*, is shown. It is the vertical distance at amidships from the baseline (upper surface of the keel plate) to the top of the main deck beams (or the under side of the deck plating, which comprises the molded line of the deck) at the side of the ship. A ship's depth, like the beam, is not constant throughout the length of the ship, because decks (especially weather decks) are rarely flat. A deck line is usually curved longitudinally, reaching its highest point at the bow, its lowest point at or somewhat aft of amidships, and again rising toward the stern. This curvature is known as *sheer*, the measure of the sheer at any point on the deck being defined as the vertical distance of that point above (positive sheer) or below (negative sheer) the molded deck line at amidships. The deck may also have transverse curvature, called *camber* or *round of beam* or *round down*, as shown in the figure, such that it arches upward from the deck at side to the centerline. The amount of camber is the height of the molded deck line at centerline above that at side.

If the ship's bottom is not flat, but slopes upward toward the sides, the bottom is said to have *deadrise,* or *rise of bottom,* or *rise of floor.* A flat plate keel, running along centerline, normally has no deadrise, and the half-width of such a keel is called the *half-siding.* From the outer edge of the half-siding the deadrise begins, its measure being the distance above the baseline at which the molded line of the bottom plating intersects the vertical molded line of maximum beam amidships. Bottom and side are joined smoothly by a circular arc or other fair curve at the *turn of the bilge,* the radius of the arc thus formed being called the *bilge radius.* As illustrated in (c) of Figure 2-3, the sides of some ship sections, particularly in small vessels such as tugboats and yachts, curve inward from their maximum breadth to the point at which they join the deck. This characteristic is known as *tumblehome,* measured by the horizontal distance from maximum breadth to breadth at deck. The opposite kind of curvature, outward as the deck is approached, is called *flare,* shown in the same figure. Sections with flare are common at the bow of most ships, as this shape tends to throw water aside and keep it off the deck if the ship should plunge deeply into a wave in heavy seas.

Two of the dimensions shown in Figure 2-3, namely the draft and the freeboard, are characteristics that depend on ship loading as well as ship geometry. *Draft (T)* is the vertical measurement from the waterline at any point on the hull to the bottom of the ship. A ship with a draft of 20 feet is also said to be "drawing" 20 feet. The *design draft* shown on a lines plan to the design waterline is a *molded draft,* measured to the molded baseline. For navigational purposes, it is more important to determine drafts to the bottom of the keel, which are called *keel drafts.* Draft marks at a ship's bow and stern indicate keel drafts. Drafts of a ship can vary from bow to stern and from voyage to voyage, depending on the amount and distribution of the cargo and other weights aboard. *Freeboard* is the vertical distance from the waterline to the deck at side, or the difference between the depth at side and the draft at any point along the ship. Since freeboard is an important measure of the safety of a ship, every ship is assigned a minimum acceptable freeboard at amidships, which is its *statutory freeboard,* determined by internationally adopted regulations.

Table of Offsets

A finished and faired lines plan is the source of the information necessary to proceed with two essential tasks of the naval architect: determining the hydrostatic properties of the ship on which calculations of flotation and stability depend, and preparing working drawings for ship construction. In either case, the ship form information documented in the lines plan must be expressed numerically and at full ship scale. The numerical equivalent of a lines plan is a table of offsets.

To define the three-dimensional ship form numerically, three coordinates of selected points on the molded hull must be specified:

- Longitudinal distance from the FP, AP, or ⊗.
- *Half breadth,* from the centerline plane.
- *Height* above the baseline plane.

In a table of offsets to be used for ship form calculations, the longitudinal coordinates of the selected points are the stations defined on the lines plan. For ship construction purposes, additional cross sections are defined, one at each structural transverse framing member. At each station or frame, the other two coordinates, called in general "offsets," are determined by measuring them from the faired lines. *Half-breadth offsets* are measured at each of the waterlines shown on the lines, and also (for ship construction offsets) at each deck, tank top, and structural flat or platform. *Heights above baseline,* or *buttock height offsets,* at each station or frame are scaled off at each of the buttocks shown on the lines plan. A typical table of offsets at stations is shown in Figure 2-4.

A finished table of offsets must represent the lines faired at full ship scale to the nearest eighth of an inch or millimeter, an accuracy that cannot be achieved in offsets scaled from the small scale lines drawing, which is typically at a scale of one hundredth to one fiftieth of full size. If the lines drawing has been faired manually by a draftsman, it must be refaired at full scale by mold loftsmen in a procedure known as "laying off" the lines to full ship scale on the mold loft floor. This time-consuming process has been largely replaced in modern shipyards by computer programs that can fair a set of lines using as input a set of approximate offsets from a small lines drawing. Computer faired lines are equally as fair at full size as they are at small scale.

For the purpose of making ship form calculations, offsets in the format shown in Figure 2-4, with longitudinal coordinates at ten or twenty stations, are used. Only the half breadths at each station and waterline are necessary. Heights are not needed to make hull form calculations. The half breadths should be in feet or meters, in either case expressed in decimal form. For ship construction purposes, the table is much more comprehensive, since offsets are required at every frame rather than at every station. There may be several hundred frames, spaced at two- to four-foot intervals, throughout the ship length. The units used to express the offsets in the U.S. measurement system are chosen for the convenience of the shipyard workers who make measurements with steel tapes—they are tabulated in feet, inches, and eighths of an inch, instead of in decimal form.

HULL FORM CHARACTERISTICS OR HYDROSTATICS

The description of a ship's form does not end with a faired lines plan and a table of offsets. The naval architect must also furnish the ship operator with information that will enable him or her to determine the drafts, freeboard, and stability

(Dimensions given in feet, inches and eighths)

Half breadths at waterlines and decks

Station (FWD End)	0'-0" WL	4'-0" WL	8'-0" WL	16'-0" WL	24'-0" WL	32'-0" WL	40'-0" WL	Fo'c'sle deck	Main deck	Second deck	Third deck	Station
0-FP	—	—	—	1-0-1	0-2-5	0-5-0	1-8-6	11-1-2	7-0-3	1-11-0	0-2-3	0
½	—	3-0-5	2-8-2	4-3-5	4-0-6	5-0-6	7-8-3	20-4-2	14-6-0	7-7-2	4-1-4	½
1	—	4-8-7	5-0-7	8-6-1	9-3-0	11-1-0	14-3-5	20-5-2	20-6-5	13-9-4	9-4-7	1
2	2-4-0	6-10-2	7-11-6	19-0-0	21-10-2	24-1-7	26-11-6		30-3-6	26-0-2	21-0-2	2
3	9-2-0	13-7-6	16-9-3	30-10-3	32-8-1	34-0-5	35-2-7		36-1-1	34-8-4	32-0-2	3
4	21-1-0	24-1-0	27-7-1	37-2-2	37-7-5	37-10-4	37-11-6		38-0-0	37-11-2	37-5-7	4
⊠5	28-5-6	33-3-0	35-8-1	38-0-0	38-0-0	38-0-0	38-0-0		38-0-0	38-0-0	38-0-0	5
6	24-10-4	36-2-6	37-10-3	37-11-7	38-0-0	38-0-0	38-0-0		38-0-0	38-0-0	37-11-6	6
7	13-6-4	34-8-1	37-0-7	35-8-1	37-5-2	37-11-6	38-0-0		38-0-0	38-0-0	36-0-1	7
8	4-5-2	27-1-4	31-2-6	26-2-2	31-9-2	35-7-3	37-4-2		37-8-1	36-9-7	26-11-4	8
9	—	15-0-3	19-3-3	11-2-4	18-1-0	26-8-5	31-3-1		33-8-7	29-6-3	11-10-1	9
9½	—	6-0-7	7-9-4	4-3-0	8-8-3	17-10-1	25-3-4		29-5-0	23-2-7	4-5-1	9½
10	—	2-8-7	3-4-1	—	—	8-6-3	16-11-6		22-5-2	—	—	10

Heights above molded baseline

Station (FWD End)	4-ft buttock	8-ft buttock	16-ft buttock	24-ft buttock	Forecastle deck Ht at ₵	Forecastle deck Ht at side	Main deck Ht at ₵	Main deck Ht at side	Second deck Height	Third deck Height	Station
0-FP	47-5-6	57-0-7	—	—	63-11-0	63-11-0	55-8-0	55-8-0	40-10-2	25-0-0	0
½	2-3-0	40-8-3	55-4-4	—	63-4-5	63-3-1	55-0-2	54-11-5	40-8-6	↕	½
1	0-11-7	8-0-7	43-4-1	50-8-1	62-2-6	61-0-5	53-2-5	53-0-0	39-9-1	25-0-0	1
2	0-0-4	0-8-7	6-9-2	31-5-7	61-0-6	60-4-1	51-8-2	51-3-0	38-10-4	20-10-0	2
3	—	0-0-0	0-6-7	3-11-1	—	—	49-3-5	48-4-1	37-5-1	↕	3
4		—	0-0-0	0-1-1			47-7-3	46-3-0	36-4-4		4
5				0-0-0			46-5-5	44-11-5	35-8-7	20-10-0	5
6	—	—	0-0-6	0-0-0			40-0-0	44-6-0	35-6-0	17-0-0	6
7	0-0-0	0-3-5	4-9-6	2-1-0			46-1-0	44-7-0	35-7-0	↕	7
8	1-2-1	8-6-6	21-11-3	13-3-3			46-4-5	44-10-5	35-10-4		8
9	14-3-4	23-4-1	30-3-6	30-0-4			46-9-7	45-4-3	36-4-4	17-0-0	9
9½	28-5-0	31-6-5	38-11-3	38-4-5			47-3-3	46-1-1	37-0-6	↕	9½
Aft end	37-1-4	41-7-0	44-1-6	—			47-6-1	46-7-3	37-5-7		10
							47-8-0	47-2-4	—		
							47-11-6	47-11-6			

Figure 2-4. Table of offsets, Mariner class.

for any loading condition or distribution of weights carried throughout the life of the ship. These are critical matters affecting the safe operation of a ship that will be treated in later chapters. For the present, we define those geometrical properties of the immersed form of the hull that will be shown later to be essential to the officer charged with the important task of assuring that the ship is operated in a safe and efficient manner.

The properties in question are called *hull form characteristics* or *hydrostatic properties* because they pertain to the underwater form of the hull. When these properties are displayed in graphical form, the set of curves is referred to as the *hydrostatic curves* or *curves of form*. Since the drafts of a ship can vary considerably with changes in loading, the hydrostatic properties are calculated and plotted for a series of drafts. Therefore the calculations described here must be repeated for a number of drafts, from the light ship draft (ship as built with no cargo, fuel, personnel, or loading of any kind) to well above the full load draft corresponding to the design waterline.

Fundamental Hull Form Characteristics
All of the hydrostatic properties to be calculated are derived from the following fundamental characteristics of the immersed hull form at each given even keel waterline.

Properties of the Waterplane. Four properties of each waterplane are required:

1. *Area of the waterplane (A_W).* The waterplane area is required to determine the change in mean draft when small weights are loaded or discharged, as shown in Chapter 3 and Chapter 5. Units: feet^2 or meters^2.

2. *Center of flotation (CF).* The CF is the centroid of the waterplane, also called the center of area or center of gravity of the waterplane. It is required for the calculation of changes in draft at bow and stern as a result of loading, discharging, or shifting weights aboard ship, as shown in Chapter 5. The CF is located on centerline because of the symmetry of the waterplane. Its longitudinal position with respect to the midship section (or the FP or AP if preferred as reference planes) must be calculated. The distance so determined is called the *longitudinal center of flotation (LCF)*. Units of LCF: feet or meters from reference plane.

3. *Longitudinal moment of inertia (I_L).* This property of the waterplane is its second moment of area about a transverse axis passing through the center of flotation. It is required for the longitudinal stability and trim (difference between forward and aft drafts) calculations described in Chapter 5. Units: feet^4 or meters^4.

HULL FORM 33

4. *Transverse moment of inertia (I_T)*. I_T is the second moment of area of the waterplane about its centerline. It is required in the calculation of initial transverse stability as shown in Chapter 3. Units: feet^4 or meters^4.

Properties of the Immersed Volume of the Hull. Three quantities associated with the immersed volume must be determined:

1. *Volume of displacement (∇)*. This is the immersed volume itself, called the volume of displacement because it is a measure of the volume of fluid displaced by the floating ship. As shown in Chapter 3, it is a fundamental property of hull form because the weight and mass of the ship are equal respectively to the weight and mass of the water displaced. The *molded volume* is calculated directly from the offsets of the molded form. Volumes of the shell and appendages like bilge keels, rudder, etc., are then added to determine the *total displacement* at each draft. Units of ∇: feet^3 or meters^3.

Two more properties of ∇ are required to locate the *center of buoyancy, B*, which is variously called the center of volume or center of gravity of the displaced water. The position of B affects the stability and trim of a ship in ways that are described in Chapters 3 and 5. Since the volume ∇ is three-dimensional, locating its centroid requires three coordinates. A ship's symmetry about its centerline plane puts the center of buoyancy on centerline when the ship floats upright, so only two coordinates of B require calculation. They are:

2. *Longitudinal center of buoyancy (LCB)*. This is the distance of B from a specified transverse reference plane, usually the midship section. Or LCB may be measured from FP or AP, so long as the reference axis is clearly stated. Units: feet or meters.
3. *Vertical center of buoyancy (KB)*. KB is the height of the center of buoyancy above the baseline or keel. Units: feet or meters.

Properties of the Stations. The last of the fundamental hull form characteristics required to prepare the hydrostatic curves are the *immersed station areas (A_S)*. The cross-sectional area of each station shown in the body plan up to the waterline in question is determined for input into the calculation of the volume of displacement. When plotted against ship length, the immersed station areas form a *sectional area curve,* whose shape represents the "fullness" or "fineness" of the ship form, an important consideration in ship resistance and powering. Units: feet^2 or meters^2.

Hull Form Shorthand—the Coefficients of Form

Designers of ships rely on their experience and the proven performance of successful ships of the past when they decide what the characteristics of a new design should be. When they compare the hull forms of several ships, they use as a starting point a number of simple dimensionless coefficients whose numerical values are, to an experienced naval architect, a kind of hull form shorthand that can quantify the notions of fullness or fineness of the various hulls, independent of their differences in size. The *coefficients of form* are also useful tools for making first estimates of a ship's resistance, powering, and seagoing performance at an early point in the design of a new ship. The most commonly used coefficients of form are defined below. In all cases the coefficients refer to the immersed portion of the hull. Ship dimensions used to compute form coefficients may be either molded or maximum dimensions, and the length may be LBP or LWL, depending on the type of hull and the designer's preference. It is therefore important that in each case the specific convention used should be clearly stated. In the following definitions, generic terms and symbols are used for the dimensions. The coefficients can be calculated for any draft of a ship, but the typical numerical values quoted here refer to ships floating at their respective design waterlines.

Overall, or "whole hull," fullness is characterized by the *block coefficient* (C_B), defined as the ratio of the immersed volume of the hull (∇) to the volume of a rectangular block whose length, width, and height are equal respectively to the length, beam, and draft of the ship, as shown by Figure 2-5.

$$C_B = \frac{\nabla}{LBT} \qquad (1)$$

The block coefficients of typical ships may vary from as low as 0.45 for a high-speed combatant ship like a destroyer or fast frigate to as full as 0.85 or more for a very large crude oil tanker. It conveys some idea of a ship's load-carrying capa-

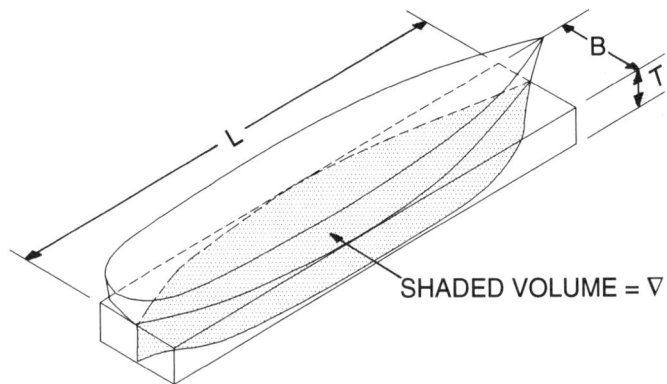

Figure 2-5. Block coefficient.

HULL FORM   35

bility compared to other ships of the same principal dimensions, but it does not tell how hull fineness is distributed longitudinally, that is, whether it is fined down primarily at the ends, around amidships, or both.

The fullness of the midship portion of a hull is described by the *midship section coefficient (C_M)*, which is the ratio of the immersed midship section area to the area of its circumscribing rectangle (see Figure 2-6).

$$C_M = \frac{A_M}{BT} \quad (2)$$

Very fine hulls typical of destroyers might have a C_M of 0.75 or less, but most large merchant ships have vertical flat sides and a flat bottom at amidships, the section departing from a rectangle only by virtue of rounded bilges, so their midship section coefficients are more like 0.95 to 0.995.

The coefficient that describes the fineness of the ends (bow and stern) of a hull without being influenced by its midship fineness is the *prismatic coefficient (C_P)*, also called the longitudinal coefficient, illustrated in Figure 2-7. It is defined as the ratio of the volume of displacement (∇) to the volume of a prism whose cross section is shaped like the immersed midship section, and whose length is the length of the ship. Thus

$$C_P = \frac{\nabla}{A_M L} \quad (3)$$

Typical values range from about 0.57 for a high-speed, fine-ended ship to 0.85 for a large bulk carrier or tanker.

Since we have defined independently the fineness of the ends (C_P) and the middle (C_M) of a hull, it is no surprise to find that its overall fineness (C_B) is mathematically related to them. The relationship is

$$\frac{\nabla}{LBT} = \left(\frac{\nabla}{A_M L}\right)\left(\frac{A_M}{BT}\right)$$

or

$$C_B = C_P \times C_M \quad (4)$$

Waterplane fullness or fineness may be quantified by defining the *waterplane coefficient (C_W or C_{WP})*, which is the ratio of the area of the waterplane (A_W) to the area of its circumscribing rectangle, as shown in Figure 2-8.

$$C_W = \frac{A_W}{LB} \quad (5)$$

Typical values of C_W at the design waterline vary from 0.67 to 0.92.

Basic Hull Form Integrals

The hull form characteristics described above are geometrical properties of stations, waterplanes, and the displaced volume, all of which are bounded by curves

36   APPLIED NAVAL ARCHITECTURE

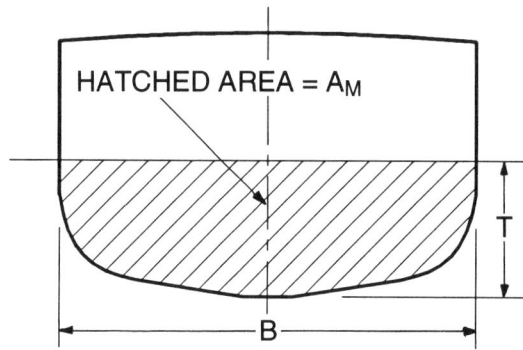

Figure 2-6. Midship section coefficient.

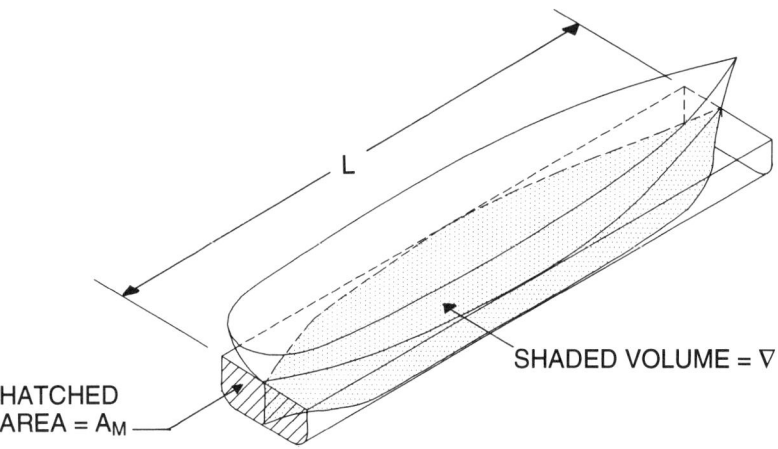

Figure 2-7. Prismatic coefficient.

or curved surfaces. This suggests that the areas and volumes and their moments can be calculated by representing them as *definite integrals,* that is, as infinite sums of infinitesimally small rectangular elements located between specified limits. Applications of this mathematical process of the integral calculus to calculations of the required properties of ship form are described below.

Suppose the shape of half of a ship's waterplane is given by Figure 2-9, shown superimposed on a Cartesian coordinate system whose origin 0 is at amidships and on centerline. The x-axis coincides with the ship's centerline and x is positive forward, and the y-axis is in the transverse direction (vertical in the fig-

HULL FORM 37

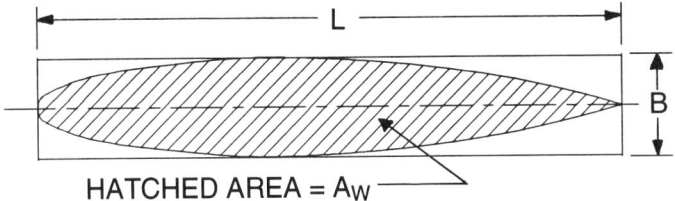

Figure 2-8. Waterplane coefficient.

ure). Specific values of the ordinate y, such as y_i, are the half-breadths of the waterplane, as shown on the lines plan. The limiting values of abscissa x are, as shown in the figure, $-1/2$ at the after perpendicular and $+1/2$ at the forward perpendicular, where L is the LBP of the ship. If the area shown is imagined to be subdivided into a large number of elemental areas, each of small width δx and height y_i like the shaded area in the figure, leaving no gaps, the waterplane's half area is the sum of all of the elemental areas. Since δx is very small, the elements are very nearly rectangles of area $\delta A_i = y_i \, \delta x$. Thus the area shown, which is the half area of the waterplane, is

$$A = \Sigma \delta A_i = \sum_{i=1}^{i=N} y_i \, \delta x \qquad (6)$$

where Σ represents the sum of all such elemental areas, from the first one at the stern (i = 1) to the last one at the bow (i = N). This area is not exact, because the assumed rectangular elemental areas do not trace the smooth curve exactly, but approximate it as a "stair step" function.

The accuracy of the calculated area can be improved by making the elemental width δx smaller, which increases the number of elemental areas that must be summed and produces smaller "stair steps" which converge ever more closely on the smooth curve as δx decreases. If δx is reduced to an infinitesimal (approach-

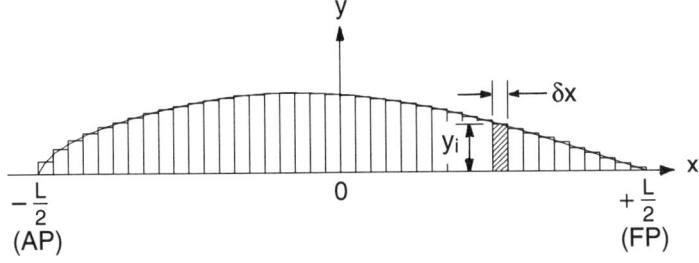

Figure 2-9. Approximate waterplane area.

ing zero) width so that N increases without limit, the expression for the area is written as the *infinite sum*

$$A = \lim_{\delta x \to 0} \sum_{x=-\frac{L}{2}}^{x=+\frac{L}{2}} y(x)\, \delta x \qquad (7)$$

Since this equation has, in principle, an infinite number of terms on the right side because δx approaches zero, the elemental areas are not countable, and the limits of the summation are therefore set in terms of x, from $-\frac{L}{2}$ to $+\frac{L}{2}$. Ordinates y are represented by y(x), or "y as a function of x," rather than by the counter i. The formal mathematical name of an infinite summation like equation (7) is the *definite integral,* and it is written in standard form by replacing the summation symbol with the integral symbol and the discrete increment δx with the differential dx, as follows:

$$A = \int_{-\frac{L}{2}}^{+\frac{L}{2}} y\, dx \qquad (8)$$

which is read, "A equals the integral of y dx between the limits of x from $-\frac{L}{2}$ to $+\frac{L}{2}$." The variable ordinate y is understood to be a function of x. It is often written as y(x) to make that clear. This is a definite integral because definite limits have been set on the variable of integration, x. So long as numerical values are taken for the limits of integration, the solution to a definite integral is always a number (not a mathematical expression). In this case it is a measure of the area of Figure 2-9.

When the function y(x) is given as an analytical expression in terms of the variable x, the integration represented by equation (8) can be obtained using the formal rules and methods of integration described in textbooks on calculus. Ship's lines, however, are rarely defined analytically, because they have been created by the fairing process rather than by mathematics. In such cases, definite integrals must be obtained by numerical integration methods, using numerical values of measurements taken from the lines drawing. The table of offsets described previously provides the required measurements.

Integrals for Waterplane Properties. The waterplane of Figure 2-9 is redrawn in Figure 2-10 using the symbolism of equation (8). Because waterplanes are symmetrical about their centerlines, all integrations are performed for the half waterplane, and the results are multiplied by two. The integral for the half area of the waterplane has already been derived, so we can write waterplane area as

HULL FORM

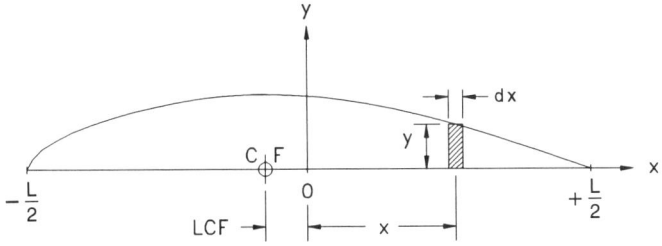

Figure 2-10. Waterplane area integration.

$$A_W = 2 \int_{-\frac{L}{2}}^{+\frac{L}{2}} y \, dx \qquad (9)$$

To locate the center of flotation, or centroid of the waterplane area, we must first calculate its first moment of area with respect to some reference axis. Using the y-axis or amidships as the axis of moments, the first moment of elemental area y dx is the area times its distance from the axis (i.e., its moment arm).

$$dm = x \, dA = x \, (y \, dx)$$

where dA = elemental area = y dx
 x = moment arm from amidships (+ fwd, –aft)
 dm = first moment of elemental area about amidships

Thus the first moment of the entire waterplane area about amidships is

$$M_\otimes = 2 \int_{-\frac{L}{2}}^{+\frac{L}{2}} x \, y \, dx \qquad (10)$$

where the multiplication by two is necessary, as before, because y values are half-breadths. The subscript ⊗ refers to amidships, the axis of moments. The centroid of the area (center of flotation) is the point at which the waterplane would balance if it were imagined to be a thin homogeneous plate. Its longitudinal distance from the axis of moments (amidships in the figure), called $\bar{x}$ in general, or LCF for a ship's waterplane, is determined by dividing the first moment of area by the area itself:

$$LCF = \bar{x} = \frac{M_\otimes}{A_W} \qquad (11)$$

The center of flotation (CF in Figure 2-10) will be forward of amidships if $M_\otimes$ is positive, or aft if it is negative. While amidships is the most common choice for the axis of moments in routine hydrostatic calculations, any point may be used, resulting in different values for the arms and moments and LCF distance; but the position of CF in the waterplane will be the same regardless of the moment axis chosen. In fact, if the moment axis had been chosen by chance such that it passed through the centroid, the moment calculated by equation (10) would be zero because forward (positive) moments would exactly cancel aft (negative) moments of the elemental areas. This "zero-moment" condition is characteristic with respect to any axis passing through the centroid, thus validating the balance point analogy.

We next examine the longitudinal moment of inertia of the waterplane (I_L). Moment of inertia is a second moment of area, so called because it is the (first) moment of area multiplied again by the moment arm. In Figure 2-10, if $di_\otimes$ represents the moment of inertia about amidships of the element shown,

$$di_\otimes = x^2 \, dA = x^2 \, y \, dx$$

Integrating over ship length and including both sides of the waterplane, we have

$$I_\otimes = 2 \int_{-\frac{L}{2}}^{+\frac{L}{2}} x^2 \, y \, dx \qquad (12)$$

where $I_\otimes$ is the moment of inertia of the waterplane about amidships as the moment axis. The moment of inertia needed to evaluate longitudinal stability (see Chapter 5), however, must be the one determined about a transverse axis passing through the center of flotation. This special moment of inertia is called the *longitudinal moment of inertia of the waterplane (I_L)*. The parallel axis theorem of mechanics states that two moments of inertia of an area about two parallel axes, one of which passes through the centroid of the area, are related as follows:

$$I_{par} = I_{cent} + A \, h^2 \qquad (13)$$

where I_{cent} = I of the area about axis through centroid
A = area
h = distance of parallel axis from centroidal axis
I_{par} = I of the area about the given parallel axis

This shows that the I of an area about an axis passing through its centroid is smaller than that about any other axis parallel to it. Applying the parallel axis theorem to the waterplane of Figure 2-10, we can determine I_L from $I_\otimes$ as follows:

HULL FORM   41

$$I_L = I_\infty - A_W (LCF)^2 \qquad (14)$$

The last of the waterplane properties that must be calculated is the transverse moment of inertia (I_T), which is its second moment of area about the longitudinal axis passing through the center of flotation, that is about the centerline of the waterplane. Now the moment of inertia of a rectangle of height h and width b about its base as an axis is $\frac{1}{3} b h^3$. Thus the elemental area shown in the figure has a moment of inertia about its base of

$$di_T = \tfrac{1}{3} y^3 \, dx$$

which, when doubled for both sides and integrated over the length of the waterplane gives

$$I_T = \tfrac{2}{3} \int_{-\frac{L}{2}}^{+\frac{L}{2}} y^3 \, dx \qquad (15)$$

Integral for Immersed Station Area. Each station of the ship, as drawn in the body plan of the lines drawing, can be represented as shown in Figure 2-11. The

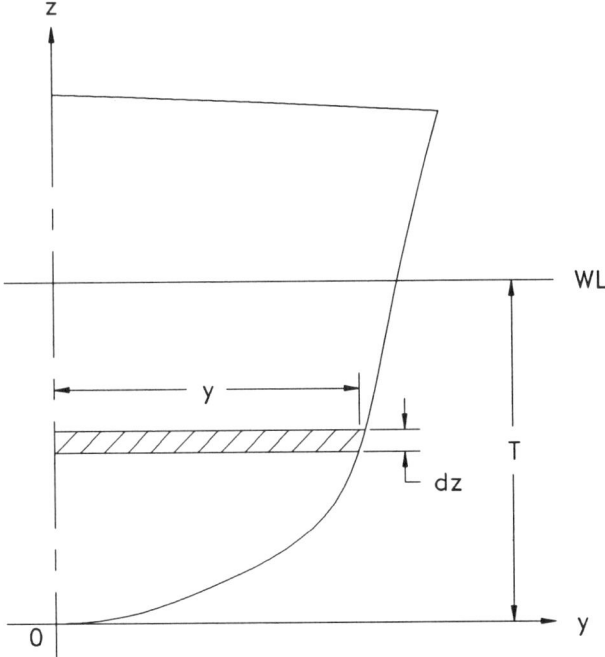

Figure 2-11. Station area integration.

intersection of baseline (y-axis) and centerline (z-axis) is taken as the origin of a coordinate system for the station. Elements of the half area of the station are given by

$$dA_S = y(z)\,dz$$

Integrating from baseline to waterline and including the other symmetrical half, the immersed station area is

$$A_S = 2\int_O^T y\,dz \quad (16)$$

where A_S = immersed station area, or *sectional area*
  y = half breadths of station
  T = draft at station

Integrals for Properties of the Immersed Volume of the Hull. The immersed volume of the hull, or *volume of displacement*, may be represented in integral form by defining elemental volumes of the immersed form as shown in Figure 2-12. The volume shown is only that up to the waterline at which the ship is floating. The volume integral is analogous to the integration for waterplane area, except that station areas A_S replace the half breadths y. Thus the elemental volume (dv) shown in the figure is written

$$dv = A_S dx$$

and the integration is performed over the length of the ship to give:

$$\nabla = \int_{-\frac{L}{2}}^{+\frac{L}{2}} A_S dx \quad (17)$$

where ∇ = volume of displacement of the ship
  A_S = sectional areas to ship's waterline, both sides

It should be observed that there is no multiplication by two in equation (17) because the sectional areas A_S are assumed to be full (both sides) station areas as determined from equation (16).

The analogy between the calculations of ∇ and A_W can be extended to the calculations of their longitudinal moments and centroids. Thus, the volume element shown in Figure 2-12 is at distance x from the midship axis of moments, just like the area element in Figure 2-10, so the longitudinal moment of volume ($M\nabla$)

HULL FORM 43

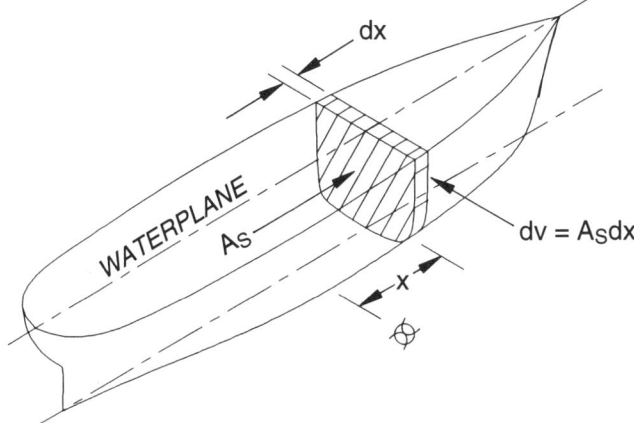

Figure 2-12. Volume of displacement by longitudinal integration.

and longitudinal center of buoyancy (LCB) are calculated like the moment of area and LCF. The elemental moment of volume about amidships is

$$dm = x\, A_S dx$$

and its integral is the moment of the entire immersed volume about amidships:

$$M_{\nabla\otimes} = \int_{-\frac{L}{2}}^{+\frac{L}{2}} x\, A_S dx \qquad (18)$$

Dividing moment of volume by the volume, we get the longitudinal distance of the center of buoyancy from the axis of moments, namely

$$LCB = \frac{M_{\nabla\otimes}}{\nabla} \qquad (19)$$

where LCB = longitudinal center of buoyancy (distance from amidships)
 $M_{\nabla\otimes}$ = moment of volume ∇ about amidships
 ∇ = immersed hull volume, or volume of displacement

The center of buoyancy (B) must also be located vertically; that is, its height above keel or baseline must be determined. This measurement is called KB (for height of B above the keel). The procedure is analogous to the above calculations of ∇, $M_{\nabla\otimes}$, and LCB, but the direction of integration is vertical from keel to waterline, and the elemental volumes are represented as infinitesimally thin horizontal layers, each with the shape of a waterplane lying between the baseline and the limiting waterline at which the ship is floating. This is illustrated in Figure 2-13. It is clear that an elemental volume is equal to A_W dz, and thus the volume of displacement is

$$\nabla = \int_0^T A_W dz \quad (20)$$

Although ∇ has been calculated previously by integrating sectional areas longitudinally using equation (17), calculating it again by integrating waterplane areas vertically serves as a useful check on the previous calculation. Moments of the elemental displaced volumes about the keel are written as $z\, A_w\, dz$, and the total moment is therefore

$$M_{\nabla K} = \int_0^T z\, A_W dz \quad (21)$$

Finally, KB is determined by dividing moment by volume:

$$KB = \frac{M_{\nabla K}}{\nabla} \quad (22)$$

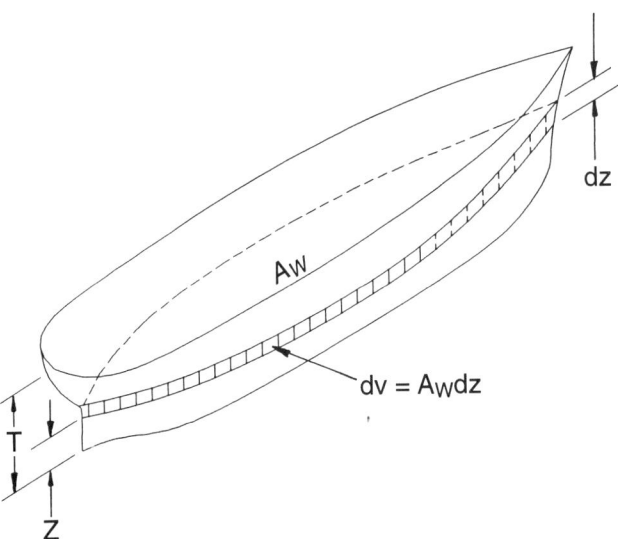

Figure 2-13. Volume of displacement by vertical integration.

HULL FORM

Hydrostatic Properties Derived from the Integrations
From the quantities calculated using the definite integrals defined above, all of the hydrostatic properties of a ship that will be used in our later studies can be determined. At this point, they will be defined without substantiation or proof of their validity, but only to gather in one place the description of those properties of ship form which, when plotted (or tabulated) against the draft of the ship, are known collectively as the *hydrostatic curves,* or *curves of form,* or *displacement and other curves*. In later chapters dealing with ship stability, trim, strength, resistance, and dynamics, the formulas listed here will be derived and analyzed as they are needed.

Practice varies as to how many hydrostatic properties are included in a set of curves of form, depending on whether the curves are intended for use by the naval architect for detailed ship design calculations or for use by the ship's officers to assess the effects of loading and ship operation on stability and trim of the vessel. For the latter purpose, fewer relationships are needed.

Properties Always Included in Hydrostatic Curves. The six hydrostatic properties listed below are always included in curves of form furnished to shipboard personnel. Nomograms or numerical values from computer printouts may be substituted for plotted curves.

1. *Displacement (Δ or Δ_m).* In the U.S. measurement system, displacement is the weight of the ship and its contents. In the SI system, displacement is expressed as a mass. It is equal to the volume of displacement times the density (weight or mass density as required) of the water in which the ship floats.

$$\Delta = \rho \, g \, \nabla \quad \text{(U.S. units)} \tag{23}$$

$$\Delta_m = \rho \, \nabla \quad \text{(SI units)} \tag{23a}$$

where Δ = displacement (weight)
Δ_m = displacement (mass)
ρ = mass density of water
g = acceleration of gravity
∇ = volume of displacement

Sometimes two displacement curves are plotted, one for seawater and one for fresh water. See Chapter 3 for a detailed derivation.

2. *Longitudinal center of buoyancy (LCB).* The LCB is determined as shown in equations (17) through (19) above. It is used in trim calculations described in Chapter 5.

3. *Height of transverse metacenter above keel (KM)*. This quantity is needed to calculate initial transverse stability. It is defined in Chapter 3 as follows:

$$KM = KB + BM = KB + \frac{I_T}{\nabla} \qquad (24)$$

where KB = height of center of buoyancy above keel, as shown by equations (20) through (22) above
BM = transverse metacentric radius (see Chapter 3)
I_T = transverse moment of inertia of waterplane, integral equation (15) above
∇ = volume of displacement, integral equation (17) above

4. *Tons per inch immersion (TPI)*, and *tons per centimeter immersion (TPC)*. These quantities are defined in Chapter 3, where it is shown that

$$TPI = \frac{A_W}{420} \quad \text{(U.S. units)} \qquad (25)$$

$$TPC = \frac{A_W}{97.56} \quad \text{(SI units)} \qquad (25a)$$

where A_W = waterplane area, integral equation (9) above.

5. *Longitudinal center of flotation (LCF)*. This is the centroid of the waterplane, determined as shown in equations (9) through (11).

6. *Moment to change trim one inch (or one centimeter) (MT1, MTcm)*. In their exact form, these quantities depend slightly on loading as well as on ship form, but approximate values which are functions of form alone are accurate enough for trim calculations, and they are included in the hydrostatic curves. They are defined as

$$MT1 = \frac{\Delta\, BM_L}{12\, L} = \frac{I_L}{420\, L} \quad \text{(U.S. units)} \qquad (26)$$

$$MTcm = \frac{\Delta_m\, BM_L}{100\, L} = \frac{0.01025\, I_L}{L} \quad \text{(SI units)} \qquad (26a)$$

where BM_L = longitudinal metacentric radius (Chapter 5)
I_L = longitudinal moment of inertia of waterplane, from equation (14)

Derivation and applications of these properties of form are given in Chapter 5.

HULL FORM   47

Additional Properties Sometimes Included in Hydrostatic Curves.

7. *Change in displacement per inch (or centimeter) trim by stern ($d\Delta PI$, $d\Delta PC$)*. The displacement curve is plotted only for the even keel condition. If a very accurate determination of displacement is required for a ship in a trimmed condition, that is, with different drafts forward and aft, this property is a convenient way to determine the corrected displacement. It is defined as

$$d\Delta PI = \frac{TPI \cdot LCF}{L} \quad \text{(U.S. units)} \quad (27)$$

$$d\Delta PC = \frac{TPC \cdot LCF}{L} \quad \text{(SI units)} \quad (27a)$$

where the LCF must be measured from amidships. Details are given in Chapter 5.

8. *Vertical center of buoyancy (KB)*. The KB (or VCB) is determined from basic hull form equations (20) through (22).

9. *Height of longitudinal metacenter above keel (KM_L)*. This is shown in Chapter 5 to be

$$KM_L = KB + BM_L = KB + \frac{I_L}{\nabla} \quad (28)$$

All of these quantities have been defined previously.

10-13. *Coefficients of form*. The four principal coefficients of form, namely C_B, C_M, C_P, and C_W are sometimes plotted against draft in the curves of form. They have been defined in equations (1) through (5).

14. *Wetted surface (WS)*. As its name implies, the wetted surface is the total area of the shell of the ship that is wetted by the water in which it floats for any given draft. It is needed by the naval architect to determine the frictional resistance of the ship as described in Chapter 7. The calculation involves measuring wetted "half girths" of the stations from the body plan, that is, the lengths of the lines tracing each station from waterline to centerline, then applying special integration techniques to them over the ship's length. Wetted surface is not needed by ship's officers for any operational calculations, thus it is often omitted from the published hydrostatic curves.

CHAPTER THREE

STATIC EQUILIBRIUM AND STABILITY

A body at rest is said to be in static equilibrium. All bodies on earth have forces acting on them (their own weight, for example). The state of static equilibrium describes a condition in which all of the externally applied forces are balanced. Such a condition, first enunciated by Newton as part of his first law of motion, is the basis of the classical study of the statics part of the mechanics of rigid bodies. Here, we will apply the principles of statics to the special situation of a floating body.

ARCHIMEDES' PRINCIPLE AND THE LAW OF FLOTATION

To understand the nature of and conditions for static equilibrium of floating bodies, another physical law must be recalled. It is the *law of flotation,* which is a special case of Archimedes' principle. Simply put, Archimedes' principle states that a body immersed in a liquid is subject to an upward vertical force equal to the weight of the liquid displaced by the body. This upward force is called a *buoyant force,* or simply the *buoyancy* of the body.

Imagine a solid homogeneous body, suspended from a spring balance, being gradually lowered into a body of still water. Keeping in mind the conditions of static equilibrium and Archimedes' principle, we examine the force recorded by the spring balance at each stage of immersion of the body, as shown in Figure 3-1. Before immersion, there is no buoyant force, so equilibrium requires that the force measured by the spring balance is equal to the weight of the body, as at (b) in the figure. During partial immersion, at (c), there are three forces: the body's weight (W) acting downward, the buoyant force (B) acting upward, and, for equilibrium, the spring balance force (F) equal to W–B and acting upward. As

STATIC EQUILIBRIUM AND STABILITY

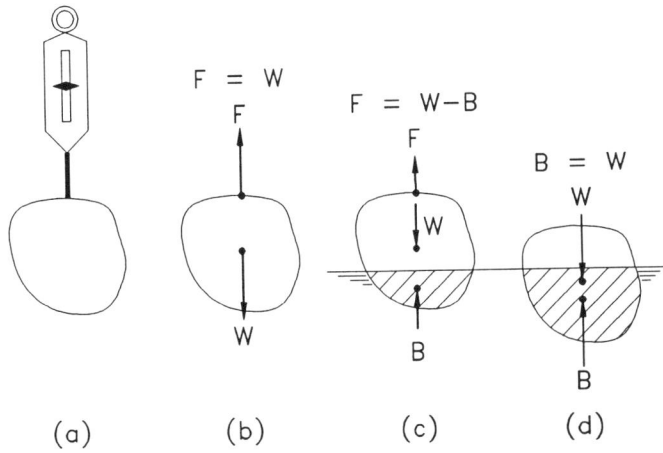

Figure 3-1. Archimedes' Principle.

immersion increases, so does the volume of water displaced and hence the buoyant force. Thus W–B decreases, since the weight is constant. Will the body float? That depends on how large B becomes before the body is totally immersed. If the increasing buoyant force becomes equal to the weight, as shown in (d) of the figure, the spring balance force will become zero, and the body will be in equilibrium with only the weight and buoyancy forces acting. That is, it will float. A very dense object (steel, stone) does not float because even when it is totally immersed the buoyant force is smaller than the weight.

It is now clear what the condition is for flotation in a particular liquid. It is that the body weigh less than the weight of a volume of the liquid that equals the volume of the body. If the body is homogeneous, it can be even more simply put. A homogeneous body floats in any liquid more dense (weight per unit volume) than the body, and sinks in a liquid less dense than itself. If the body, like a ship, is not homogeneous, the extent of immersion at which it will float (the ship's draft) can only be determined by knowing exactly the "below the waterline" volume of the hull (called *volume of displacement,* ∇, or immersed volume) for any draft, the ship's weight (called its *displacement,* Δ), and the density of the water. The fundamental relationship between draft and displacement for a given ship in seawater is thus reduced to a geometric determination of the volume of the hull up to any waterline. The hydrostatic calculations that determine this relationship have been described in Chapter 2.

Put into equation form, and using the symbol Δ instead of W for the displacement (weight) of the floating body, the above concepts can be expressed as follows:

50   APPLIED NAVAL ARCHITECTURE

$$\Delta = \rho g \nabla = \frac{\nabla}{\delta} \quad (1)$$

where Δ = weight of the body
        ρ = mass density of the water
        g = acceleration of gravity
        ∇ = volume of displaced water
        $\delta = \dfrac{1}{\rho g}$ = specific volume of the water
           = reciprocal weight density

Note that this single equation combines the two principles described above. The buoyant force acting on a partially immersed body is, according to Archimedes,

$$\text{buoyant force} = \rho g \nabla$$

while the requirement of static equilibrium of a floating body is

$$\text{buoyant force} = \text{gravitational force (weight)}$$

Equation (1) is thus an expression of a static balance of vertical forces. In this book, the symbol ρ represents the mass density of water (or any homogeneous substance) or mass per unit volume. The weight density is then expressed by ρg. The buoyant force, or weight of water displaced, is thus equal to the volume of displacement (∇) times the weight density (ρg). An alternative way to express the density is the specific volume, which is the reciprocal of the weight density, for which the symbol δ is employed in this book.

It is just as correct to describe this equilibrium condition as a balance of masses rather than weights. That is, just as the weight of the floating body equals the weight of water displaced, so the mass of the floating body equals the mass of water displaced. The term "displacement," which has a long history in the English language as a name for the weight of a ship, could as well be used to describe its mass. That is, in fact, the convention adopted in the International System of Units, called here the SI system. To keep things straight while gaining proficiency in solving problems in both the SI system and the traditional U.S. system (also called the "English" system), we adopt the following nomenclature in this book:

Δ will always refer to a displacement *weight*.
Δ_m will always refer to a displacement *mass*.

While it is possible to express displacement as either a weight or a mass in either the U.S. or SI measurement systems, and to convert units from one system to the

STATIC EQUILIBRIUM AND STABILITY

other, we shall for the most part keep to the traditional interpretations when dealing with each system. That is, ship mass will be used in connection with SI units and ship weight when dealing with U.S. units. Thus the SI equivalent to equation (1) is written:

$$\Delta_m = \rho \nabla \tag{1a}$$

where Δ_m = mass of the body
 ρ = mass density of the water
 ∇ = volume of displaced water

The units of measurement of the quantities described in equations (1) and (1a) are as follows:

U.S. units
 Δ = displacement (weight) in long tons (2,240 pounds)
 ρg = weight density in long tons per cubic foot
 δ = reciprocal weight density in cubic feet per long ton
 ∇ = displaced volume in cubic feet

SI units
 Δ_m = displacement (mass) in metric tons (1,000 kg)
 ρ = mass density in metric tons per cubic meter
 ∇ = displaced volume in cubic meters

Values of the densities of fresh and seawater in both systems of measurement are given in Table 3-1.

Table 3-1. Density of Seawater and Fresh Water

	U.S. Units ($g = 32.17\,ft/sec^2$)		SI Units ($g = 9.807\,m/sec^2$)	
	Fresh	Sea	Fresh	Sea
Mass Density	1.94	1.99 slugs/ft^3 (lb-sec^2/ft^4)	1000 1.000	1025 kg/m^3 1.025 MT/m^3
Weight Density	62.4	64.0 lb/ft^3	9.807	10.052 kN/m^3
Reciprocal Wt. Dens.	36*	35 ft^3/LT	0.1020	0.0995 m^3/kN
Specific Gravity	1.000	1.025	1.000	1.025

* 35.9 is more exact, but 36 is often used instead, as an easily remembered and acceptable approximation.

Tons Per Inch and Tons Per Centimeter Immersion

The relationships expressed by equations (1) and (1a) apply to portions of the immersed volume of a ship as well as to the entire volume of displacement. For example, when a weight is loaded on a ship, the buoyant force must increase by exactly the amount of the added weight in order that equilibrium be maintained. Also, the buoyant volume must increase by exactly the volume of water whose weight is equal to the weight added, thus satisfying the law of flotation. This notion is employed to define a particularly useful hydrostatic property, the *tons per inch immersion* or *tons per centimeter immersion*. These quantities are used to determine small changes in draft of a ship that result from loading or discharging relatively small weights.

If a weight is loaded so that the increase in draft is the same throughout the length of the ship, the condition is known as *parallel sinkage*. Parallel sinkage is illustrated in Figure 3-2. In order for an added weight to produce parallel sinkage, the weight must be loaded so that the resultant of the additional buoyancy created is collinear with the added weight as shown in the figure. For a thin parallel sinkage layer, the buoyant force (b) shown in the figure will be at the center of flotation of the ship's waterplane. The volume of the layer, or volume of added buoyancy, is

$$v = A_W t \qquad (2)$$

where A_W = area of the waterplane, ft² or m²
t = thickness of the layer, feet or meters

The buoyant force created by this layer of buoyancy is

$$b = \rho g v$$

and it must be equal to the added weight (w). That is,

$$w = b = \rho g v = \rho g A_W t \qquad (3)$$

The weight in equation (3) is the weight that will increase the draft of the ship by t feet.

Equation (2) is exact only if the ship's sides in the region of the layer are perfectly vertical. The ship is then said to be *wall-sided*. Typical ships are not wall-sided, but if the layer is relatively thin, say not more that a few feet on a ship of ordinary cargo ship size, the approximation of equation (2) is acceptable. If it is very thin, say one inch or one centimeter, the result is quite accurate. Equating the added weight to the buoyant force of a parallel sinkage layer of unit thickness

STATIC EQUILIBRIUM AND STABILITY 53

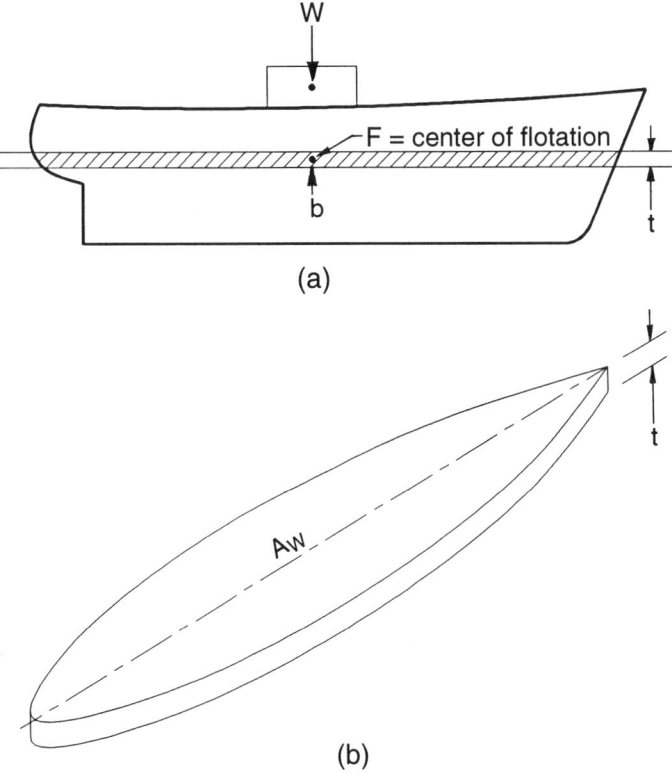

Figure 3-2. The parallel sinkage layer.

defines a quantity that might be called "tons per unit of immersion," which is useful to ships' officers during cargo loading or planning. In U.S. units, the traditional unit of immersion is the inch, and weights are expressed in long tons. Introducing these units into equation (2) defines the quantity called *tons per inch (TPI)*. Let t = 1 inch = 1/12 foot and, for seawater, $\rho g = 1/35$ tons/ft^3:

$$w = \text{TPI} = \rho g A_w t = \left(\frac{A_w}{35}\right)\left(\frac{1}{12}\right) \qquad (4)$$

$$\text{TPI} = \frac{A_w}{420}$$

In SI units, immersion is expressed in centimeters, and mass in metric tons is substituted for weight. Hence t = 1/100 meter and $\rho = 1.025$ MT/m^3. Thus *tons per centimeter (TPC)* is written:

54   APPLIED NAVAL ARCHITECTURE

$$\text{TPC} = \rho A_W t = (1.025 A_W)(\frac{1}{100}) \qquad (4a)$$

$$\text{TPC} = \frac{1.025 A_W}{100} = \frac{A_W}{97.56}$$

with A_W given in square meters. In the hydrostatic curves or charts provided with oceangoing ships, the seawater quantities defined above are given for any draft. The fresh water equivalents may be deduced by dividing TPI or TPC by 1.025. For brackish water, TPI or TPC must be divided by the water's specific gravity.

Effect of Water Density on Draft

Example 3-1
The barge shown in Figure 3-3 weighs 350 LT (long tons), and floats upright and on an even keel as indicated. At what draft (T) will it float in seawater? In fresh water?

The weight-volume relationship is given in equation (1):

$$\Delta = \rho g \nabla$$

in which Δ = 350 LT, and reciprocal density $\delta = 1/\rho g = 35$ ft^3/LT for seawater. Hence,

$$350 = \frac{\nabla}{35}$$

or
$$\nabla = 35 \times 350 = 12{,}250 \text{ ft}^3,$$
the volume of displacement in seawater

Geometrically, this volume consists of a triangular prism lower part plus a rectangular prism upper part, as follows:

Volume of triangular part = $(\frac{1}{2}) \times 6 \times 16 \times 100 = 4{,}800$ ft^3
Volume of rectangular upper part (to the waterline) = $(T - 6)(16 \times 100)$
= $1{,}600 (T - 6)$ ft^3

Thus $1{,}600 (T - 6) + 4{,}800 = 12{,}250$
and $T_S = 10.66$ feet in seawater.

In fresh water the reciprocal weight density is 35.9 ft^3/LT, hence

$$\nabla = 35.9 \times 350 = 12{,}565 \text{ ft}^3$$
and $T_F = 10.85$ feet in fresh water.

STATIC EQUILIBRIUM AND STABILITY

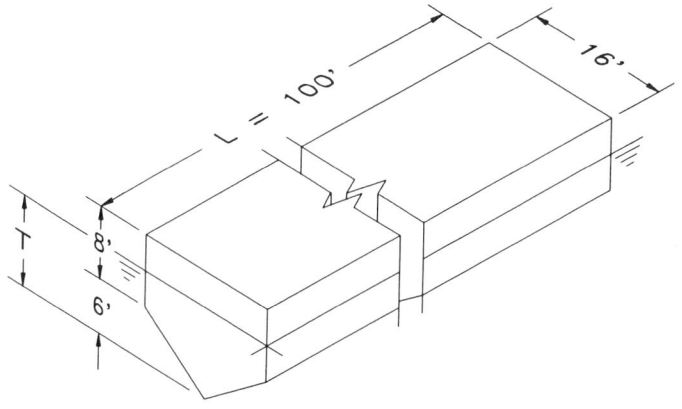

Figure 3-3.

This worked example illustrates a number of important things about the application of the principle of flotation. It shows, for example, that the barge floats at a deeper draft in fresh water than it does in seawater. This is to be expected, since the fresh water is less dense than the seawater, thus it requires a larger volume of fresh water to produce the 350 tons of buoyant force. The relationship between water density and draft is of very practical significance in ship operation. Maximum allowable drafts are assigned to every merchant ship in accordance with the regulations of the International Convention on Load Lines. The basic statutory load line marks the limit of the ship's draft in seawater. If the ship loads cargo in a fresh-water port, or where the water density is between that of fresh and seawater, it may legally be loaded so as to immerse the seawater load line so that upon entering sea water it will rise to the allowable waterline.

Determining the difference in drafts of the example barge in sea- and fresh water was done simply by repeating the volume calculation for each type of water. For a vessel of more complicated geometry, the procedure would not be so simple. The calculation is generalized by observing that the entire change in immersed volume when passing from seawater into fresh water takes place within a thin layer above the sea waterline. The geometry of that layer is approximated as a prism shaped like the waterplane on its top and bottom, and having vertical sides, like the parallel sinkage layer shown in Figure 3-2. We define the following:

∇_S = immersed volume in seawater
∇_F = immersed volume in fresh water
ρ_S = mass density of seawater

ρ_F = mass density of fresh water
A_W = area of the waterplane
t = thickness of the layer, or increase in draft

The volume of the layer may be expressed in two ways: as the fresh-water volume minus the seawater volume, and as the volume of a prism of area A_W and thickness t. Equating the two expressions:

$$\nabla_F - \nabla_S = A_W t \qquad (5)$$

Since the weight (or mass) of the barge remains constant,

$$\rho_F g \nabla_F = \rho_S g \nabla_S \qquad (6)$$

or

$$\nabla_F = \nabla_S \left(\frac{\rho_S}{\rho_F}\right) \qquad (7)$$

Substituting (7) into (5),

$$\nabla_S \left(\frac{\rho_S}{\rho_F}\right) - \nabla_S = A_W t$$

$$t = \frac{\nabla_S}{A_W}\left(\frac{\rho_S}{\rho_F} - 1\right) \qquad (8)$$

This equation is valid in either measurement system, giving a draft change in feet for the U.S. system and in meters for the SI system. Note also that since the densities appear only in the form of a ratio, either mass or weight densities may be used, or specific gravities, so long as they are consistent in numerator and denominator.

Equation (8) may be generalized to apply to any two water densities. Let subscript H refer to the water of higher density and subscript L to the lower-density water. The change in draft is:

$$t = \frac{\nabla_H}{A_W}\left(\frac{\rho_H}{\rho_L} - 1\right) \qquad (9)$$

Example 3-2
Verify the draft in fresh water of the barge of Example 3-1 by calculating its change in draft on moving from sea- to fresh water, using equation (8).

$$A_W = 100 \times 16 = 1{,}600 \text{ ft}^2$$
$$t = (12{,}250/1{,}600)(1.025 - 1) = 0.19 \text{ ft.}$$

Note that the seawater specific gravity has been used as the density ratio in this case, since the denominator density represents fresh water. Now add the draft increase to the seawater draft:

STATIC EQUILIBRIUM AND STABILITY

$$T_F = T_S + t = 10.66 + 0.19$$
$$T_F = 10.85 \text{ ft}$$

Example 3-3
A cargo liner 161 m long by 23.2 m beam floats at a pier where the mass density of the water is 1,010 kg/m³. It is desired to load this ship at the pier to a draft such that upon entering seawater the draft will be 8.75 m. The ship's hydrostatic curves (for seawater) show that, at a draft of 8.75 m, the displacement is 19,420 MT and the TPC is 27.62 MT/cm. To what draft should this ship be loaded at the pier?

Use subscript H for the seawater conditions and subscript L for the conditions at the pier. Then

$$\rho_H = 1.025 \text{ MT}/\text{m}^3 \qquad \text{[Table 3-1]}$$

$$\rho_L = 1,010 \text{ kg}/\text{m}^3 = 1.010 \text{ MT}/\text{m}^3 \qquad \text{[given]}$$

$$\nabla_H = \frac{\Delta_{mH}}{\rho_H} = \frac{19,420}{1.025} = 18,946 \text{ m}^3 \qquad \text{[Equation (1a)]}$$

$$A_W = 97.56 \times \text{TPC} = 97.56 \times 27.62 = 2,694.6 \text{ m}^2 \qquad \text{[Equation (4a)]}$$

$$t = \frac{\nabla_H}{A_W}\left(\frac{\rho_H}{\rho_L} - 1\right) \qquad \text{[Equation (9)]}$$

$$= \frac{18,946}{2,694.6}\left(\frac{1.025}{1.010} - 1\right)$$

$$= 0.104 \text{ m, say } 0.10 \text{ m}$$

That is, the draft at the pier may be 0.10 m more than the required seawater draft, so the ship may be loaded to a draft of

$$T \text{ at the pier} = 8.75 + 0.10 = 8.85 \text{ meters.}$$

EQUILIBRIUM OF A FLOATING BODY

In the preceding calculations, static balance of the floating vessels was achieved by equating the buoyant force acting upward to the weight force acting downward. That is, the resultant of the forces acting on the body was zero. For the body to be in equilibrium, it must also be shown that the resultant moment of the forces acting on the body is zero. Therefore we must know the positions of the lines of action of the forces as well as their magnitudes and directions. The two forces are completely described as follows:

58   APPLIED NAVAL ARCHITECTURE

1. The *weight* force acting *vertically downward* through the *center of gravity (G)* of the body.
2. The *buoyant* force acting *vertically upward* through the *center of buoyancy (B)* of the body.

The two conditions for equilibrium are:

1. The buoyant force is equal to the weight force. (Resultant force equals zero.)
2. The center of gravity and the center of buoyancy are in the same vertical line. (Resultant moment equals zero.)

The center of buoyancy is, as defined in Chapter 2, the centroid of the volume of displacement. Its position depends only on the geometry of that volume, so that its coordinates KB (height of B above keel) and LCB (longitudinal distance of B from reference station) are uniquely determined as hydrostatic properties. Because of the starboard-port symmetry of the hull, the TCB (transverse distance of B from centerline plane) of a ship when floating upright is zero.

A ship's center of gravity, on the other hand, cannot be determined from the hull geometry. The loaded ship consists of many discrete items (structure, machinery, equipment, cargo, fuel, supplies, etc.), each with its own weight and position in the ship. The center of gravity of such a system of weights is defined as that point through which the total weight of the system (sum of all individual weights) may be assumed to act for any determination of the static behavior of the system. It is also the center of mass of the ship. The unique property of the center of gravity is that the moment of weight (or the sum of the moments of each individual weight of the system) about any axis passing through it is zero. Like the center of buoyancy, the center of gravity is completely determined by its three coordinates with respect to the ship's principal planes:

KG = height of G above baseline, or keel
LCG = longitudinal distance of G from reference station
TCG = transverse distance of G from centerline plane

Procedures for calculating KG, LCG, and TCG of a ship are described later. For the special case of a homogeneous floating object such as a solid wood block, the center of gravity is at the geometric centroid of the object.

Figure 3-4 depicts a homogeneous rectangular block of wood 10 units long, 4 units wide, and 2 units thick of specific gravity 0.50. How many positions of floating equilibrium are there for this block if it is set afloat in fresh water? Without elaborate calculation, it is clear that since the block weighs just half as

STATIC EQUILIBRIUM AND STABILITY 59

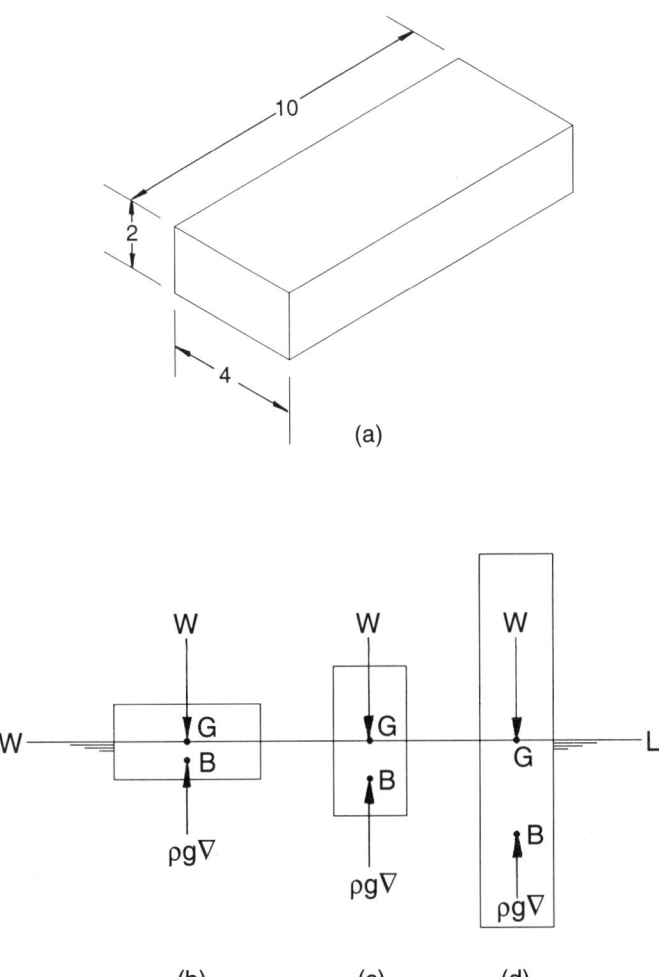

Figure 3-4. Equilibrium of half-immersed wood block.

much as an equal volume of fresh water, it must float exactly half immersed. The buoyant force ($\rho g \nabla$) will then be equal to the gravitational force or weight (Δ) and the net vertical force on the system will be zero. But we must also show that there are no net moments acting on the block. That is, of the infinite number of possible attitudes of half immersion, only those in which the center of gravity and center of buoyancy are in perfect vertical alignment will be equilibrium conditions. The center of gravity is in the geometric center of the block of wood, and

60   APPLIED NAVAL ARCHITECTURE

the center of buoyancy is in the geometric center of the immersed volume. A little thought will verify that there are six positions of the block that satisfy the "no-moments" requirement. They are illustrated in Figure 3-4(b,c,d), in which each of the attitudes shown represents two cases, since each may be inverted without changing the force relationships.

Although all six positions represent static equilibrium, they are not all attainable in nature. Our own experience and intuition about the way this simple board floats tells us that only the two cases shown in Figure 3-4(b) are possible. To understand why, we must examine the behavior of the body when it is disturbed slightly from each of its equilibrium positions.

Stable, Neutral, and Unstable Equilibrium
Static equilibrium conditions are classified into three states of equilibrium, depending on how the body reacts when it is displaced slightly by an external disturbance, which changes the forces acting on it. The equilibrium condition is said to be:

- *Stable* — if the new forces acting on it tend to return it to its original position.
- *Unstable* — if the new forces acting on it tend to increase the magnitude of the disturbed position.
- *Neutral* — if it is still in equilibrium in the disturbed position.

TRANSVERSE STABILITY

Our observations of how this board will float thus suggest that the position shown in Figure 3-4(b) is stable, while those shown in 3-4(c) and (d) are unstable. To verify that this is the case, we examine the forces involved as the body is disturbed from equilibrium. In this study, the disturbance is assumed to cause a small rotation of the body about a longitudinal axis. The associated stability is called *initial transverse statical stability*. The significance of the terms is:

- *Initial* — very small angle of rotation, body remains nearly upright.
- *Transverse* — rotation is in the transverse direction, that is, about the longitudinal axis. Such a rotation is known as "heel."
- *Statical* — velocity of rotation and inertial effects are ignored. Only the direction and magnitude of moments are of concern.
- *Stability* — tendency of the body to return to its initial equilibrium position.

Consider first the block in the equilibrium condition shown in Figure 3-5, that is, floating with the broadest surface (4 × 10) horizontal. Because of the symme-

STATIC EQUILIBRIUM AND STABILITY 61

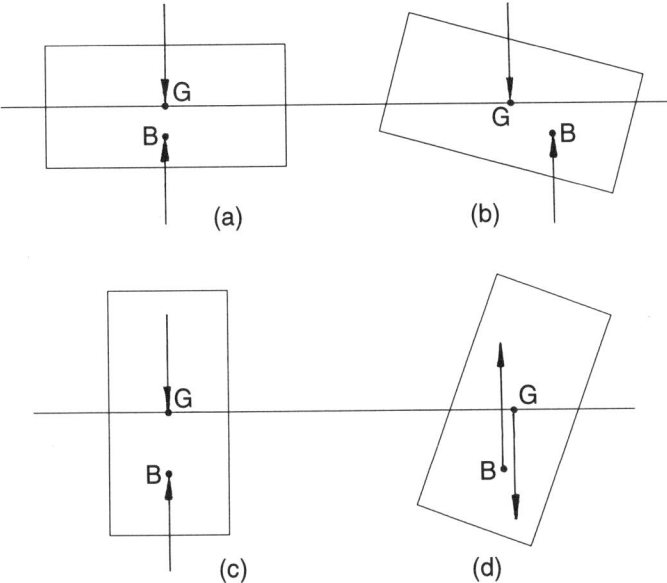

Figure 3-5. Test for stability.

try and homogeneity of the body, the centers of gravity and buoyancy are easily located at the geometric centers of the block and the immersed portion, respectively. When the block is given a small angle of clockwise heel, as in (b) of the figure, the center of buoyancy shifts toward the low side because of the changed geometry of the immersed volume from rectangular to trapezoidal. The center of gravity remains fixed. Both forces are unchanged in magnitude and direction, but the lines of action are not coincident, and a moment or couple is formed. The moment is, in this case, called a *righting moment* because its sense is opposed to the rotation imparted on the block by the initial disturbance, that is, it is counterclockwise. As soon as the disturbing force is removed, the block will return to its original equilibrium position, which is therefore called a state of stable equilibrium.

If the same test is made on the block floating with the broad side vertical, as in (c) and (d), the moment formed by the weight and buoyancy forces is in the same direction as the direction of initial heel—clockwise in the figure. It is thus called a *heeling moment,* and it tends to rotate the body away from the initial position and toward a new equilibrium position. The original position is thus an unstable equilibrium position. An unstable equilibrium position cannot be maintained in practice because it is quite impossible in nature to avoid even the slightest disturbance which will create a heeling moment. The unstable state applies to the block in the four equilibrium conditions depicted by Figure 3-4 (c) and (d).

62   APPLIED NAVAL ARCHITECTURE

A similar test of stability could have been made on this block by rotating it ever so slightly in the longitudinal direction, that is, by pushing down slightly on the end (rather than on the long edge) of the block when it was in the stable condition. It should be obvious that it will return to the original position when the force is removed. That is, the block is stable with regard to angular disturbances in the longitudinal direction as well. The experiment in that case would be a test of longitudinal stability, a subject to be addressed in Chapter 5.

A special stability situation arises in the case of a homogeneous circular cylinder—a dowel stick or telephone pole. As shown in Figure 3-6, the weight and buoyancy forces are collinear whether the cylinder is upright, with the spot marked top at 12 o'clock, or heeled, with the top spot at any other position. No amount of rotation will produce a heeling or righting moment, hence this object is in a state of *neutral equilibrium,* or *indifferent stability.* This is also the case for a homogeneous sphere—a croquet ball, for example. This cannot be demonstrated experimentally, as the experiment on the wood block can, because it is quite impossible to shape a cylinder or sphere with such precision that the center of gravity is in the exact center, and it is unlikely that the material would be perfectly homogeneous.

The Measure of Initial Transverse Stability

Having demonstrated the existence of stable and unstable states, we must now address the question of how much stability or instability a floating body possesses. Since a moment (righting or heeling) is produced in the heeled condition, the answer to this question lies in the magnitude of that moment. The displacement (weight) of the body is constant as it is heeled, therefore the moment depends on the distance separating the lines of action of the weight and buoyancy forces, hence ultimately on the path taken by the center of buoyancy as the body is heeled away from the upright condition.

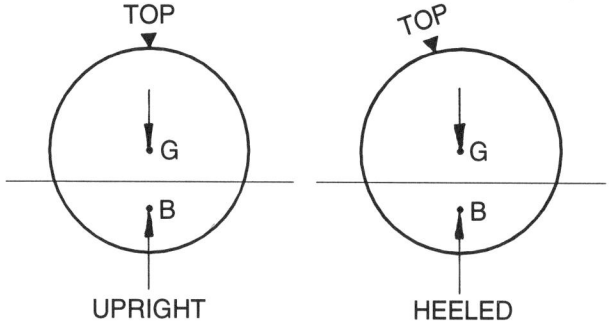

Figure 3-6. Neutral equilibrium.

STATIC EQUILIBRIUM AND STABILITY

To understand what takes place, imagine a ship floating in equilibrium as shown in Figure 3-7, with centers of gravity and buoyancy as indicated. It is then heeled to a small angle φ by an external force, that is, without changing the displacement or position of the center of gravity. The result, shown in (b), is that while the volume of displacement remains unchanged, its shape changes, and the center of buoyancy shifts in the direction that the ship was heeled. In the heeled condition, a horizontal line is drawn from G to intersect the line of action of the buoyant force at Z. This distance, GZ, is called the *righting arm,* and the moment $\Delta \times GZ$ is the *righting moment.* The ship was in stable equilibrium because the moment tends to right it. The path taken by the center of buoyancy from B to B′ as the ship heels is, for very small angles of heel, approximated by a circular arc. This is exactly true only if the angle is vanishingly small, but the approximation is satisfactory for normal ship forms up to angles of heel of about 10 degrees. For the study of initial transverse stability, we restrict the angle of heel to within that range, and the analysis that follows pertains only to small angles of heel. In the accompanying diagrams, exaggerated angles are drawn for clarity of the diagram.

The Metacenter and Metacentric Height

Note the geometry of the diagram in Figure 3-7(b). The ship's centerline, perpendicular to upright waterline WL, represents the line of action of the buoyant force in upright condition (a). In heeled condition (b), the buoyant force upward through new center of buoyancy B′ is perpendicular to inclined waterline W′L′. The two lines must therefore intersect at the point labeled M, and the angle between them is φ, the (small) angle of heel. The point M, called the *transverse metacenter,* is defined as the intersection of the upright condition buoyancy vector and the buoyancy vector for a very small angle of heel, the rotation having taken place at constant displacement. So long as the path of the center of buoyancy for successive small angles of heel is a circular arc, it is clear that all such

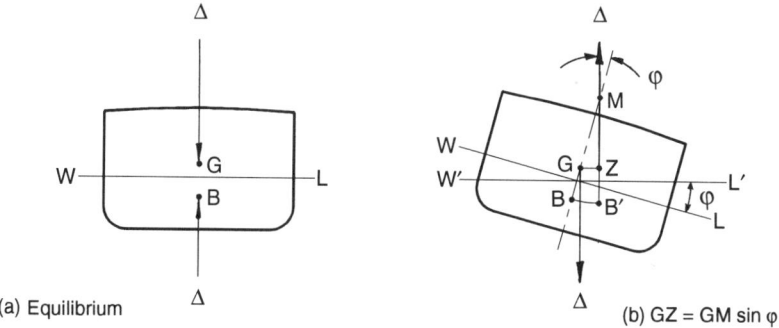

(a) Equilibrium　　　　　　　　　　　　　(b) GZ = GM sin φ

Figure 3-7. The righting arm (GZ).

buoyancy vectors will intersect at the point M, which is thus the center of the circular path traced by the center of buoyancy as the ship heels through small angles. The radius of the circle is BM, called the *metacentric radius*. The relative positions of G and M determine the magnitude of the righting arm, since

$$GZ = GM \sin \varphi \qquad [\varphi \to 0] \qquad (10)$$

in which GM is called the *metacentric height*. As the geometry of the figure shows, the righting arm, and hence the stability of the ship, depends directly on the metacentric height. If G is below M, as in the figure, GZ is a righting arm. If G and M coincide, GZ is zero, and the ship is in a condition of neutral stability. If G is above M, GZ is a heeling, or capsizing arm, and the ship is in an unstable condition.

In order to evaluate a ship's initial stability numerically, we must know the locations of both G and M. The location of G depends on the distribution of weights throughout the ship, that is, on the loading of the ship. The calculation of its location will be described later. The location of the metacenter (M) depends on the form, or geometry of the ship, since that is what influences the path taken by the center of buoyancy as the ship heels. An examination of the mechanics of the transference of buoyancy due to heeling reveals how the position of M is determined.

When the ship shown in Figure 3-8(a) is heeled to small angle φ, a long thin wedge-shaped volume emerges from the water on the "high side," while a similarly shaped volume immerses into the water on the "low side." The buoyancy associated with the emerged wedge is transferred from the centroid of that wedge to the centroid of the immersed wedge. This shift of buoyancy causes a shift in the whole ship's center of buoyancy. Assuming that the transfer takes place with no change in displacement, that the angle of heel is very small, and that the sides of the ship are wall-sided throughout the small thickness of the wedges, a typical cross section of the ship can be depicted as in (b) of the figure. We define:

$$\begin{aligned}
\text{WL} &= \text{the waterline when upright} \\
\text{W'L'} &= \text{the waterline when heeled} \\
y &= \text{the half-breadth of the ship's waterplane} \\
b_e &= \text{the centroid of the emerged wedge} \\
b_i &= \text{the centroid of the immersed wedge}
\end{aligned}$$

The other symbols are as previously defined. Since the waterplane half breadths vary with position in the ship, we must first determine the characteristics of an elemental slice of volume of length dx, then integrate over the ship's length. The elemental wedge volumes are equal to the product of wedge area and length:

STATIC EQUILIBRIUM AND STABILITY

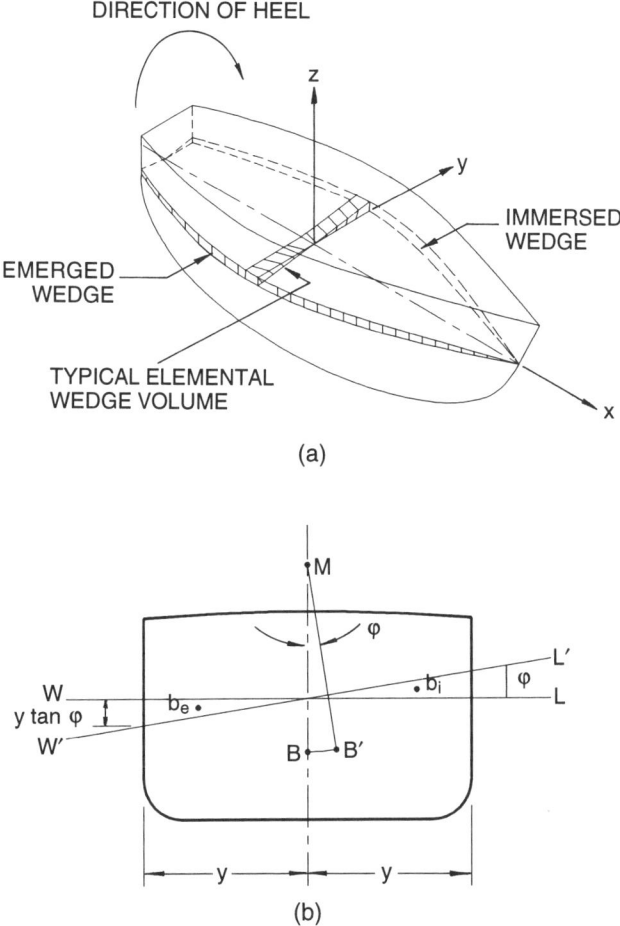

Figure 3-8. The mechanics of transference of buoyancy due to heeling.

$$dv_e = dv_i = \tfrac{1}{2} y \, (y \tan \varphi) \, dx \tag{11}$$

This buoyant volume is transferred from point b_e to b_i, which is, in the limit as φ approaches zero, perpendicular to the ship's centerline, and equal to

$$\overline{b_e b_i} = \tfrac{4}{3} y \tag{12}$$

Thus the moment of transference of buoyant volume from the emerged to the immersed side is

66   APPLIED NAVAL ARCHITECTURE

$$dm = \overline{b_e b_i} \; dv = (\tfrac{4}{3} y)(\tfrac{1}{2} y^2 \tan\varphi \, dx) = \tfrac{2}{3} y^3 \tan\varphi \, dx \tag{13}$$

For the whole ship moment of transference, this moment of elemental volume must be integrated over the length of the ship:

$$v \, \overline{b_e b_i} = (\tfrac{2}{3}) \int_0^L y^3 \, dx \, (\tan\varphi) \tag{14}$$

It was shown in Chapter 2 that $(\tfrac{2}{3}) \int_0^L y^3 \, dx$ is the transverse moment of inertia of the ship's waterplane (I_T). Therefore the moment of transference of buoyancy may be written:

$$v \, \overline{b_e b_i} = I_T \tan\varphi \tag{15}$$

In response to this moment, the whole ship's center of buoyancy, that is, the centroid of the volume of displacement (∇) shifts a distance $\overline{BB'}$ in a direction parallel to $\overline{b_e b_i}$ such that the response moment equals the moment of transference. That is,

$$\nabla \, \overline{BB'} = v \, \overline{b_e b_i} = I_T \tan\varphi \tag{16}$$

and from Figure 3-8(b),

$$\overline{BB'} = BM \tan\varphi \tag{17}$$

Thus:

$$\nabla \, BM \tan\varphi = I_T \tan\varphi$$

$$BM = \frac{I_T}{\nabla} \tag{18}$$

In words, equation (18) states that the transverse metacentric radius equals the transverse moment of inertia of the waterplane divided by the volume of displacement.

A number of important inferences may be drawn from this relationship. It shows that the location of the transverse metacenter of a ship is entirely determined by the geometry, or form, of the immersed hull. For a given ship, the height of the metacenter above the keel (KM) is:

$$KM = KB + BM \tag{19}$$

and since both KB and BM are properties of ship form, KM is also. Because the metacenter represents the limiting position of the ship's center of gravity for the

STATIC EQUILIBRIUM AND STABILITY

ship to be stable, knowing the KM at any draft gives the operator of the ship guidance as to how one must load the ship to keep its center of gravity below the metacenter. We note also that the moment of inertia of the waterplane governs, in large part, the height of the metacenter, and therefore the stability of the ship. Because the beam of the ship has such a strong influence on I_T (it is proportional to the *cube* of the beam), it now comes as no surprise that a beamy, or wide, floating object is more stable than a narrow one. Equation (11) thus verifies our intuitive notion that a board, for example, will float with its broad side horizontal because it has then maximized its waterplane moment of inertia, and thus also the height of its metacenter and the magnitude of its stability.

In the following example, the stability of the wood block previously examined is calculated.

Example 3-4

Show that the $2 \times 4 \times 10$-inch board of specific gravity 0.50 shown in Figure 3-4 is stable in condition (b) and unstable in condition (c) of that figure. To do so, calculate the metacentric height (GM) for each condition.

Using ship nomenclature for our floating board, note that in condition (b) we have L = 10, B = 4, D = 2, and T = 1 inch for the length, beam, depth, and draft, respectively. In condition (c), the corresponding dimensions are L = 10, B = 2, D = 4, and T = 2. Keel point K, by definition at the intersection of the baseline and centerline, is the reference point for measuring vertical dimensions. For case (b), we have:

$$I_T = \frac{LB^3}{12} = \frac{10 \times 4^3}{12} = 53.33 \text{ in}^4$$

Note that this simple formula for I_T applies only to the rectangular waterplane. It can be verified by formal integration (see equation [14]), putting y = B/2, a constant, for the half breadth. That is,

$$I_T = \tfrac{2}{3}\int_0^L y^3 \, dx = \tfrac{2}{3}(\tfrac{B}{2})^3 \int_0^L dx = \tfrac{1}{12} B^3 L = \frac{LB^3}{12}$$

$$\nabla = LBT = 10 \times 4 \times 1 = 40 \text{ in}^3 \quad \text{(rectangular "hull")}$$

$$BM = \frac{I_T}{\nabla} = \frac{53.33}{40} = 1.33 \text{ in}$$

Of course, for this simple geometry, we can derive a special formula for BM, namely:

$$BM = \frac{I_T}{\nabla} = \frac{\tfrac{1}{12}LB^3}{LBT} = \frac{B^2}{12\,T}$$

Caution should be exercised in the use of this attractively simple formula for BM. It is valid only for the special case of a hull (or block of wood) that is completely rectangular. BM for a ship's hull must be determined by calculating I_T and ∇ separately by numerical integration as shown in Chapter 2. Similarly, the height of the center of buoyancy is calculated by simple formula for this rectangular hull,

$$KB = \frac{T}{2} = \frac{1}{2} = 0.5 \text{ in}$$

Because the block is homogeneous, the center of gravity is at its geometric center:

$$KG = \frac{D}{2} = 1.0 \text{ in}$$

The metacentric height can now be calculated.

$$GM = KM - KG = KB + BM - KG = 0.50 + 1.33 - 1.00 = 0.83 \text{ in}$$

Since G is below M, the GM is positive, and the block is stable. For condition (c), the calculations are as follows:

$$BM = \frac{B^2}{12\,T} = \frac{2^2}{12 \times 2} = 0.17 \text{ in}$$
$$KB = \frac{T}{2} = \frac{2}{2} = 1.00 \text{ in}$$
$$KG = \frac{D}{2} = \frac{4}{2} = 2.00 \text{ in}$$
$$GM = KB + BM - KG = 1.00 + 0.17 - 2.00 = -0.83 \text{ in}$$

Thus the GM is negative, M is below G, and the block is unstable in this condition. Direct comparison of the corresponding quantities in the two conditions shows why the results are so different.

Case (b)		Case (c)
0.50	KB	1.00
1.33	BM	0.17
1.83	KM	1.17
1.00	KG	2.00
+ 0.83	GM	− 0.83

Note that the metacentric radius (BM) is much smaller in case (c). The narrow "beam" of the block is the culprit. Had we calculated I_T separately, we would get

STATIC EQUILIBRIUM AND STABILITY

$$I_T = \left(\frac{1}{12}\right) \times 10 \times 2^3 = 6.67 \text{ in}^4$$

compared to 53.33 in^4 for condition (b). The volume of displacement is the same for both conditions, so the beam *at the waterline* is seen to be very influential in maintaining adequate transverse stability.

The above example demonstrates the influence of certain dimensions and proportions on stability, without requiring extensive numerical calculations to determine the basic elements of stability—KB, BM, and KG. A ship, by contrast, has a more complicated form, so the geometric elements (KB, I_T, ∇, BM, and KM) cannot be determined from simple equations. They are, instead, precalculated as described in Chapter 2 for all likely drafts of the ship and are displayed as hydrostatic properties, in either graphical or numerical form. Having a plot of KM against draft obviates much of the calculation in the above example, and replaces it with a single readout at the given draft.

The Center of Gravity

Although the position of the metacenter is fixed by the ship's geometry, the center of gravity is not. Its position depends on the distribution of weights aboard the ship, including the ship itself and everything it carries. Since the ship and its loading are not homogeneous, its center of gravity is not determined from its geometry as it was for the piece of wood. Calculating the center of gravity of a system of discrete weights requires that we know two things about each weight:

- the magnitude of the weight
- the location of the center of gravity of the individual weight within the system of weights (the ship)

To locate each weight in the ship requires three coordinates of its center of gravity with respect to the ship's principal planes, as previously defined for the whole ship center of gravity. That is, each weight has a vertical, longitudinal, and transverse position within the ship. For the present purpose of determining the KG of the ship, we need only to identify the vertical coordinate of each weight, that is, its kg (lowercase letters will mean individual weights, uppercase means the whole ship). The longitudinal coordinates will be needed later for trim calculations, and the transverse coordinates for calculations of list caused by off-center loading.

Once the weight (w) and vertical center of gravity (kg) of each item is known, calculating the whole ship weight (Δ) and vertical center of gravity (KG) is done simply by equating the moment of Δ to the sum of the moments of all of the w's about the keel, or baseplane.

$$\Delta \times KG = \Sigma (w_i \times kg_i) \qquad (20)$$

$$KG = \frac{\Sigma (w_i \times kg_i)}{\Delta} \qquad (21)$$

where Δ = total weight (ship displacement) = Σw_i
 i = 1, 2, 3, . . . to total number of individual weights in the system (ship + loading)

If the ship displacement and the items to be loaded are known as masses, as in SI system measurements, the same formulas apply by substituting Δ_m for Δ and individual masses for the w_i's. The calculations required to execute equation (21) are very simple but quite voluminous, because of the enormous number of items that make up a ship and its loading. To systematize this computation, the weights are classified into groups according to type, and standard forms and computer program files have been devised to assist the person who must do this tedious repetitive task.

Classification of Weights
During the process of ship design, weights and centers of the components of the ship must be estimated so that the dimensions and proportions of the ship can be adjusted as necessary to provide adequate stability when the ship is built and loaded to its designed capacity. Standard weight classifications are usually followed by the ship designer. A typical classification system for merchant ships is published by the U.S. Maritime Administration, for example. Other agencies and ship designers devise and use similar standard formats. Broad weight classifications are defined, and they are then broken down into many subcategories, right down to the individual structural elements and items of machinery and equipment to be installed. The broadest classification of weight groups for the ship itself is:

- *Hull steel* — includes all ship structural elements.
- *Propulsion machinery* — engines, gears, shaft, propeller, and all auxiliary machinery and equipment associated with the propulsion plant.
- *Outfit* — equipment for hotel services, navigation, cargo handling, etc., furnishing, joiner work, deck coverings, paint, for example.

These three categories contain all of the items that are a permanent part of the ship exclusive of loading of any kind, such as cargo, fuel, personnel, fresh water, stores, seawater ballast, etc. The condition of the ship without loading is called the *light* condition, and its weight is called the *light ship weight,* or *light ship displacement.* Its mass is the *light mass.*

STATIC EQUILIBRIUM AND STABILITY

Example 3-5
A cargo ship 150 meters long has masses and vertical centers as follows:

Steel	4,402 MT at	7.7 m above keel
Machinery	889 MT at	5.7 m above keel
Outfit	1,859 MT at	12.8 m above keel

Determine this ship's light mass and KG.

The solution is best organized into a table to calculate and sum the masses and their moments about the keel:

	Mass	KG	Moment
Steel	4,402	7.7	33,895.4
Machinery	889	5.7	5,067.3
Outfit	1,859	12.8	23,795.2
Light ship	7,150		62,757.9

In this table, each moment is the product of the corresponding mass and its kg. Thus the sum of the moments, 62,757.9 MT-m, is equal to the moment of the light ship about the keel. The sum of the mass column is clearly the light mass of the ship.

$$\text{Light mass} = 7{,}150 \text{ MT}$$

$$KG = \frac{62{,}757.9 \text{ MT-m}}{7{,}150 \text{ MT}} = 8.78 \text{ m}$$

Given any set of discrete masses or weights, the calculation of total mass or weight and the position of its center of gravity is done like the above example. In fact, the mass and kg of each of the groups above (steel, machinery, outfit) was determined by a similar calculation into which the masses and kg's of hundreds (sometimes thousands) of individual items listed in each group were entered.

The light ship displacement and KG are never changed unless the ship undergoes a conversion that involves adding, removing, or moving about any of the items that are parts of the light ship. Light ship displacement and KG are therefore treated as constants when calculating the displacement and KG of a loaded ship.

The weights carried by the ship, on the other hand, can vary considerably, thereby changing the ship's stability. Fuel and fresh water are consumed continuously, and cargo weights change from voyage to voyage, while some weights, like crew and stores, change very little or very infrequently. It is the responsibility of the ship's officer in charge of cargo to calculate the stability of the ship and to plan its loading so that sufficient stability will be maintained as each voyage progresses. To assist in this task, the ship designer prepares a *capacity plan* and a *trim and stability booklet* which are placed aboard ship. Among other things, these documents list in tables and show in drawings the essential characteristics of all compartments, tanks, and spaces into which items of ship loading may be placed. These characteristics include the name, location, and cubic capacity of each space, plus the vertical and longitudinal coordinates of the geometric centroid of the space.

Sample pages from the trim and stability booklet of a Mariner class dry cargo vessel, designated as a C4-S-1a by the U.S. Maritime Administration, are included in Appendix A. Some of the worked example problems and practice problems following each chapter refer to Mariner class ships. The solution of many of those problems requires the use of the hydrostatic data and loading table data given in Appendix A.

Using these documents and lists of cargo consignments and fuel to be loaded, the officer classifies the weights into categories, plans where each item will be loaded, and calculates the resulting weight or mass and KG (also LCG, to be treated later) of the loaded ship. The broad categories of variable loads carried are as follows:

- *Cargo* — also called cargo deadweight or "payload." This category is frequently subdivided by type, such as general dry cargo, liquid cargo, container cargo, refrigerated cargo, etc., according to the particular design of the ship.
- *Fuel oil*
- *Fresh water*
- *Salt-water ballast* — this is the variable ballast, carried only when the ship is so lightly loaded with cargo that a deeper draft is needed. Fixed (not removable) ballast, installed in some ships to provide adequate stability, is included in the light ship weight.
- *Crew and effects*
- *Stores*

The sum of all of the above variable weights is properly called the *total deadweight*. Total deadweight may also be defined as the difference between the load displacement and the light ship displacement, in either weight or mass units. It is often referred to as simply "the deadweight," a practice that should be avoided because the same term is often used to mean cargo deadweight, or payload. Nei-

STATIC EQUILIBRIUM AND STABILITY 73

ther should the terms cargo capacity and cargo deadweight be used as equivalent terms, because cargo capacity is a measure of the volume (cubic feet, cubic meters) available for cargo, while cargo deadweight is measured in tons.

Calculation of Metacentric Height

The following example illustrates how the metacentric height is determined for a ship in a particular loading condition. Note that the calculation of KG involves considerable effort, while KM is determined simply by reading its value from the hydrostatic curves at the final displacement or draft. This is in contrast to the calculations described in Example 3-4 for the homogeneous wood block, in which KG was determined by inspection, but KM required considerable calculation because hydrostatic curves for the piece of wood had not been worked out in advance.

Example 3-6

A Mariner class ship has a lightship displacement of 7,675 tons and KG of 31.5 feet. On a particular voyage, a summary of the loading of the ship is given in the following table. Using this information and the data from the Mariner trim and stability booklet in Appendix A, determine the ship's mean seawater draft, KM, KG, and GM.

Items loaded	Tons	Feet above keel
General dry cargo	8,520	27.9
Reefer cargo	355	26.6
Fuel oil	2,456	5.4
Fresh water	257	22.9
Crew, stores, lube oil	63	40.0

The loaded displacement and KG are determined by a "weights and moments" calculation that combines the loaded items with the light ship, as follows:

Item	Weight	KG	Moment
Light ship	7,675	31.5	241,763
Dry cargo	8,520	27.9	237,708
Reefer cargo	355	26.6	9,443
Fuel	2,456	5.4	13,262
Fresh water	257	22.9	5,885
Crew, etc.	63	40.0	2,520
	19,326		510,581

Loaded displacement = Δ = 19,326 tons
KG = 510,581 / 19,326 = 26.42 feet

74   APPLIED NAVAL ARCHITECTURE

The mean draft in seawater is determined by entering the hydrostatic properties (Appendix A) at the displacement of 19,326 tons. The draft at that displacement is seen to be 27'8". From the same diagram at this draft, the KM is 31.15 feet. The metacentric height in the loaded condition is therefore:

$$GM = KM - KG = 31.15 - 26.42 = 4.73 \text{ feet}$$

HOW CHANGES IN LOADING AFFECT STABILITY

One of the outcomes of the relationship between moments of individual weights and that of a system of weights described by equation (21) is that if any weight is shifted, or added, or removed from a system in equilibrium, the equilibrium will be disrupted and the system will adjust itself to a new and different equilibrium condition. If cargo (or any other weight) is shifted about, loaded on, or discharged from a ship, some change *must* take place in the ship's weight or the position of G, or both, as the ship regains equilibrium in the new loading condition.

Shifting a weight causes the center of gravity to shift in a direction parallel to the direction of the weight shift, with no change in the displacement of the ship. The shift in G will be such that the moment of the shift in the *ship's* weight equals the moment caused by the *shifted* weight.

$$\Delta \times GG' = w \times d$$

$$GG' = \frac{w\,d}{\Delta} \qquad (22)$$

where: w = weight (or mass) of the shifted item
        d = distance it is shifted
        Δ = weight or mass of ship (includes w)
        GG' = distance moved by ship's G

This expression of regained equilibrium is valid for any direction of movement of the weight. The direction of movement of G is always parallel to "d" and in the same sense. For now, we are concerned only with vertical movement. Note the similarity of this expression to equation (16), which dealt with the shift in the ship's center of buoyancy as a result of a shift in a small buoyant volume across the ship.

Example 3-7
A ship floats in seawater at a draft of 21'6". At this draft, the hydrostatic curves give a displacement of 13,620 long tons and KM = 30.6 feet. KG is 28.3 feet. Deck cargo weighing 220 tons is shifted from the deck to the lower hold, a dis-

STATIC EQUILIBRIUM AND STABILITY

tance of 28 feet downward. Determine the draft, displacement, KG, and GM after the shift.

$$GG' = \frac{w\,d}{\Delta} = \frac{220 \times 28}{13,620} = 0.45 \text{ ft downward}$$

$$KG = 28.30 - 0.45 = 27.85 \text{ ft}$$

Since no weight was loaded or discharged, the draft, displacement, and KM remain unchanged. The final metacentric height is

$$GM = KM - KG = 30.60 - 27.85 = 2.75 \text{ ft}$$

Loading or discharging a weight involves changes in the displacement as well as the KG. Therefore the draft and the KM will also change. When a single weight is loaded or discharged, however, equation (22) can be used to determine the resulting shift of G by applying the following interpretations:

- Δ is the displacement of the ship *after* loading or discharging the weight.
- d is the vertical distance from the original G to the weight loaded or discharged.
- G moves *towards* an added weight and *away from* a discharged weight.

Example 3-8
After the cargo shift on the ship of Example 3-7, cargo weighing 640 tons is loaded on the deck, at 40.5 feet above the keel. What is the resulting KG?

$$w = 640 \text{ tons}$$
$$d = 40.5 - 27.85 = 12.65 \text{ ft (weight above G)}$$
$$\Delta = 13,620 + 640 = 14,260 \text{ tons } (after \text{ loading w})$$
$$GG' = \frac{640 \times 12.65}{14,260} = 0.57 \text{ ft upward}$$

The shift is upward, *toward* the added weight. Therefore the final KG is

$$KG = 27.85 + 0.57 = 28.42 \text{ ft}$$

While equation (22) can be used as in the above example to determine a new KG after loading or discharging a single weight, it is not recommended if more than one weight is involved. Nor is it the only approach for a single weight. The alternative, and completely general, procedure for any number of weights is the relationship expressed by equation (21), treating the ship before loading or discharging cargo as a single weight. The problem in Example 3-8 could have been done as follows, for example.

Example 3-9
Repeat the KG calculation of Example 3-8, using the general principle of moments expressed by equation (21).
The solution in tabular form is:

	Weight	KG	Moment
Ship before loading	13,620	27.85	379,317
Cargo loaded	640	40.5	25,920
Loaded ship	14,260		405,237

$$\text{Loaded KG} = \frac{405,237}{14,260} = 28.42 \text{ ft}$$

Note that any number of weights could have been loaded or discharged by extending the above table to include a line for each weight. Discharged weights would be entered as negative weights, and would therefore have negative moments as well.

THE EFFECT OF OFF-CENTER WEIGHTS

If, on a ship floating in stable equilibrium and perfectly upright, a weight is shifted laterally, the center of gravity of the ship will shift laterally in accordance with equation (22). The shift in the center of gravity in a lateral direction will not change the metacentric height, as a vertical shift does. But the equilibrium of the system will be altered, and the ship must respond to regain equilibrium. What is the nature of the ship's response?

Figure 3-9 shows what happens. The lateral shift of the weight causes the center of gravity of the ship to move perpendicular to the ship's centerline from G to G', as shown at (a). The distance moved is GG' = wd/Δ, as in equation (22). This disturbs the ship's equilibrium, as shown in (b). The ship responds by heeling to an angle φ such that the center of buoyancy moves to position B', vertically below G'; see Figure 3-9(c). So long as the angle of heel is small, say less than about 10 degrees, the path taken by the center of buoyancy is a circular arc as described in the analysis of the expression for the metacentric radius. The buoyant force vector then passes through the metacenter, as in (d), and B'G'M form a straight vertical line. As shown in (d), the angle of heel can be written:

STATIC EQUILIBRIUM AND STABILITY

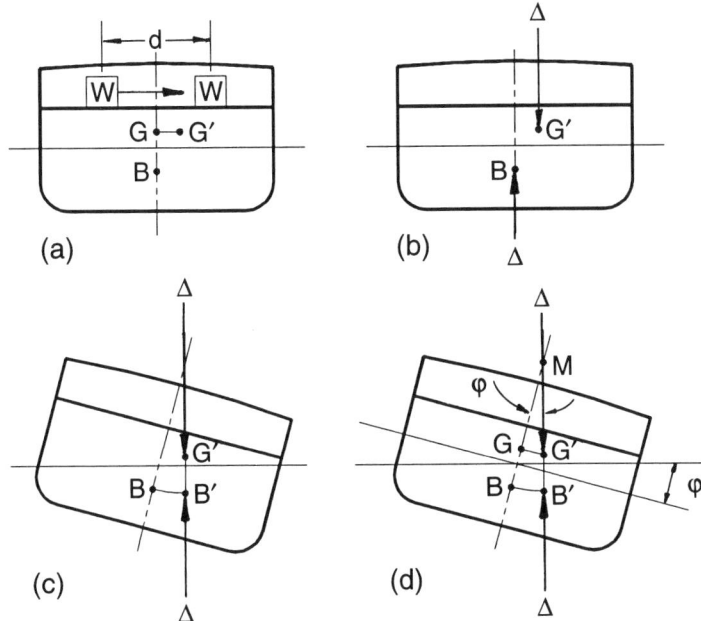

Figure 3-9. Effect of lateral shift of weight.

$$\tan \varphi = \frac{GG'}{GM} = \frac{wd}{\Delta\, GM} \quad [\varphi \text{ small}] \qquad (23)$$

This relationship establishes the angle of heel caused by weights shifted laterally on a ship. It is also applicable to weights loaded at or discharged from off-center positions on the ship, so long as the effects of the loaded or discharged weights on the displacement (Δ), the KG, KM, and GM are determined first before solving equation (23). That is, Δ and GM in equation (23) apply to the ship after loading is completed.

Example 3-10

A ship displacing 28,000 tons has KG = 36.0 feet and KM = 42.0 feet. A weight of 360 tons is loaded on deck at 62 feet above the keel and 21 feet to starboard of centerline. Determine the angle of heel caused by the off-center weight. Assume KM remains unchanged as the weight is loaded.

As shown in Figure 3-10, the loading is modeled as if the weight had been (a) loaded at the original center of gravity, then (b) shifted upward by distance d_1 to 62 feet above the keel on centerline, then (c) shifted to starboard by distance d_2 to its final position.

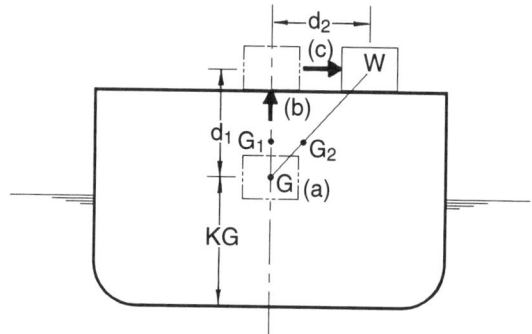

Figure 3-10.

(a) $\Delta = \Delta_{orig} + w = 28{,}000 + 360 = 28{,}360$ tons

(b) $GG_1 = \dfrac{wd_1}{\Delta} = \dfrac{360\,(62-36)}{28{,}360} = 0.33$ ft (up)

$KG_1 = KG + GG_1 = 36.00 + 0.33 = 36.33$ ft

$G_1M = KM - KG_1 = 42.00 - 36.33 = 5.67$ ft

(c) $\tan \varphi = \dfrac{wd_2}{(\Delta\ G_1M)} = \dfrac{360 \times 21}{28{,}360 \times 5.67} = 0.0470$

$\varphi = 2.69$ degrees

Several points are worthy of note. The calculation of KG_1 could have been done alternatively in tabular form, as in Example 3-9. Also, it should be noted that the center of gravity moved precisely toward the position of the added weight; that is, G_2 lies on the line joining G and w. This can be verified by calculation:

$$G_1 G_2 = \dfrac{wd_2}{\Delta} = \dfrac{360 \times 21}{28{,}360} = 0.267\,\text{ft}$$

Then show that triangle GG_1G_2 is geometrically similar to the triangle formed by the three positions of the weight:

$$\dfrac{GG_1}{G_1G_2} = \dfrac{0.330}{0.267} = 1.24$$

$$\dfrac{d_1}{d_2} = \dfrac{26}{21} = 1.24$$

Thus the principle that the center of gravity of a system of weights moves directly toward an added weight is verified.

STATIC EQUILIBRIUM AND STABILITY

The relationship between moments produced by laterally shifted weights and angle of heel described by equation (23) is accurate only for angles of heel small enough that the buoyancy vector passes through the metacenter as the ship heels. If the angle is very small, say less than two degrees, the results are quite accurate. Because of this, equation (23) in modified form is put to use for a very important measurement at the completion of construction of every new ship. It is used to determine very precisely the position of the center of gravity of the ship.

The Inclining Experiment

The procedure is known as the inclining experiment, and it is required by international regulations that every passenger and cargo ship be inclined when its construction is completed to determine the light ship weight (displacement) and position of its center of gravity. To that end, equation (23) is solved for GM, and written:

$$\text{GM} = \frac{wd}{\Delta \tan \varphi} \qquad (24)$$

In this form, it is known as the *inclining experiment formula*. In order to apply it to measurements taken on the actual ship, the procedure, illustrated in Figure 3-11, consists of placing onboard a special inclining weight (w), shifting it in a direction precisely laterally (perpendicular to the ship's centerline plane) through a carefully measured distance (d), allowing the ship to heel (incline) and regain equilibrium in the inclined condition, then measuring the angle of inclination (φ) as precisely as possible.

A simple, reliable device for measuring the angle of inclination is a pendulum, which is indicated diagrammatically in the figure. The pendulum consists of a plumb line, that is, a filament or light wire with a weight tied on one end. The line is suspended so that it hangs freely. A horizontal board called a batten is erected so that the pendulum line hangs close to it. When the pendulum swings as the ship is heeled, the initial and final positions of the line are marked on the batten so that the pendulum deflection, distance (a) in the figure, can be measured. The pendulum length (l), from the point of suspension to the upper edge of the batten, is also measured. Thus the triangle traced by the pendulum and batten is geometrically similar to the triangle GG'M. That is, $\tan \varphi = a/l$.

Draft measurements and water density measurements taken at the time of the inclining experiment enable an accurate determination of the displacement of the ship (Δ). The metacentric height (GM) determined in this manner is called the "as-inclined" metacentric height. It is subtracted from the "as-inclined" KM, determined from the ship's hydrostatics or calculated by integration of the displaced hull form, to determine the "as-inclined" KG. The longitudinal center of gravity (LCG) is also calculated, using procedures to be described in Chapter 5.

80   APPLIED NAVAL ARCHITECTURE

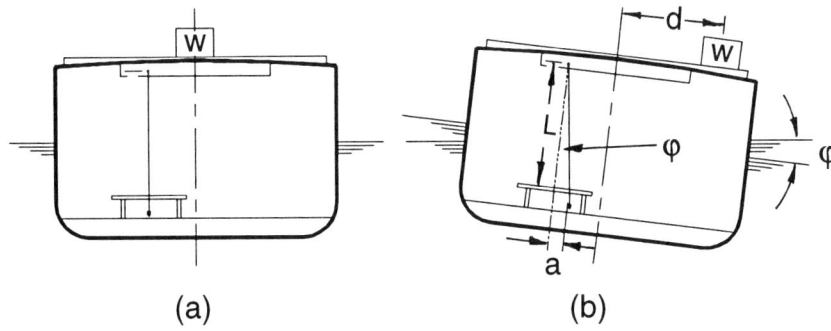

Figure 3-11. Inclining experiment.

Example 3-11
During an inclining experiment of a 465-foot-long ship floating in a graving dock containing water of specific gravity 1.011, an inclining weight of 18.51 tons is shifted 32 feet to starboard, causing a pendulum 18.33 feet long to deflect by 2.20 inches. The hydrostatic curves for this ship at the drafts during the inclining experiment show that the KM is 28.83 feet and the displacement in seawater (35 cubic feet per ton) is 12,999 tons. Determine the ship's as-inclined GM and KG.

The ship's displacement (weight) is not 12,999 tons, because the dock water is not seawater. The corrected displacement is determined by multiplying the reference seawater displacement by the ratio of the densities (or specific gravities) of the dock water and seawater.

$$\Delta = 12{,}999 \left(\frac{1.011}{1.025}\right) = 12{,}821 \text{ tons}$$

The angle of inclination is determined from the pendulum deflection and length. Both must be expressed in the same units to calculate the tangent of the angle of inclination. Therefore the length is multiplied by 12 to express it in inches.

$$\tan \varphi = \frac{a}{l} = \frac{2.20}{12 \times 18.33} = 0.01000$$

Although it is not necessary to determine the angle of inclination in degrees in order to calculate the GM, it is instructive to do so to make sure that it is a small angle:

$$\varphi = \tan^{-1} 0.01000 = 0.573 \text{ degrees}$$

At less than one degree, the small angle assumption is certainly acceptable. The as-inclined metacentric height is:

$$GM = \frac{wd}{\Delta \tan \varphi} = \frac{18.51 \times 32}{12{,}821 \times 0.01000} = 4.62 \text{ ft}$$

Finally, the as-inclined KG is:

STATIC EQUILIBRIUM AND STABILITY 81

$$KG = KM - GM = 28.83 - 4.62 = 24.21 \text{ ft}$$

Since the purpose of the measurement is to determine the light ship characteristics, the final step is to convert, by weight and moment calculations, the as-inclined condition to the light ship condition. This requires recording the magnitudes and locations of all weights aboard that are not properly part of the light ship (the inclining weight, equipment, and personnel, for example) as well as all weights of items that are part of light ship but are not yet on board.

Example 3-12

Listed below are the weights and heights above baseline of the items to be deducted and added in order to convert the as-inclined condition to the light ship condition of the ship whose as-inclined KG was calculated in Example 3-11. Determine the ship's light ship displacement and KG.

Weights to deduct:
    Inclining weight: 18.51 tons from 62 ft above keel
    Track for car: 5.80 tons from 59 ft above keel
    Personnel on board: 1.21 tons from 55 ft above keel

Weights to add to complete vessel:
    Joiner work: 26.50 tons at 68 ft above keel
    Davits: 33.73 tons at 57 ft above keel
    Lockers: 1.34 tons at 49 ft above keel

The calculation is best done in tabular form as shown.

	Weight, tons	Feet above keel	Moment, ft-tons
As-inclined ship	12,821	24.21	310,396
Weights to deduct			
Inclining weight	− 18.51	62	− 1,148
Track	− 5.80	59	− 342
Personnel	− 1.21	55	− 67
Weights to complete			
Joiner work	+ 26.50	68	+ 1,802
Davits	+ 33.73	57	+ 1,923
Lockers	+ 1.34	49	+ 66
Light ship	12,857		312,630

Thus the light ship displacement is 12,857 tons, and the KG light ship = 312,630/12,857 = 24.32 feet.

The importance of a properly conducted inclining experiment to establish with great accuracy the light ship displacement, KG, and LCG cannot be overemphasized. The values thus obtained are recorded in the ship's documents, in particular in the trim and stability booklet. They form the baseline values that are the point of departure for all stability and trim calculations that ship's officers will do throughout the ship's lifetime. Thus errors must be avoided at all costs. The inclining experiment described in the sample calculation above has been very much simplified compared to an actual experiment. For example, it is standard practice to use three pendulums and several weights, and to make as many as a dozen or more trials, or weight shifts, reading the deflections of all three pendulums at every trial. The resulting moments and tangents are averaged graphically. Also, a number of corrections not described here are usually made, such as correcting the displacement for trim and the angle of inclination for the effect of the wind if it is significant.

Detailed procedures, calculation forms, descriptions of equipment, requirements for preparing a ship for inclining, and reporting regulations are promulgated by the approving authority designated by each government to which the international regulations apply. In the United States, the U.S. Coast Guard publishes such regulations and participates in all merchant ship inclinings to ensure that the procedures are followed. For naval vessels, the U.S. Navy issues detailed instructions and reporting forms.

THE EFFECT OF FREELY SUSPENDED WEIGHTS

In all preceding discussions involving the calculation of the position of a ship's center of gravity, each individual item aboard a ship, whether part of the ship itself or of its loading, was treated as if its weight or mass were concentrated at its own center of gravity. If the ship heeled or moved in any direction, the position of the individual item's center of gravity was assumed to remain fixed within the ship frame of reference. These assumptions are realistic for most weights actually placed aboard ships. But if any weights are not "tied down" to the ship, but are free to move about as the ship heels, it follows that the ship's center of gravity will also move as described by equation (22).

An example of such a weight is a piece of cargo being loaded or discharged using the ship's own derricks or booms. How such a weight affects the ship compared to a fixed weight is shown in Figure 3-12. The figure represents diagrammatically five angles of heel of the ship superimposed: two to port, one upright, and two to starboard. A fixed weight in the lower hold is shown, with its five corresponding vertical weight vectors, each perpendicular to one of the waterlines. Each vector passes through a common point within the weight itself, and that point is, by definition, the center of gravity of the weight. The weight's cen-

STATIC EQUILIBRIUM AND STABILITY 83

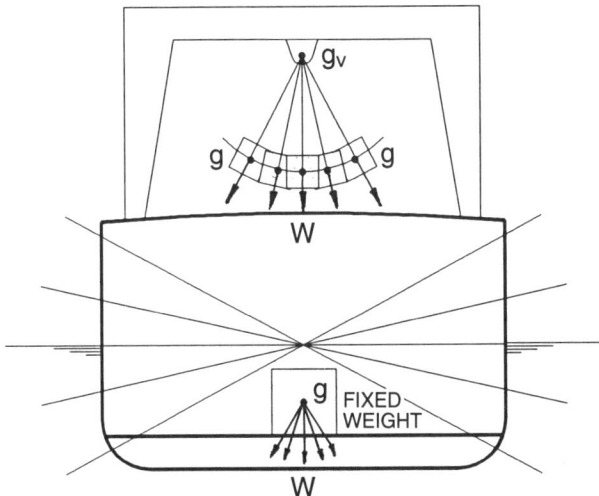

Figure 3-12. Freely suspended weight.

ter of gravity is in a fixed position within the weight, and also with respect to the ship, since it does not move about as the ship heels.

Consider now, in the same figure, the weight suspended from the gantry crane and free to swing. It also has its center of gravity within; but as the ship heels, the suspended weight swings, the weight always remaining vertically below the point of suspension. Thus for all angles of heel, the weight vector passes through the point of suspension, which can therefore be considered the virtual center of gravity (g_v) of the weight so long as it remains suspended. The effect of a suspended weight on ship stability may be accounted for simply by calculating the loading of the weight as if its center of gravity were fixed, but at its point of suspension. Regardless of how high or how little the weight is lifted by the wire, its virtual center of gravity is always above the actual weight, thus reducing the stability of the ship. The rise in the ship's center of gravity caused by lifting a weight already aboard by ship's gear can be determined by equation (22), with distance (d) defined as the vertical distance from the weight's initially fixed position to the point of suspension.

For most ordinary cargo-loading situations, it is not necessary to make such corrections to the ship's center of gravity. That is because the typical crane capacity, and thus maximum weight that can be lifted, is so small compared to the ship's weight that the quantity $GG' = wd/\Delta$ is negligibly small. In cases of extraordinarily heavy suspended weights, such as are possible on special heavy-lift ships and crane barges, the suspended weight effect should not be ignored in making stability estimates when designing the vessel.

84   APPLIED NAVAL ARCHITECTURE

THE EFFECT OF LIQUIDS WITH FREE SURFACES

As the suspended weight analysis demonstrates, stability can be adversely affected if some of the weights aboard a ship are free to shift spontaneously as the vessel changes its attitude in the water. For a ship at sea, the most common type of freely shifting weight is in the form of liquids that do not completely fill the tanks in which they are carried. A partially filled tank is said to be *slack* and the liquid is said to have a *free surface*.

Inside every partially filled tank, some liquid shifts toward the low side as the ship heels over, resulting in a shift of the ship's center of gravity toward the low side and hence a reduction in the righting arm. Determining the extent of this reduction in righting arm for all angles of heel, for every tank, and for any degree of fullness of each tank requires voluminous calculations based on a detailed study of the geometry of each tank. On the other hand, the influence of free surface on initial stability (GM), valid for small angles of heel, is independent of the quantity of liquid in the tank and of the angle of inclination, so long as it is small.

As illustrated in Figure 3-13, when a ship inclines to a small angle of heel, the transference of buoyancy from the emerged to the immersed wedge causes the center of buoyancy to shift to B' on the low side, the upward force of buoyancy Δ then passing through the transverse metacenter (M). The mechanics of this transference of buoyancy has been shown (Figure 3-8) to be summarized by equation (18):

$$BM = \frac{I_T}{\nabla} \qquad (18)$$

Within the tank partially filled with a liquid in Figure 3-13 it is clear that an analogous shift of a wedge-shaped quantity of liquid takes place, such that at any small angle of heel the center of gravity of the liquid traces path gg', and the line of action of its weight vector always passes through a point "m," the metacenter of the liquid. Note that the liquid metacenter is a point fixed in the ship that satisfies the definition of a virtual center of gravity, as in the case of a suspended weight. It is therefore more often referred to as g_v. The location of g_v is determined by direct analogy to the expression for the metacentric radius.

$$gg_v = \frac{i_t}{v} \qquad (25)$$

where: gg_v = virtual rise in center of gravity of the liquid inside the tank
  i_t = transverse moment of inertia of the surface of the liquid in the tank
  v = volume of the liquid in the tank

The Free Surface Correction

The virtual rise of the weight of the liquid produces a (virtual) *moment of free surface* equal to the product of the weight of liquid in the tank (w) and its virtual

STATIC EQUILIBRIUM AND STABILITY

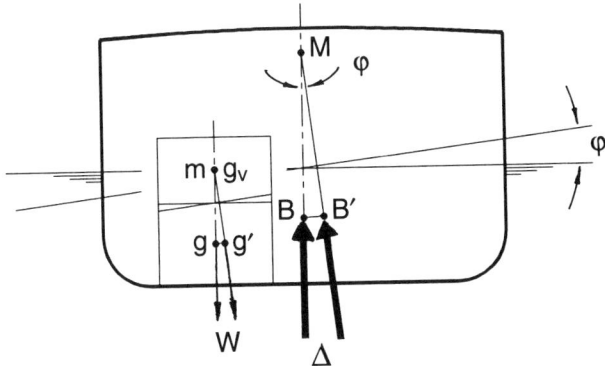

Figure 3-13. Free surface effect.

rise (gg_v). The effect of this moment of free surface on the stability of the ship must be determined. If a weight is actually shifted upward causing a moment of w × d foot-tons, the upward shift in the ship's center of gravity is determined by dividing the moment w × d by the ship's displacement, as shown by equation (22). Similarly, the virtual rise of a weight of liquid causes a virtual rise in the ship's center of gravity to G_V, as follows:

$$GG_v = \frac{w\, gg_v}{\Delta} = \frac{w\, (i_t/v)}{\Delta} \qquad (26)$$

Both the weight and the volume of the liquid appear in (26), so it can be put into a more convenient form by substituting the weight density of the liquid in the tank ($\rho_t g$) for w/v, or its reciprocal weight density (δ_t) for v/w, giving:

$$GG_v = \frac{\rho_t\, g\, i_t}{\Delta} = \frac{(i_t/\delta_t)}{\Delta} \qquad (27)$$

This form is convenient when U.S. units are used, with displacement Δ as a weight. For use with SI units and displacement as a mass, it is better to write

$$GG_v = \frac{\rho_t\, i_t}{\Delta_m} \qquad (27a)$$

Noting that the ship displacement in equation (27) can also be expressed as $\rho_s g V$, where $\rho_s g$ is the weight density of the water in which the ship is floating, and letting δ_s be the specific volume of the ship water, other useful forms of the expression for the free surface effect are:

$$GG_v = \frac{i_t}{V}\left(\frac{\delta_s}{\delta_t}\right) = \frac{i_t}{V}\left(\frac{\rho_t}{\rho_s}\right) \qquad (28)$$

86   APPLIED NAVAL ARCHITECTURE

Equation (28) is correct in both U.S. and SI units.

The tank moment of inertia (i_t) in all of the above equations refers to the transverse moment of inertia of the liquid surface inside the tank. For a tank that is rectangular in plan view, the moment of inertia is

$$i_t = \frac{l\,b^3}{12} \text{ ft}^4 \text{ or m}^4 \tag{29}$$

where: l = length (longitudinal in ship) of the tank, ft or m
         b = beam or width (transverse in ship) of the tank, ft or m

For tanks of more complicated geometry, such as those bounded by the sloping or curved sides of the ship, i_t must be calculated by numerical integration during ship design. For all of the tanks on an actual ship, either the tank moments of inertia or the moments of free surface for the liquids appropriate to each tank are worked out and provided in tables in the trim and stability booklet for the convenience of the ship's officers.

Example 3-13

Determine the free surface effect, or virtual rise in the ship center of gravity, caused by free liquids in the following tanks:

(a) Seawater ballast in a tank 60 ft long and 20 ft wide on a ship of 18,000 long tons displacement in seawater.

(b) Like (a) except the tank is "turned sideways" so that it is 20 ft long and 60 ft wide.

(c) Fresh water in a tank 5 m long and 3 m wide on a ship displacing 25,000 MT in seawater.

(d) Fuel oil with a specific volume of δ_t = 38 ft^3/ton in a tank 40 ft long and 18 ft wide on a rectangular barge 120 ft long and 18 ft wide floating at a draft of 6 ft in fresh water.

(a) $$i_t = \frac{l\,b^3}{12} = \frac{60 \times 20^3}{12} = 40{,}000 \text{ ft}^4 \quad [\text{Eqn. (29)}]$$

For U.S. units, equation (27) is used, with δ_t = 35 ft^3/ton, since the tank contains seawater.

$$GG_V = \frac{(i_t/\delta_t)}{\Delta} = \frac{(40{,}000/35)}{18{,}000} = 0.063 \text{ ft}$$

(b) $$i_t = \frac{20 \times 60^3}{12} = 360{,}000 \text{ ft}^4$$

STATIC EQUILIBRIUM AND STABILITY 87

$$GG_V = \frac{360,000}{(35 \times 18,000)} = 0.571 \text{ ft}$$

Note the enormous increase in the tank moment of inertia and the GG_V when the width dimension is large. A few tanks like this one with free surfaces could easily render a ship unstable and cause a dangerous condition. Wide tanks should be avoided in ships. This is the reason that the cargo spaces in tankers and the double bottom spaces in other ships are normally subdivided longitudinally into two, three, or four tanks athwartships.

(c) $$i_t = \frac{5 \times 3^3}{12} = 11.25 \text{ m}^4$$ [Table 3-1]

$$\rho_t = 1.000 \text{ MT/m}^3$$

Equation (27a) is the choice when SI units are specified.

$$GG_V = \frac{\rho_t i_t}{\Delta_m} = \frac{1.000 \times 11.25}{25,000} = 0.00045 \text{ m}$$

Free surface in a small tank like this one has a negligible effect on ship stability. This is why tanks such as potable water tanks and fuel oil settling tanks, which are always slack because their contents are continuously being drawn upon, are always made small, especially in the width dimension.

(d) $$i_t = \frac{40 \times 18^3}{12} = 19,440 \text{ ft}^4$$

$\delta_t = 38 \text{ ft}^3/\text{ton}$, and $\delta_s = 36 \text{ ft}^3/\text{ton}$ (fresh water)

Equation (28) is the form suggested in this case.

$$GG_V = \frac{i_t}{\nabla}(\frac{\delta_s}{\delta_t}) = \frac{19,440}{12,960}(\frac{36}{38}) = 1.421 \text{ ft}$$

If several tanks have free surface, the total GG_V is simply the sum of the GG_V's for each tank. Since the displacement is fixed, the simplest way to calculate the total GG_V is to add the separate moments of free surface for each tank before dividing by the displacement. That is, equations (27) and (27a) may be written to express the cumulative effect of all tanks having free surface as follows:

$$\text{Total } GG_V = \frac{\Sigma (i_t/\delta_t)}{\Delta} \quad (30)$$

$$\text{Total } GG_V = \frac{\Sigma (\rho_t i_t)}{\Delta_m} \quad (30a)$$

In practice, the summation is determined by adding the tabulated moments of free surface for all slack tanks in a given condition of loading.

The Effect of Cornering. The effect of free surface on ship stability derived above is valid so long as the path traced by the liquid center of gravity within the tank traces a circular arc as depicted in Figure 3-13. Only then is there a fixed position of the virtual center of gravity of the liquid (g_v). The circular path assumption is valid only for small angles of heel, and only if the surface of the liquid, upon rising or falling on the sides of the tank, does not go past the corner of the tank at the top or bottom.

Figure 3-14 shows the cornering of the liquids in various tanks. The resulting changes in stability depend on the shape of the tank, particularly its depth-to-breadth ratio, and on the degree of fullness of the tank. As shown in the figure, any tank that is almost full or almost empty will corner at a small angle of heel. If the liquid level in the tank is not close to either the top or bottom, it will corner at a small angle only if it is very shallow compared to its width (like a double bottom tank, for example). A deep tank that is about half full may not corner at any typical angle of heel or roll.

When cornering takes place, an exact determination of how free surface changes a ship's stability requires calculations of the actual shift of position of each tank's center of gravity for any angle of heel and percent fullness of the tanks. Voluminous calculations of this sort are rarely done, but simpler approximations are often made in the design office, and the results are tabulated in the trim and stability booklet. The approximation is expressed in the form of a fictitious reduced moment of free surface for each tank when the tank is nearly full but not pressed full (hard against the tank top and into the vent pipe). Tanks carrying liquids like fuel oil and cargo oils cannot be pressed full because thermal expansion of the oil can cause overflow out of the vent pipe. When tanks are taking on such liquids, pumping is stopped when the tank is from 95 to 98 percent full. For nearly full tanks, therefore, reduced i_t / δ_t values (moments of free surface) are calculated and tabulated along with the ones for slack tanks. When making routine free surface calculations, the "95%" or "98%" values of i_t / δ_t are chosen for those tanks in which "full" means "leaving space for expansion." There are then three possible conditions for each tank as regards the free surface effect:

1. Pressed full or empty, in which case $i_t / \delta_t = 0$, since there is no free surface. Tanks containing ballast water are normally pressed full, because water does not expand as the temperature rises.
2. "Full" of oil, which is normally between 95 percent and 98 percent full, in which case the reduced value of i_t / δ_t recorded in the table of tank capacities and free surface moments is assumed.

STATIC EQUILIBRIUM AND STABILITY 89

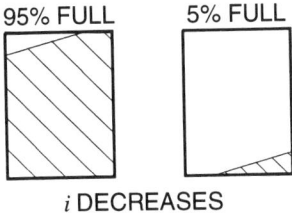

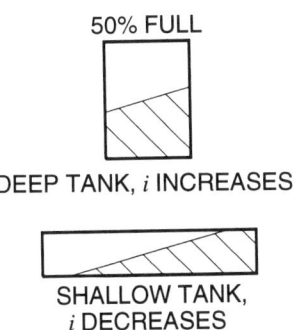

Figure 3-14. Cornering of liquids with a free surface.
Angle of heel is 15 degrees in this illustration.

3. Slack, meaning other than conditions 1 or 2 defined above, in which case the full value of i_t / δ_t, also given in the table, is assumed. In normal operating conditions, all fresh-water tanks, all fuel oil settling tanks, and one pair of matching port and starboard fuel oil storage tanks (or a single one, if it is on centerline) are assumed to be slack.

Taking on Liquids. It has been noted above that the effect of free surface on the ship (GG_V) is independent of the quantity of liquid in the tank. This does not mean that the quantity of liquid has no influence on the stability of the ship. The change in the ship's stability caused by the weight and centroid of the liquid in each tank must be calculated when determining Δ and KG. The following worked example illustrates how taking liquids aboard affects the displacement, KG, GM, KM, and angle of list of a vessel.

Example 3-14
A rectangular barge 200 feet long, 50 feet wide, and 20 feet deep weighs 1,850 tons and has KG = 9.0 feet when it is empty. It is divided into two tanks of equal

90   APPLIED NAVAL ARCHITECTURE

size by a single longitudinal bulkhead on centerline. If 600 tons of oil having a specific volume of 40 cubic feet per long ton are loaded into the starboard tank, determine the resulting stability and angle of list of the barge. The barge is floating in seawater.

Calculation of displacement and KG: The depth of the oil in the tank, and hence the height of its center of gravity, must first be determined by expressing the volume of oil in terms of both its specific volume and of the dimensions of the oil in the rectangular tank.

$$\text{volume of oil} = lbd = \delta w$$

where l = length of tank = 200 ft
b = breadth of tank = 50/2 = 25 ft
d = depth of oil in tank, to be determined
δ = specific volume of oil = 40 ft^3/ton
w = weight of oil = 600 tons

$$200 \times 25 \times d = 40 \times 600$$
$$d = 4.80 \text{ ft}$$

The center of gravity of the oil is at half its depth, hence it is 4.80/2 = 2.40 feet above the baseline.

Item	Weight	VCG	Moment about keel
Barge	1,850	9.00	16,650
Oil	600	2.40	1,440
	2,450		18,090

$$\Delta = 2{,}450 \text{ tons}$$
$$KG = \frac{18{,}090}{2{,}450} = 7.38 \text{ ft}$$

Calculation of GM uncorrected for free surface: The draft of the loaded barge, as if it were on even keel, is determined from:

$$LBT = 35\Delta$$

$$T = \frac{35\,\Delta}{LB} = \frac{35 \times 2{,}450}{200 \times 50} = 8.58 \text{ ft}$$

The KM of a rectangular barge is determined from simple formulas for KB and BM:

$$KB = \frac{T}{2} = \frac{8.58}{2} = 4.29 \text{ ft}$$

$$BM = \frac{B^2}{12\,T} = \frac{50^2}{12 \times 8.58} = 24.28 \text{ ft}$$

$$KM = KB + BM = 4.29 + 24.28 = 28.57 \text{ ft}$$

The uncorrected metacentric height is:

$$GM = KM - KG = 28.57 - 7.38 = 21.19 \text{ ft}$$

The free surface correction:

$$i_t = \frac{l\,b^3}{12} = \frac{200 \times 25^3}{12} = 260{,}417 \text{ ft}^4$$

$$GG_V = \frac{(i_t/\delta_t)}{\Delta} = \frac{260{,}417}{40 \times 2{,}450} = 2.66 \text{ ft}$$

$$G_V M = GM - GG_V = 21.19 - 2.66 = 18.53 \text{ ft}$$

Angle of list: Loading the oil in the starboard tank is equivalent to loading it on centerline, then "shifting" it to starboard. The distance of this imagined shift of weight is the lateral distance from the centerline of the barge to the centroid of the oil as stowed, that is, to the midpoint of the starboard tank. Since the tank is 25 feet wide, its midpoint is 25/2 = 12.5 feet to starboard of the barge centerline. Therefore the angle of list to starboard is given by:

$$\tan \varphi = \frac{wd}{\Delta\,G_V M} = \frac{600 \times 12.5}{2{,}450 \times 18.53} = 0.1652$$

$$\varphi = 9.4°$$

PROBLEMS

1. A ship has the following principal dimensions:
   LBP = 700 ft
   B = 101 ft
   D = 60 ft to main deck

When floating at load draft of 39 feet in seawater, the hydrostatic curves give:

Displacement = 54,200 tons
TPI = 126
KM = 40.7 ft
KB = 21.1 ft

Calculate the following:
Block coefficient
Waterplane coefficient
Volume of displacement
Transverse moment of inertia of waterplane

2. Given the following ship characteristics:
Length = 620.0 ft
Beam = 78.0 ft
Draft = 21'6" in seawater
Waterplane area = 35,448 ft^2
Transverse moment of inertia of waterplane = 19.86×10^6 ft^4
Midship section coefficient = 0.972
Block coefficient = 0.675

Determine:
a. Displacement (long tons) in seawater
b. Midship section area
c. Prismatic coefficient
d. Waterplane coefficient
e. Metacentric radius
f. Tons per inch immersion

3. A ship has the following characteristics when floating in seawater:
Δ = 44,500 tons C_B = .689 KM_T = 59.35 ft
L = 720 ft C_W = .770   KB = 19.00 ft
B = 105.8 ft C_M = .968

Calculate the following and indicate units:
a. Draft, seawater
b. Waterplane area
c. Midship section area
d. Prismatic coefficient
e. Transverse moment of inertia of waterplane
f. Tons per inch immersion

STATIC EQUILIBRIUM AND STABILITY 93

4. A cargo ship having the characteristics labeled *ORIG* below is to be jumbo-ized by adding a new 96-foot-long section of parallel body having the shape of the original midship section. Calculate the items left blank in each list.

	Orig	New
L, ft	650	_____
B, ft	95	_____
T, ft	34.5	34.5
C_B	0.675	_____
C_M	0.970	_____
C_W	0.750	_____
C_P	_____	_____
TPI	_____	_____
Δ, tons seawater	_____	_____

5. A shallow-draft river vessel 310 feet long and 42 feet on the beam has the following characteristics at a draft of 8.0 feet in fresh water with nothing aboard (light ship):

$$C_B = 0.750 \qquad C_M = 0.920 \qquad C_W = 0.697$$

The ship is drydocked to install a rubbing strip on the outer bottom. The wood strip is 220 feet long, 8 feet wide, and 1.5 feet thick, and it weighs 36 tons including all fasteners. Determine the vessel's draft from the bottom of the strip when it is refloated in fresh water after the strip has been installed.

6. A ship loads to its summer load line in a river where the reciprocal density of the water is 35.8 cubic feet per long ton. It then moves downriver to a port where the water is at 35.4 cubic feet per long ton, consuming 70 tons of fuel and water during the passage. After loading 300 tons of cargo at the new port, it is noted that the ship is again at its summer load line.

What is the summer load displacement (long tons) in seawater at 35.0 cubic feet per ton?

7. A rectangular barge 320 feet long, 42 feet wide, and 12 feet deep is floating in seawater at a draft of 5.5 feet. At what draft would it float in fresh water if the loading remains unchanged? How many tons of cargo would be removed to make the fresh-water draft equal to what it was in seawater?

8. Show that a homogeneous "log" of square cross section and specific gravity 0.50 will float in fresh water with its corners at the top and bottom and the waterplane across the diagonal of the cross section. To prove this, calculate its GM as a function of the length of its side (S) in two attitudes: with the flat sides up (show that GM is negative) and with the corners up (positive GM).

9. A cylinder is placed in water with its axis vertical. Show that, if the center of gravity is in the waterplane, the cylinder will just float upright (GM = 0) if the radius divided by the draft is greater than $\sqrt{2}$.

   Hint: Moment of inertia of a circle about its diameter is $\frac{1}{4}AR^2$ where A = area of circle, and R = radius.

10. A Mariner class ship floats at a mean draft of 23'0" in seawater with a GM of 2.20 feet. The following loading then takes place:

    Discharge 230 tons cargo from 40.3 ft above keel
    Load 480 tons cargo at 28.3 ft above keel
    Load 240 tons fuel at 2.5 ft above keel

    Determine the draft, displacement, and GM after loading is complete.

11. A Mariner floats at a draft of 18'0" in seawater with GM = 6.40 feet. It then loads as follows:

    800 tons loaded at 20 ft above keel
    1,000 tons loaded at 15 ft above keel
    1,400 tons loaded at 25 ft above keel
    900 tons shifted from 7 ft above keel to 36 ft above keel

    Calculate the final draft, KG, and GM.

12. The only bridge across the river within a hundred kilometers has collapsed. A local entrepreneur has a rectangular barge 22 meters long, 4.4 meters wide, and 2.4 meters deep (keel to deck), which floats at a draft of 1.0 meter and has KG = 1.2 meters when nothing is aboard. The barge owner offers (for a fee) to carry a large truck across the river on the deck of the barge. The truck mass is 35 metric tons and its center of gravity is 2.0 meters above the ground. Calculate the initial stability (GM) of the truck-barge combination. Would you advise the trucker to accept the offer?

13. A Mariner vessel, floating at a draft of 23'6", has a GM of 1.5 feet, which does not meet the required GM standard. How far above the keel must 1,400 tons be loaded to increase the GM to 2.0 feet?

STATIC EQUILIBRIUM AND STABILITY 95

14. A Mariner loading cargo displaces 17,000 tons, with KG = 27.0 feet. It is to make a 12-day passage, consuming 75 tons of fuel per day from its double bottom tanks (VCG = 2.6 feet). How much deck cargo may be loaded (at a VCG of 46.5 feet) if the GM on arrival at the destination is to be not less than 2.0 feet?

15. A rectangular barge 60 feet long and 20 feet wide weighs 300 tons and its center of gravity is 8.5 feet above the keel. Show that when floating upright in seawater it is in unstable equilibrium. Fifty tons of solid ballast are now placed in the barge at 2 feet above the keel. Calculate the resulting transverse metacentric height. How far could the ballast be raised (above its original installed position) to reduce the metacentric height to 0.2 feet?

16. A ship of 10,500 metric tons light ship displacement is inclined by shifting a 30-ton mass through 11.3 meters. A pendulum 5.8 meters long deflects by 21.0 centimeters. KM_T is 8.32 meters. Calculate the as-inclined KG.

17. A barge of constant triangular cross section is 180 feet long and has a beam of 24 feet at the waterline when floating in *fresh* water at 9 foot draft. A weight of 15 tons (already on board) is shifted 10 feet across the deck, causing the barge to heel 2.3 degrees.
    Determine the KG of the barge:

    a. As inclined
    b. After removing the inclining weight, if it was 20 feet above keel

18. A barge 250 feet long and 35 feet wide floats in seawater with a displacement of 1,500 tons and KG of 7.5 feet. Then 150 tons is loaded at 12 feet above the keel and 8 feet to starboard of centerline. What angle of list will result?

19. A rectangular barge 120 feet long, 24 feet wide, and 10 feet deep to the deck floats at a draft of 5.3 feet in seawater and has KG = 5.7 feet. A locomotive weighing 90 tons is then loaded on deck so that its center of gravity is 6 feet above the barge's deck and 1.2 feet to starboard of the barge's centerline. What angle of heel (or list) will this cause?

20. An empty recangular barge 140 feet long and 30 feet wide weighs 504 tons and has KG = 6.5 feet. It floats in seawater. The following loading then takes place:

    230 tons loaded at 4 feet above keel
    350 tons loaded at 10 feet above keel
    120 tons loaded at 6 feet above keel
     30 tons loaded at 13 feet above keel

    The 30-ton weight listed above is placed 12 feet off centerline to port. Calculate the draft, displacement, metacentric height, and angle of list of this barge after loading is complete.

21. A Mariner arrives in a seawater port at a draft of 23 feet and GM = 3.2 feet. It then loads as follows:

    600 tons discharged from 13.3 feet above keel on centerline
    80 tons loaded at 50 feet above keel and 26 feet off centerline

    What will be its angle of list when loading is completed?

22. A Mariner floats in seawater at a draft of 20′6″ even keel with no list and KG = 28.5 feet. The following weights are then loaded:

    350 tons at 15 ft above keel, on centerline
    140 tons at 32 ft above keel, 25 ft to port of centerline
    90 tons at 28 ft above keel, 12 ft to starboard of centerline

    a. Determine the draft, GM, and angle of list of the loaded ship.
    b. How many tons of fuel should be shifted from a port to a starboard double-bottom tank (a distance of 48 feet, centroid to centroid) to eliminate the list?

23. A rectangular barge 67 meters long, 14 meters wide, and 6 meters deep (to main deck) floats at a mean draft of 3 meters in seawater, with KG = 3.7 meters and with a list to starboard of 6 degrees. If 250 tons of cargo are to be loaded on the main deck at 7 meters above the baseline, how far to port of centerline should it be loaded so that the barge will complete loading with no list?

24. A Mariner floating at a mean draft of 22′3″ and with KG = 28.0 feet has a list of 5 degrees to port. How far to starboard of centerline should 200 tons of cargo be loaded at 28 feet above the keel in order to correct this list?

STATIC EQUILIBRIUM AND STABILITY   97

25. A Mariner ship fitted with heavy-lift gear is floating in seawater at 25′6″ draft with KG = 28.5 feet and no list. It then lifts 50 tons from the pier, the point of suspension of the heavy-lift boom being 48 feet above the main deck and 46 feet from the ship's centerline. The main deck is 44.5 feet above the keel. Determine the ship's GM_T and angle of heel at three stages during loading:

   a. When the center of gravity of the 50 tons is 5 ft above the deck and 46 ft from centerline.
   b. When it has been swung to centerline, still suspended and still 5 ft above the deck.
   c. When it has been placed and secured on the deck on centerline, with its center of gravity 2 ft above the deck.

26. An empty rectangular barge 140 feet long and 25 feet wide floats at a draft of 4′8″ even keel in seawater and has a KG = 6.0 feet. The following loading then takes place:

   260 tons of cargo loaded at 4 ft above the keel.
   320 tons of cargo loaded at 10 ft above the keel.
   100 tons of fresh water loaded at 6 ft above the keel.

   The fresh water has a free surface that is 40 feet long and 13 feet wide. Calculate the draft, displacement, and metacentric height corrected for free surface of the loaded barge.

27. A Mariner enters port with 17,700 tons displacement, KG = 28.0 feet, and GG_v = 1.05 feet. A piece of heavy machinery which weighs 180 tons is lifted by shore-based cranes and placed on deck so that its center of gravity is 47 feet above the ship's keel and 19 feet to starboard of centerline. What will be the final mean draft, G_vM, and angle of list when loading is complete?

28. A ship of 9,000 tons displacement floats at a draft of 23 feet and has an upper deck 34 feet above the keel. A sea breaking over the bulwarks floods a portion of the upper deck with water to a depth of 3 feet. The flooded area is 60 feet long and 40 feet wide. If KG before taking on water was 20.0 feet, calculate the *reduction* in GM, assuming that KM remains constant. The effects of both the added weight and free surface must be determined.

29. A rectangular barge 55 meters long, 11.2 meters wide and 4.25 meters deep to the deck floats in fresh water at a draft of 1.50 meters when empty, and has a KG of 2.00 meters. Diesel oil (specific gravity = 0.90) is then loaded into a central tank 24.5 meters long and 11.2 m wide to a depth of 3.6 meters. The bottom of the tank is the bottom of the barge. Determine the resulting draft, displacement, and metacentric height corrected for free surface.

30. A tanker displacing 36,000 tons in seawater begins loading oil (36.25 cubic feet per ton) into an empty center tank that is 58 feet long, 40 feet wide, and 50 feet deep. Before loading began, the ship's GM was 6.5 feet and there was no free surface. Calculate its GM corrected for free surface when the tank is three-quarters full. The KM_T = 37.0 feet may be assumed to be unchanged during loading. Bottom of the tank is at the keel.

31. A rectangular barge 120 feet long, 30 feet wide, and 13 feet deep floats in seawater at a draft of 7 feet when the only weights aboard are those for the inclining experiment. It is then inclined and a transverse shift of 5 tons through 20 feet heels the barge 2 degrees. After the barge is returned to upright (the inclining equipment remains aboard), 480 tons of fuel oil having a specific gravity of 0.90 are loaded equally into three double-bottom tanks, each 10 feet wide and 120 feet long, formed by two equally spaced longitudinal bulkheads. Free surface is created in each of the tanks. Calculate the virtual metacentric height (G_VM) in this condition.

32. A ship displacing 9,200 tons has a tank (42 feet long, 30 feet wide) half filled with oil (specific volume 42 cubic feet per ton). While in this condition the ship is inclined by shifting 45 tons 18 feet to starboard, and a 28-foot-long pendulum deflects 10 inches. Determine the virtual metacentric height (G_VM) and the real metacentric height (GM) of this ship in the as-inclined condition.

33. A tanker floats at a displacement of 28,000 tons in seawater with a KM of 27.5 feet and KG of 22.0 feet. Then deballasting of a centerline tank full of seawater begins. The tank is 68 feet long, 30 feet wide, and 36 feet deep. What will be the G_VM of the ship when the tank is exactly one-third full? Assume that KM remains constant as deballasting proceeds.

STATIC EQUILIBRIUM AND STABILITY 99

34. A rectangular barge 240 feet long, 35 feet wide, and 16 feet deep floats in seawater at a draft of 4.0 feet with KG = 7.0 feet when empty. It is divided into two equal compartments by a longitudinal bulkhead on centerline. Bulk cargo stowing at 78 cubic feet per ton is loaded evenly in the starboard compartment to a depth of 13 feet, and 1,050 tons of *fresh* water is loaded in the port compartment. Calculate the final mean draft, virtual metacentric height, and angle of heel.

35. A large tanker displacing 145,000 tons in seawater has a GM corrected for free surface of 15.0 feet, with a center tank 100 feet long and 68 feet wide containing oil (38 cubic feet per ton) to a depth of 70 feet. A cross-connecting valve is opened from this tank to the adjacent starboard wing tank (100 feet long, 42 feet wide) until the oil reaches equal levels in center and wing tanks, then the valve is closed. Calculate the final GM and angle of heel in this condition. (Tanks may be assumed rectangular—ignore bilge radius.)

CHAPTER FOUR

STABILITY AT LARGE ANGLES

Reliance on the metacentric height as a measure of transverse stability is limited, as described in Chapter 3, to situations in which the ship heels to small angles from the upright, typically less than about 10 degrees. If the upsetting forces that act on ships in service, such as those caused by wind, waves, cargo handling, and turning, could not produce inclinations larger than a few degrees, the study of metacentric, or initial transverse statical, stability would be sufficient for both ship designer and operator. Clearly this is not the case, however. Ships can and do heel and roll to larger angles under the influence of large heeling moments. To ensure proper design and safe operation we must know how a ship behaves when heeled to large angles.

RIGHTING ARM AND RIGHTING MOMENT

Whatever the angle of heel, the proper measure of a ship's ability to return to upright is the righting moment, equal to the product of the ship's weight (Δ) and the righting arm (GZ), as shown in Figure 4-1. This figure, intended to represent a large angle of heel, should be compared to its small angle counterpart shown in Figure 3-7. The difference is that at large angles the buoyant force vector does not pass through the metacenter (M). The reason is that, as the angle of heel increases beyond a few degrees, the "B-path" departs from a circular arc of radius BM. The consequence of this departure is that the righting arm is no longer related in any simple way to the metacentric height, that is, GZ is not equal to GM sin φ, as it is in the case of very small angles of heel. In fact, no exact formula is known that relates GM to the righting arms GZ for large angles, except for the very restrictive class of hull forms for which the center of buoyancy traces a circular path when the vessel heels to any angle. This will be the case only for spheres, circular cylinders, or bodies of revolution floating with their axis of

STABILITY AT LARGE ANGLES

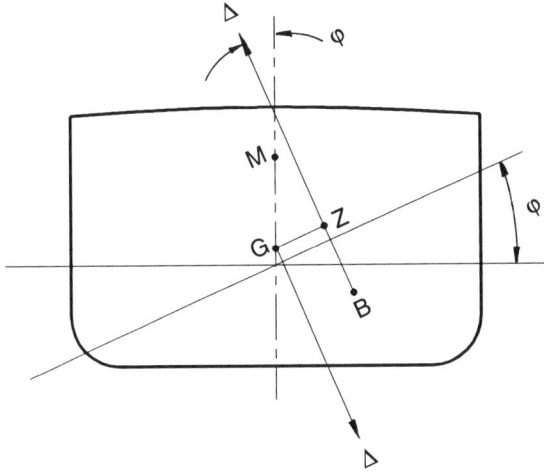

Figure 4-1. At large angle of heel, GZ is *not* GM sin φ.

symmetry parallel to the water surface. For such forms, the transverse metacenter lies on the axis of symmetry and the righting arms for all angles of heel are equal to GM sin φ. The only practical hull forms satisfying these conditions are circular section pontoons and submarines whose hull forms are essentially bodies of revolution.

STABILITY OF A BODY OF REVOLUTION

Although the body of revolution is not a realistic form for surface ships, it is instructive to examine the behavior of such a vessel when it is heeled to large angles. Imagine a vessel, all of whose sections are circular, loaded such that it floats with its axis horizontal, but with its center of gravity below its axis. It is immaterial whether all sections are of the same diameter (a circular cylinder) or not (a body of revolution, for example, shaped like a cigar or a football). Figure 4-2 depicts this body floating at various angles of heel, the upright condition (φ = 0°) representing the position of stable equilibrium, in which G is directly below the center of the circle. It is apparent, as each of the positions shown are examined, that, at any angle of heel, the buoyant force acts through the center of the circle, which is thus the transverse metacenter M. The righting arms GZ are, for all angles of heel, equal to GM sin φ. Note that the righting arms increase to a maximum value equal to GM at 90 degrees, then decrease again to zero in the inverted position, or 180 degree heel. This upside-down position is therefore one of

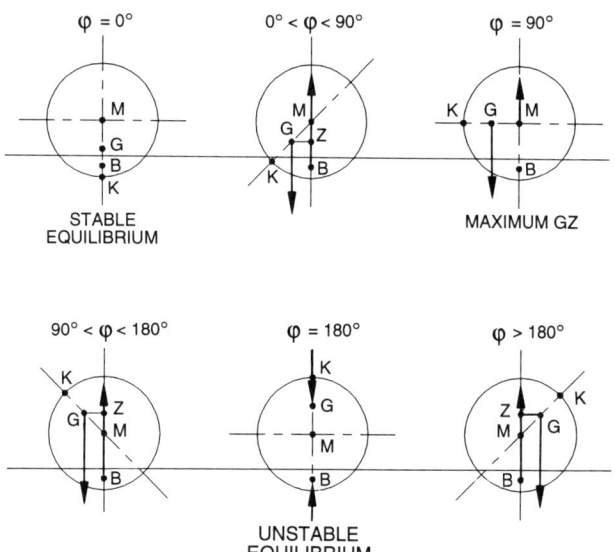

Figure 4-2. Stability of a body of revolution. GZ = GM sin φ for all angles of heel.

equilibrium (GZ = 0), but the slightest disturbance from it in either direction will cause the vessel to rotate until it returns to its original upright condition. Thus it is a position of unstable equilibrium.

The stability at large angles for this vessel can be summarized by plotting the righting arms (GZ) against the angle of heel (φ), as shown in Figure 4-3. This curve is known as a *statical stability curve*. In this special case, the statical stability curve may also be described as a sine curve of amplitude GM. Note that the same curve would result regardless of the draft (or displacement) of the cylinder because KM is equal to the radius of the cylinder whatever amount is underwater. Righting moments, of course, increase with displacement (or weight), even though righting arms do not.

This statical stability curve shows that a body of revolution has, in one complete revolution, two equilibrium angles (zero and 180 degrees), only one of which is stable (zero degrees). Thus such a body will return to the upright condition from any angle of heel imposed on it externally.

The unique feature of the circular-section vessel that enables its statical stability curve to be described by a sine curve is that at all angles of heel the immersed portion of the body is the same shape. Put in another way, we could observe, by imagining the various diagrams in Figure 4-2 to be superimposed on one another, that the path taken by center of buoyancy B describes a circle of radius BM at all angles of heel.

STABILITY AT LARGE ANGLES

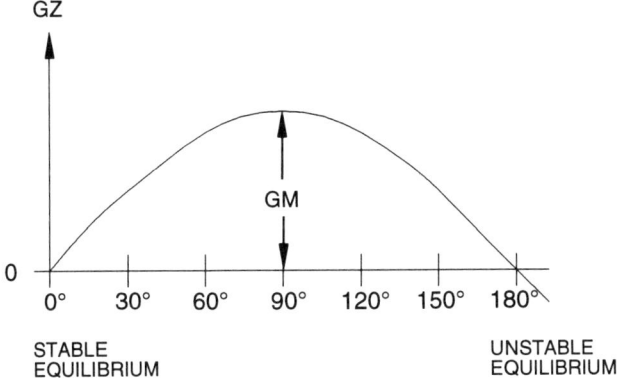

Figure 4-3. Statical stability curve for a body of revolution.

The characteristic most strongly influencing the righting arms is the position of the center of gravity, that is, KG. If the center of gravity is lowered, righting arms at all angles will get larger, and, conversely, if G moves up, all righting arms will decrease. This relationship applies to ship forms as well.

DETERMINING THE RIGHTING ARMS FOR SHIP FORMS

For hull forms other than bodies of revolution, the naval architect must determine the righting arms and displacements of the ship floating at each of numerous waterlines representing large heel angles by direct numerical integration of the hull form in the heeled condition. Reference to the general picture of a righting arm in Figure 4-1 will show that anything that changes the positions of either G or B in a specific ship will change the righting arm. The position of B depends on the displacement (or upright draft) and the angle of heel. The position of G depends on the loading of the ship, and it is characterized by its three coordinates. The above statements can be put in the form of a complete physical equation as follows:

$$GZ = f(\varphi, \Delta, KG, LCG, TCG) \qquad (1)$$

A program of calculations that would determine the variations in GZ caused by typical variations on each of the five parameters shown in this equation would be extremely voluminous. Supposing that only five different values of each of the five variables were enough to define the full range of their variability (it probably would not), a total of 3,125 righting arms would have to be calculated to account for all combinations of the variables.

The magnitude of the job can be reduced considerably by making two observations:

- Righting arms are only slightly (often negligibly) affected by typical variations in the LCG, and hence in the vessel's trim. It is usually sufficient to determine them for the even-keel (no-trim) condition only.
- The effects of variations in KG and TCG can be determined by simple correction formulas after the GZs have been determined for an arbitrarily assigned assumed position of G. The assumed G, sometimes referred to as the "pole point" for the GZ calculations, is taken at any convenient height above the keel, on the centerline; that is, the transverse position (TCG) is taken as zero.

As a result of these simplifications, the physical equation for GZ of a ship of known form is reduced to

$$GZ = f(\varphi, \Delta) \tag{2}$$

and calculations are performed for angles of heel up to 90 degrees in increments of 10 or 15 degrees, and for five to seven displacements at each angle. Thus a typical program of calculations consists of determining at least 30, and up to 60 or more, different GZs.

Even so, the calculations are tedious and repetitive, and since they consist of numerical integrations of offset measurements taken from the lines drawing, they are most often done by special computer programs written for the purpose. All comprehensive software packages that calculate hydrostatic properties include routines to determine the righting arms. The specific calculation procedures of various programs may vary considerably, some using as basic input the same offsets used in standard hydrostatic calculations (to which are added special offsets describing the weather deck and watertight spaces above it), others creating special offsets measured parallel to the heeled waterlines, or even measured radially from a selected point in the inclined waterline. The programs have the advantage, beyond the obvious one of saving an enormous amount of tedious hand calculations, of automatically adjusting the statically balanced ship to the correct trim for each heeled waterline, thus making unnecessary the assumption mentioned above that trim (or LCG) corrections are small enough to be ignored. We shall not attempt to describe individual techniques, but a brief examination of the underlying geometry and mechanics basic to all such calculations follows. The calculation programs will probably be variations of one of the two basic methods outlined below—the whole body method or the wedge method.

Whole Body Method

For the whole body method, an inclined waterline such as WL in Figure 4-4 is drawn across a body plan showing both sides and the weather deck of each station. A vertical axis, perpendicular to WL, is passed through an arbitrarily chosen pole point G_o, representing the assumed center of gravity. Two properties of each station are determined, the exact method depending on the particular program details. They are the immersed area (A_s), and the moment of that area (M_s) about the vertical axis. The areas and moments thus obtained are then integrated over the length of the ship to determine immersed volume (∇) and the moment of ∇ about the axis, respectively. The righting arm is determined by dividing the moment by the volume. This calculation must be repeated for each angle of heel and displacement to be considered.

Wedge Method

The wedge method operates on the volumes and moments of emerged and immersed wedges of buoyancy produced as a ship heels from an upright to an inclined waterline, rather than on the entire immersed volume of the ship. Although the analysis is somewhat more complicated than that of the whole body method, it requires fewer measurements to be taken from the body plan, since it is only necessary to choose enough offsets to define the wedge geometry. Because it reduced the required number of exacting measurements, this method was preferred in the past when hand calculations were necessary. A number of spe-

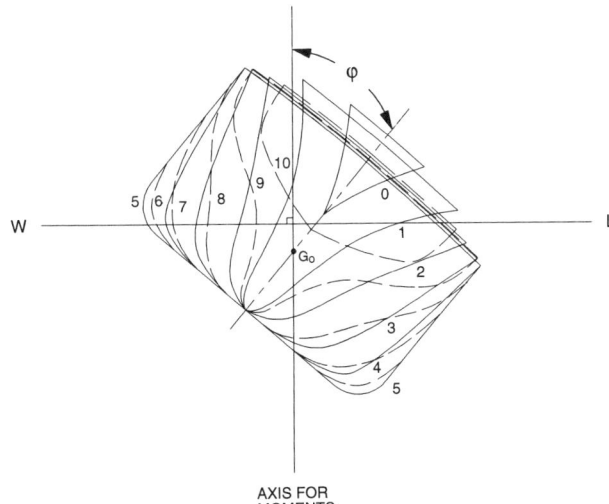

Figure 4-4. Body plan for whole body method.

106 APPLIED NAVAL ARCHITECTURE

cific procedures for defining and integrating the areas and moments of the wedge sections were devised for hand calculations.

When a ship is heeled to a large angle of inclination, the upward-acting buoyant force does not pass through the metacenter, and the upright and inclined waterlines do not intersect on the ship's centerline. This is illustrated in Figure 4-5. The following definitions apply to the figure:

φ = angle of heel imposed by an external force
WL = upright waterline
W'L' = waterline when inclined to angle φ
B, B' = centers of buoyancy of ship when upright and heeled
b, b' = centroids of volumes of emerged and immersed wedges
h, h' = projections of b and b' on inclined waterline W'L'
G = center of gravity of the ship
GZ = righting arm at angle of heel φ
Δ = weight (displacement) of the ship
$\overline{BB'}$ is parallel to $\overline{bb'}$
GZ, BR, and $\overline{hh'}$ are all parallel to W'L' and perpendicular to the force of buoyancy through B'Z

Since the ship has been inclined by externally applied forces, its displacement remains unchanged. Therefore the volumes of the emerged and immersed wedges are equal.

Let v = volume of emerged and immersed wedges
∇ = volume of displacement of ship

The ship's center of buoyancy shifts from B to B' in response to the shift of wedge buoyancy from b to b'. Expressed in terms of moments of buoyant volume,

$$v \times \overline{bb'} = \nabla \times \overline{BB'} \qquad (3)$$

A similar moment equation can be written for both volume moments resolved into the horizontal plane of the inclined waterline W'L':

$$v \times \overline{hh'} = \nabla \times BR \qquad (4)$$

Therefore
$$BR = \frac{v \, \overline{hh'}}{\nabla} \qquad (5)$$

The ship weight acts vertically downward through center of gravity G. Thus

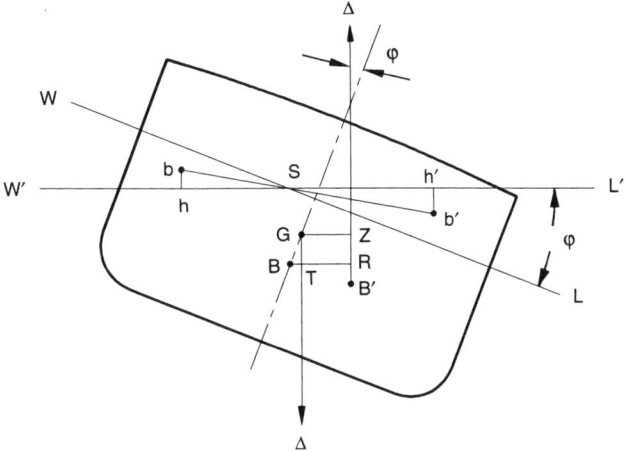

Figure 4-5. Righting arm by wedge method.

$$GZ = BR - BT \tag{6}$$

where, in triangle GBT, with angle φ at vertex G,

$$BT = GB \sin \varphi \tag{7}$$

Combining equations (5), (6), and (7):

$$GZ = \frac{v \, \overline{hh'}}{\nabla} - GB \sin \varphi \tag{8}$$

Equation (8) is the fundamental equation of stability, applicable to any angle of inclination, large or small. It appears to be very simple, but its solution in a practical case requires considerable calculation. If the hydrostatic calculations for the upright condition of a ship have already been done, and an assumed KG is chosen, it is clear that for any given upright waterline like WL in Figure 4-5, the displacement (and hence ∇), KB, KG, and GB will be known. Any angle may then be chosen. Determining GZ for the given condition then reduces to calculating wedge volume v and horizontal "arm" $\overline{hh'}$. Here, the difficulties of applying this formula begin. A trial-and-error iterative procedure must be used to determine the position of the inclined waterline such that the volumes of emerged and immersed wedges are equal. Wedge volumes must be calculated by numerical integration of the wedge cross-sectional areas over the length of the ship. Finally, moments of the wedge volumes about the point S in the figure (intersection of waterlines WL and W'L') must be calculated in order to determine b and b', and hence h and h'. The complications arising in these calculations, especially when the deck edge is immersed as in the given figure, are immediately apparent.

CROSS CURVES OF STABILITY

The results of the righting arm calculations for a ship are plotted in a set of curves known as *cross curves of stability*. The entire set of cross curves of stability apply to the ship only if its center of gravity coincides with the pole, or position of G, assumed in making the calculations. A typical set of cross curves is shown in Figure 4-6. The range of displacements over which cross curves have been determined is from the light ship displacement at the lower end to a displacement usually well above the load displacement, so that stability can be assessed at deep drafts associated with potential flooding situations.

In the preparation of cross curves of stability, certain assumptions have been made, as follows:

- The ship's center of gravity remains fixed at the pole point, or assumed center of gravity, regardless of the angle of heel.
- The ship's hull, consisting of the bottom, sides, and weather deck, is assumed to be perfectly watertight.
- Superstructures and deckhouses above the weather deck are normally assumed to be nonwatertight. Any actual watertightness of such structures, maintained by the proper closure of watertight doors, will provide a margin of safety of additional intact stability beyond that indicated by the cross curves at angles of heel that immerse the structures in question.
- Adjustments are made to account for the volumes and moments of immersed appendages such as rudder, propellers, sonar domes, etc., and freely flooding spaces like large seachests.

CONSTRUCTING A STATICAL STABILITY CURVE

Cross curves of stability are a convenient form in which to store the information necessary to determine the large-angle stability characteristics of a ship at any displacement, but at a single assumed position of the center of gravity. For the ship operator as well as for the naval architect during the design process, what is needed is a determination of the righting arms or moments of a ship heeled to any angle while in a given loading condition at one displacement and with its center of gravity at a position different from the one that was assumed in preparing the cross curves. The *statical stability curve* in which righting arms are plotted against angle of heel is the appropriate format.

The statical stability curve was introduced in Figure 4-3 as a sine curve for a body of revolution. For a ship whose cross curves are given, the statical sta-

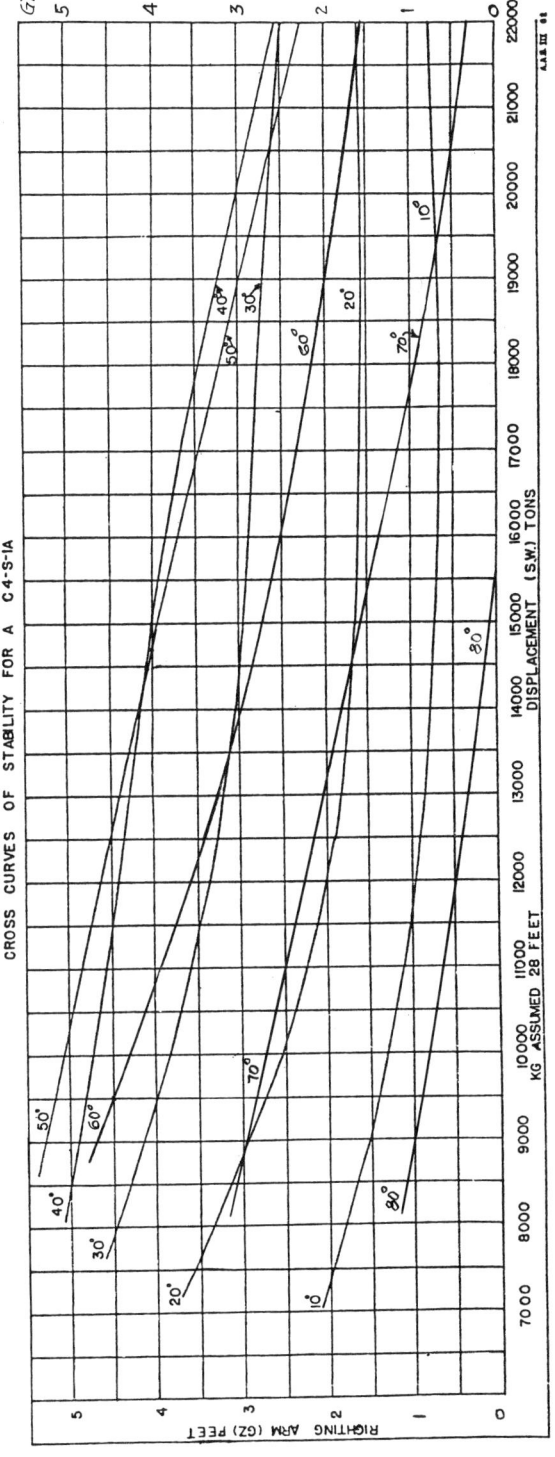

Figure 4-6. Cross curves of stability, Mariner class.

bility curve is determined by picking values of the righting arms from the cross curves at the appropriate displacement and correcting the values thus obtained to reflect the actual position of the center of gravity. Two corrections may be required: one for the height of G above the keel (KG) and one for the distance of G off centerline (TCG). The vertical correction is almost always necessary, since it would be only by gratuitous coincidence that a ship's KG in a given loading condition would be exactly at the pole point chosen arbitrarily when the cross curves were prepared. The lateral correction is, by contrast, rarely needed in routine loading conditions because good practice in loading a ship requires that the center of gravity be on centerline so that the ship will float upright.

The correction for the vertical position of G is shown in Figure 4-7. In the figure,

G_c = center of gravity assumed in preparation of the cross curves (pole point)
$G_c Z_c$ = righting arm from cross curves
G = actual center of gravity
GZ = actual righting arm
φ = angle of heel

Comparison of the actual righting arm with that read out from the cross curves shows that

$$GZ = G_c Z_c - GG_c \sin \varphi \qquad (9)$$

and it is seen that all righting arms will be smaller than those plotted in the cross curves, because the actual center of gravity is higher than the assumed center of gravity. Thus stability at large angles decreases as G rises, just as initial stability measured by GM does. Supposing that the actual center of gravity is below the assumed one, it is easy to show that the righting arms will be larger than those from the cross curves, also in accordance with equation (9). To demonstrate that, imagine that in Figure 4-7 the positions of $G_c Z_c$ and GZ are reversed. The same sine correction would result, but it would be added to $G_c Z_c$. Combining these two results, we have

$$GZ = G_c Z_c \pm GG_c \sin \varphi \qquad (10)$$

in which we must choose *plus* if G is *below* G_c, and choose *minus* if G is *above* G_c.

STABILITY AT LARGE ANGLES

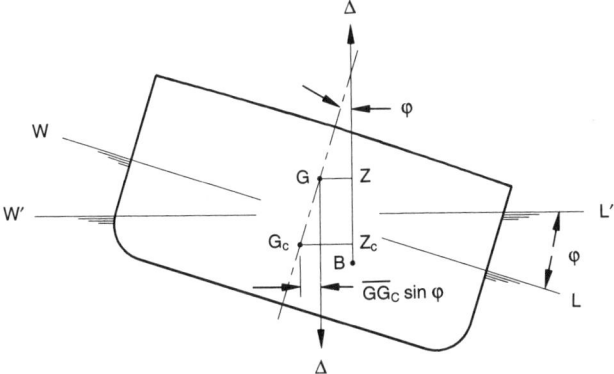

Figure 4-7. Effect of vertical shift of G on GZ.

Example 4-1

Determine statical stability curves of a Mariner class ship, whose cross curves of stability are given in Figure 4-6, at a displacement of 12,000 tons, and with KG values as follows:

(a) KG = 26 ft
(b) KG = 28 ft
(c) KG = 30 ft

Plot all three curves on the same axes.

Since the cross curves were determined for an assumed KG of 28 feet, curve (b) is determined by reading out and plotting the uncorrected values of GZ from Figure 4-6 at 12,000 tons displacement.

For curve (a),
GG_c = 28 − 26 = 2 feet
$GG_c \sin \varphi$ = 2 sin φ, to be added, since the actual G is below the assumed G.

For curve (c),
GG_c = 30 − 28 = 2 feet

and the correction 2 sin φ is to be subtracted, since the actual G is above the assumed G. The work is most conveniently arranged in tabular form, as follows.

112   APPLIED NAVAL ARCHITECTURE

"1"	"2"	"3"	"4"	"5"
$\varphi°$	G_cZ_c (KG = 28) Δ = 12,000	$GG_c \sin \varphi =$ $2.0 \sin \varphi$	GZ at KG = 26 "2" + "3"	GZ at KG = 30 "2" − "3"
0	0	0	0	0
10	0.95	0.35	1.30	0.60
20	2.00	0.68	2.68	1.32
30	3.38	1.00	4.38	2.38
40	4.42	1.29	5.71	3.13
50	4.62	1.53	6.15	3.09
60	3.60	1.73	5.33	1.87
70	2.27	1.88	4.15	0.39
80	0.50	1.97	2.47	−1.47

The three statical stability curves given by columns 2, 4, and 5 in the above table are shown in Figure 4-8. It is apparent that the righting arms at all angles of heel are reduced as the ship's center of gravity rises and KG increases.

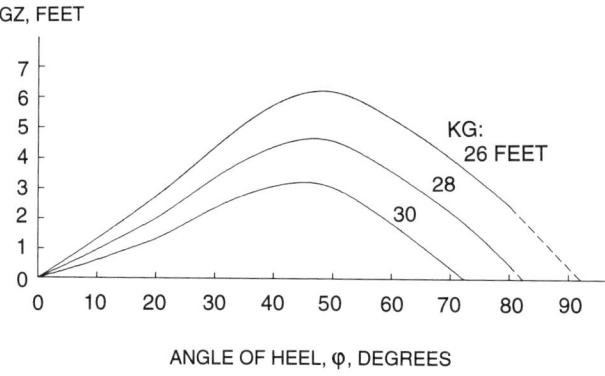

Figure 4-8.

ANATOMY OF THE STATICAL STABILITY CURVE

The conventional plotting of a statical stability curve is shown in Figure 4-9. Since the center of gravity is on centerline, the direction of assumed heel (starboard or port) is immaterial, because the centerline symmetry of the hull would cause the center of buoyancy to assume symmetrically corresponding positions on either side at any given angle of heel. Although the curve can be plotted for all angles of heel for which cross curves have been determined, it is typically terminated where the GZs become negative, that is, where righting arms change to capsizing arms.

STABILITY AT LARGE ANGLES

To have a full understanding of intact ship stability, we must know not only how a statical stability curve is determined, but also why it is shaped as shown, and what significance is to be attached to its typical features. Let us therefore examine the important features, or the "anatomy" of a typical statical stability curve.

The First Few Degrees of Heel

The validity of a statical stability curve is not restricted to large angles of heel the way the study of initial stability was restricted to small heel angles. The curve of righting arms is valid for all angles, small and large. The initial portion of the curve (say the first 10 degrees or so) must therefore be consistent with the measure of initial stability, that is, the metacentric height (GM). The relationship can be determined by recalling that for small angles,

$$GZ = GM \sin \varphi \quad [\varphi \text{ small}] \qquad (12)$$

and noting that as φ approaches zero, $\sin \varphi = \varphi$ (radians). Thus for these small angles, it is as correct to write

$$GZ = GM \times \varphi \quad [\varphi \text{ small}] \qquad (13)$$

Either of equations (12) or (13) may be plotted on the statical stability curve, the simpler of the two being the straight line represented by equation (13), which has a slope equal to the metacentric height (GM). The statical stability curve traces that straight line at the origin, departing from it very little for the first few degrees. This property of the statical stability curve, namely that *the metacentric height is a measure of the slope of a statical stability curve at the origin,* should always be used as an aid to plotting the curve, by running the curve in tangent to

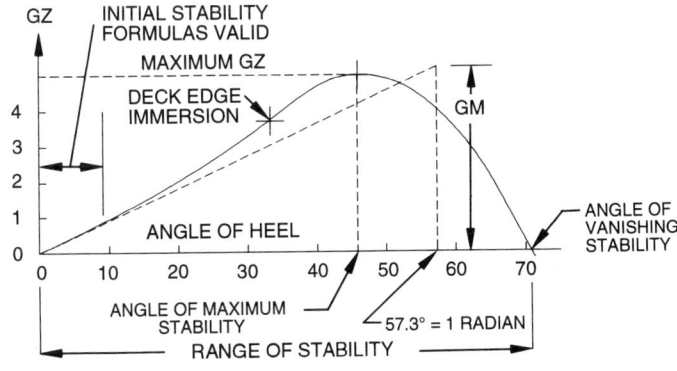

Figure 4-9. Anatomy of the statical stability curve.

the straight line at the origin. Although the straight line is meaningless beyond the first few degrees of heel, as an aid to plotting it, it is customary to establish it by noting that, at an angle of one radian (equal to $180/\pi$, or 57.3 degrees), the straight line passes through the value GZ = GM. Thus, as is shown in the figure, if GM is laid out as an ordinate at 57.3 degrees and that point is connected to the origin by a straight line, the statical stability curve will approach that line asymptotically as it approaches the origin. It should be noted that had a sine curve of amplitude GM been plotted as well, all three curves would be coincident near the origin.

Increasing Slope Beyond Initial Straight Line
For the great majority of ship hull forms, the statical stability curve departs from its initial straight path with increasing slope so that it rises above the tangent line as the angle of heel increases. The reason for this is shown diagrammatically in Figure 4-10. As the ship heels from waterline zero (WL 0) to WL 1, buoyancy is transferred from emerged wedge number 1 to immersed wedge number 1. As the shift of buoyancy is largely across the ship and only slightly upward, as shown by the broken lines in the figure, the shift in the center of buoyancy to B_1 is mostly athwartships, thus producing righting arms that increase at a relatively rapid rate. A similar shift takes place in the next incremental shift of wedge buoyancy, denoted by wedges number 2, and in number 3 as well. So long as this general trend continues, righting arms GZ rise at an ever-increasing rate.

Deck Edge Immersion
Eventually, as the angle of heel continues to increase, a point is reached at which deck edge immersion takes place, denoted in Figure 4-10 as waterline WL 3. In fact, since a ship's section shapes vary from bow to stern, deck edge immersion is not a sudden occurrence along the entire length of the ship, but takes place gradually over a range of heel angles. But the general trend, as shown in Figure 4-10, is that the shifts of buoyancy in wedges 4 and 5 (and beyond) after deck edge immersion have much smaller transverse components and much larger vertical components than previous ones, as shown by the broken lines for shift number 5. The resulting upward, rather than outward, path of the center of buoyancy is accompanied by a much-reduced growth in GZ, and thus an inflection point in the statical stability curve. This steadily decreasing slope of the statical stability curve beyond deck edge immersion leads to the peak of the curve, and ultimately to its rapid plunge beyond the peak.

This effect of deck edge immersion on large angle stability has important consequences for both the design naval architect and the ship operator. The design implications are that a ship designed to have a small freeboard (those that carry dense cargo which does not require large cargo space volume) may develop inadequate righting arms and moments at large angles because deck edge immersion and the peak of the GZ curve will occur at relatively small angles of heel.

STABILITY AT LARGE ANGLES

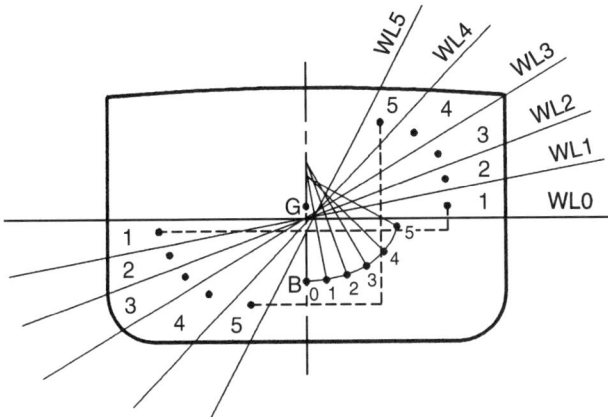

Figure 4-10. Incremental transference of buoyancy.

To avoid this problem, low-freeboard ships must be designed with relatively large metacentric heights, because the large initial slope of the statical stability curve will tend to ensure that adequate righting moments will be achieved in spite of the small angle at which the curve reaches its peak. For the ship's officer, charged with the safe operation of the ship, deck edge immersion caused by rolling or by a steadily increasing list of the ship at sea is an easily observed condition for which corrective action must be taken whenever possible, since it could well be a precursor to capsize.

The Peak of the Curve

The peak of a statical stability curve identifies two quantities that are important in evaluating the overall stability of a ship. They are the *maximum righting arm* and the *angle of maximum stability*. The importance of the maximum righting arm (GZ_{max}) is that the product of the displacement and GZ_{max} is the maximum steady heeling moment that the ship can experience without capsizing; to put it another way, if a ship is inclined to the angle of maximum stability by an externally applied constant heeling moment, the ship will capsize. More about heeling moments and their relationship to the statical stability curve will be discussed later.

Range of Stability

Beyond the angle of maximum stability, righting arms decrease, often more rapidly than they had increased up to that point. The rapid decrease ultimately leads to the point at which GZ becomes zero, and the curve recrosses the axis. The angle at which this occurs is the *angle of vanishing stability,* because thereafter the GZs are negative. That is, they are capsizing or upsetting arms, rather than righting arms. Any ship that inclines beyond its angle of vanishing stability will capsize, regardless of the cause of the inclination or its duration. The typical statical stability curve like that in Figure 4-9 crosses the horizontal axis at two angles of inclination, each of which represents a condition of static equilibrium, since GZ equals zero. The two zero crossings are, however, very unlike each other as regards the ship's behavior. An "up-crossing" of the zero axis (GZ's positive as heel increases beyond the crossing) is a stable equilibrium condition, because temporary inclinations to larger angles create righting moments that will restore the ship to the equilibrium angle when the cause of the inclination is removed. Temporary inclinations in either direction from a "down-crossing" angle (GZs negative at larger angles) such as the angle of vanishing stability, however, will cause the ship to incline away from that equilibrium condition when the cause is removed. A down-crossing angle is thus an angle of unstable equilibrium. Since a ship forced by external heeling moments to heel or roll to any angle of heel between up-crossing and down-crossing will return to the stable equilibrium attitude upon release of the heeling moment, the range of angles between the two crossings is called the *range of stability*. The range of stability should always be described by quoting both equilibrium angles, even in the most common case in which the stable equilibrium angle is zero. Thus a ship that is in stable equilibrium when upright and whose angle of vanishing stability is 80 degrees is said to have a range of stability of zero to 80 degrees. Situations in which the stable equilibrium angle is not at zero degrees will be addressed later.

As a practical matter, one must be careful not to depend too much on any ship's ability to recover from angles of inclination beyond its angle of maximum stability because the cross curves of stability were determined *on the assumption of perfect watertight integrity of the weather deck.* Except for submarines, this assumption is incorrect. Openings must be provided in the deck for cargo hatches, access to spaces below decks, ventilation, piping, and any of a host of other practical purposes. Properly designed deck penetrations are made weathertight, but few of them can be made truly watertight, and there is always the possibility of human error, that is, leaving doors and hatches open that should be tightly closed against heavy weather. At angles of heel that immerse part of the deck, the possibility always exists that water will be shipped through such openings. This event is called *downflooding,* and the smallest angle of heel at which it can occur is known as the *downflooding angle.* If large quantities of water enter the ship, the statical stability curve is no longer correct; the ship's loading has changed. Therefore at any angle of heel greater than the downflooding angle, a statical stability curve must be considered invalid. For large ships, downflooding is rarely considered to be a critical issue, but for some smaller vessels, fishing craft in particular, fish holds with hatches open on the deck combined with small freeboards pose a real threat of dangerous downflooding. The range of stability of such vessels should be assumed to be terminated at the downflooding angle rather than at the angle of vanishing stability.

STABILITY AT LARGE ANGLES

NEGATIVE INITIAL STABILITY AND THE STATICAL STABILITY CURVE

A ship with a negative metacentric height (G above M) cannot float upright since it is then in unstable equilibrium. Is it therefore destined to capsize at the slightest external disturbance? Fortunately for most ships, the answer to this question is no, although this conclusion cannot be reached by the study of initial stability alone. A look at the statical stability curve of a ship with negative GM provides the evidence to support this claim.

Example 4-2

A Mariner class ship floats in seawater at a draft of 27′0″ and with KG = 31.8 feet. Show that this ship is unstable when upright. Calculate and plot the statical stability curve for this condition of loading.

From the Mariner hydrostatic data, Appendix A, we read at a draft of 27′0″ that Δ = 18,800 tons and KM = 31.1 feet. Therefore the metacentric height is

$$GM = KM - KG = 31.1 - 31.8 = -0.7 \text{ ft}$$

Thus GM is negative, the ship is unstable when upright, and it is unable to float upright. Since the ship's center of gravity is more than 28 feet above the keel (as assumed in constructing the cross curves), righting arms from the cross curves of stability must be corrected by subtracting $GG_c \sin \varphi$, where

$$GG_c = KG - 28 = 31.8 - 28.0 = 3.8 \text{ ft}$$

Reading the GZs from the cross curves (Figure 4-6) at a displacement of 18,800 tons, the following table is constructed:

"1"	"2"	"3"	"4"
$\varphi°$	$G_c Z_c$ (KG = 28) Δ = 18,800	$GG_c \sin \varphi$ = 3.8 sin φ	GZ at KG = 31.8 "2" – "3"
0	0	0	0
10	0.63	0.66	–0.03
20	1.53	1.30	+0.23
30	2.72	1.90	+0.82
40	3.26	2.44	+0.82
50	3.02	2.91	+0.11
60	2.04	3.29	–1.25

Beyond the angle of 60 degrees, calculations are unnecessary, as all GZs will be negative.

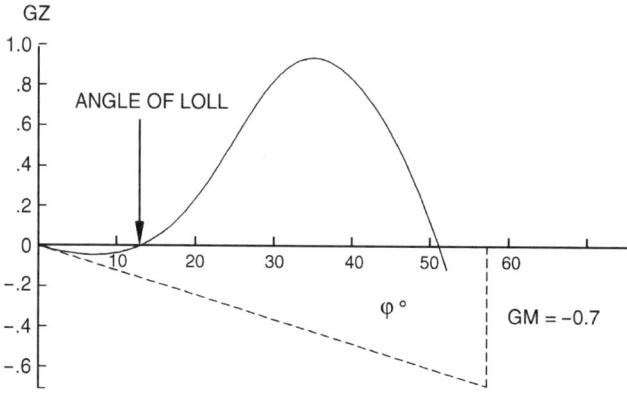

Figure 4-11.

The resulting statical stability curve, including the negative GM plotted at one radian, is shown in Figure 4-11.

Note the unique shape of the statical stability curve in Figure 4-11. There are three points of equilibrium defined by zero crossings, but two are unstable equilibrium points—the downcrossings of the zero axis. The negative metacentric height is set out below the axis, and the statical stability curve starts out at a negative slope, as expected. But since the ship form is such that the curve rises above its initial tangent line, the curve crosses the axis at $\varphi = 13$ degrees, and is thereafter positive until the angle of vanishing stability, 51 degrees in this case, is reached. The angle at which the ship regains stable equilibrium is called the *angle of loll,* a term reserved only for the special case of negative initial stability. The ship of Figure 4-11 will come to rest at the angle of loll of 13 degrees in still water. The range of stability for this ship is from the angle of loll to the angle of vanishing stability, that is, from 13 degrees to 51 degrees.

A unique feature of the angle of loll is that it exists on both sides of the upright. The ship will seek to come to rest at its angle of loll either to port or starboard, depending on the direction of the moment that caused its initial heel from the upright. To illustrate this point, a *two-sided statical stability curve* is shown in Figure 4-12. A two-sided curve is plotted by reversing the positive and negative GZ directions on the left side of the diagram. Thus, on the right side of the plot, righting and heeling arms are plotted in the usual sense, upward for righting. On the left side, righting arms are plotted downward. The choice of which sides represent starboard and port, respectively, is arbitrary. Plotted in this way, the curve is continuous and smooth through the origin, and the significance of stable upcrossings and unstable downcrossings is preserved. If left-pointing ar-

STABILITY AT LARGE ANGLES

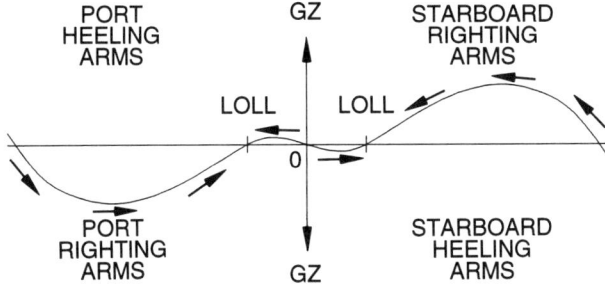

Figure 4-12. Two-sided statical stability curve, negative GM.

rows are imagined on all parts of the curve that are above the axis and right-pointing arrows on the parts below the axis, the arrows will always point toward a stable equilibrium point and away from an unstable one.

This method of plotting makes clear the nature of an angle of loll. The ship cannot remain upright, because that is an unstable equilibrium point. Instead, it will settle to the angle of loll on either side of upright. As it rolls to angles less than the angle of loll, it will return to the angle of loll and appear to be listing in that direction. A larger disturbance can cause it to roll beyond the upright, whereupon it will then settle temporarily to the angle of loll in the other direction. This produces an erratic rolling behavior, coupled with a very long rolling period and hence a slow or lazy roll. These two characteristics are the unmistakable signs of negative initial stability. Although the ship is capable of surviving this condition without capsizing so long as it remains intact, it is not an acceptable condition, and in fact can be quite dangerous if the GM continues to reduce or if the ship takes on flooding water from collision or downflooding. During normal operations, a ship's center of gravity usually rises in the course of a voyage because of the fuel consumed from the double bottom tanks. The prudent shipmaster will periodically check the GM as it decreases during the voyage. Should the GM approach zero, steps must be taken to lower the center of gravity by ballasting empty ballast tanks or by shifting liquids to lower tanks if that is possible.

HEELING MOMENTS AND THE STATICAL STABILITY CURVE

Heeling moments caused by the lateral shift of weights were dealt with in Chapter 3 by applying the inclining experiment formula. That formula gives satisfactory results in describing the relationship between small heeling moments and the small angles of list they produce. The formula fails, however, when large heeling moments capable of causing heel angles greater than 10 degrees or so are experienced. Determining the angle of heel or list in such circumstances requires resorting to the statical stability curve.

Consider a ship whose loading is such that the ship's center of gravity is not on centerline, but has shifted laterally to a position like G' in Figure 4-13. Righting arms GZ have been determined from the cross curves of stability, already corrected for the appropriate height of the center of gravity, assumed to be on centerline. We must determine the reduction in righting arm from GZ to G'Z' when the ship is heeled to angle φ. The shift of the center of gravity (GG') is assumed to be purely lateral, that is, perpendicular to the centerline or parallel to the waterline when the ship was upright. Both righting arms are measured parallel to the inclined waterline. The corrected righting arm is therefore given by

$$G'Z' = GZ - GG'\cos\varphi \qquad (14)$$

as shown in Figure 4-13. The condition shown is not the equilibrium angle assumed by the ship as a result of the shift of G. It depicts the righting arm that would develop under the influence of an externally applied moment in addition to the moment caused by the shift of the center of gravity. The correction GG' cos φ is shown as a reduction in the righting arm because the direction of the angle of heel is assumed to be toward the low side, that is, in the same direction as the shift of G. Had we examined an external moment in the opposite direction, GZ would increase by the amount GG' cos φ. In general, we are interested only in the negative correction for a transverse shift of G, because it is imperative to identify the potentially unsafe condition associated with the reduced righting arms as the ship is inclined toward the low side.

Example 4-3

A Mariner floats with no list in seawater at drafts of 22'0" forward and aft, with KG = 28.0 feet. An estimated 500 tons of cargo shifts 30 feet to port. Calculate and plot the statical stability curve for this condition and compare the angle of list from the curve to that predicted by the inclining experiment formula.

From the Mariner hydrostatic data, at a draft of 22 feet, Δ = 14,800 tons and KM = 31.4 feet. Therefore:

$$GM = KM - KG = 31.4 - 28.0 = 3.4 \text{ ft}$$

No sine correction is required in this case because the KG of the ship is equal to that assumed for the cross curves. But the shift of cargo causes a lateral shift of the ship's center of gravity as follows:

$$GG' = \frac{w\,d}{\Delta} = \frac{500 \times 30}{14,800} = 1.01 \text{ ft}$$

STABILITY AT LARGE ANGLES

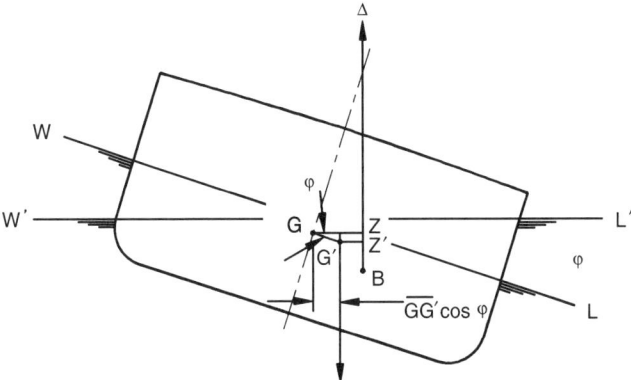

Figure 4-13. Effect on GZ of transverse shift of G.

To correct the righting arms for this shift of the center of gravity, we must subtract 1.01 cos φ from each GZ. The GZs from the cross curves and the corrections to them are shown in the following table and in Figure 4-14.

"1"	"2"	"3"	"4"
	GZ at KG = 28)	GG' cos φ =	GZ corrected
φ°	Δ = 14,800	1.01 cos φ	"2" – "3"
0	0	1.01	–1.01
10	0.81	0.99	–0.18
20	1.67	0.95	+0.72
30	2.98	0.87	2.11
40	4.00	0.77	3.23
50	3.97	0.65	3.32
60	2.82	0.51	2.31
70	1.63	0.35	1.28
80	0.10	0.18	–0.08

Note that, as expected, the statical stability curve of a ship with off-center weights shows that the ship cannot remain upright, because that is not a condition of equilibrium. The stable equilibrium point shown at 12 degrees (up-crossing) is the angle of list produced by the shifted center of gravity. The range of stability is from 12 to 79 degrees. The upright ship has a heeling arm of GG', the distance of the lateral shift of the ship's center of gravity.

The angle of list from the statical stability curve can be compared with the angle of list that would be predicted using the inclining experiment formula, equation (23) of Chapter 3:

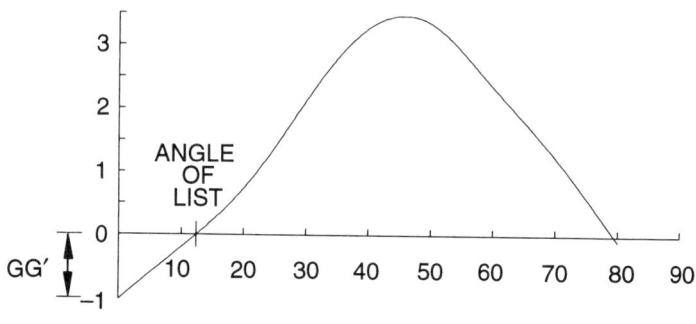

Figure 4-14.

$$\tan \varphi = \frac{w\,d}{\Delta\,GM} = \frac{GG'}{GM} = \frac{1.01}{3.4} = 0.297$$

$$\varphi = 16.5 \text{ degrees}$$

The difference in the angle of list determined by the two methods should be noted. The correct list is the angle determined from the statical stability curve, since that method is valid for any angle, large or small. The inclining experiment formula is accurate only if the angle is very small; in principle, only if it is vanishingly small. Note that the use of the formula overestimates the angle of list. That is always the case.

HEELING MOMENT AND RIGHTING MOMENT CURVES

There is an alternative method of plotting the results of Example 4-3 that is useful in discussing the general case of heeling moments applied to ships. It makes use of the fact that a statical stability curve (which shows the *righting arms* of a ship at a constant displacement) is just a convenient way to represent the *righting moments* that cause the ship to return to stable equilibrium upon removal of the moment that heeled it. Multiplying every GZ in a statical stability curve by the displacement (Δ or Δ_m) of the ship will produce its righting moment curve. Since the displacement is a constant, the shape of the curve does not change—the GZ scale may simply be converted to a righting moment scale by replacing it with the values of $\Delta \times GZ$ or $\Delta_m \times GZ$. If a curve of heeling moments is drawn on the same plot as a curve of righting moments, equilibrium conditions will be represented by the intersections of the two curves. The procedure is illustrated by the following example, using the same data as in the previous example.

STABILITY AT LARGE ANGLES

Example 4-4
Using the righting arms and shift of G given in Example 4-3, do the following:

(a) Plot the righting *moment* curve of the ship prior to the lateral shift of the center of gravity. Extend the curve to the left by 30 degrees, to show heel in the opposite direction.

(b) Plot the curve of heeling moments produced by the shift of G on the same axes as the righting moment curve.

(c) Identify the angle of list, the angle of vanishing stability, and the range of stability.

The righting moment and heeling moment values are determined by multiplying the values in columns "2" and "3" of the solution to Example 4-3 by the displacement, 14,800 tons. This produces the following table:

"1" $\varphi°$	"2" $\Delta \times GZ$ Ft – tons	"3" $\Delta\, GG' \cos \varphi$ Ft – tons
0	0	14,948
10	11,988	14,652
20	24,716	14,060
30	44,104	12,876
40	57,200	11,396
50	58,756	9,620
60	41,736	7,548
70	24,124	5,182
80	1,480	2,664

These curves are plotted in Figure 4-15. The angle of list is read from the first crossing of the two curves (12 degrees), the angle of vanishing stability from their second crossing (79 degrees). The manner of showing the range of stability is indicated on the plot. Note that extending the curves to the left of the origin shows the increase in the moment tending to cause the angle of list as the ship is heeled to starboard, that is, in the direction opposite to that of the shift of weight.

Righting and heeling moments were plotted in Figure 4-15 because moments of forces are what cause a ship to heel, list, return to equilibrium, or capsize. But since the ship's displacement does not vary with the angle of heel, a further simplification may be made to this technique for solving large-angle heeling problems. The righting arms from the statical stability curve need not have been converted to righting moments, a procedure which changed the values but not the

124   APPLIED NAVAL ARCHITECTURE

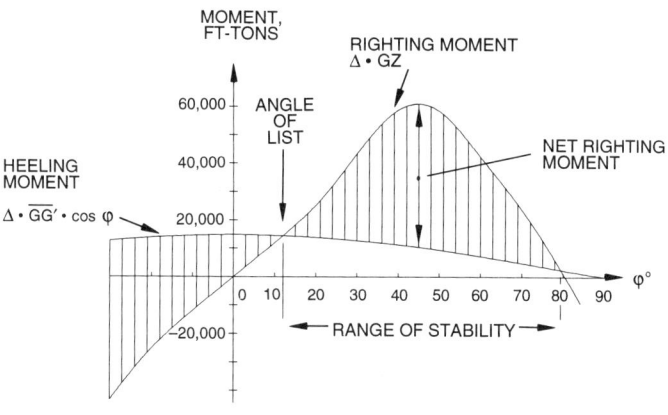

Figure 4-15.

shape of the curve. Instead, the heeling moment values ($\Delta\, GG'\cos\varphi$) may be divided by the displacement, and the resulting curve ($GG'\cos\varphi$) plotted on top of the statical stability curve. The resulting plot will have exactly the same shape, angle of list, and all of the important characteristics of Figure 4-15, except that all ordinates will be expressed as "arms" rather than moments. Any heeling or righting moment corresponding to a particular arm may then be determined by multiplying the arm by the displacement. Heeling "arms" determined in this way are not always the measure of a distance with any physical meaning. They must be understood only as heeling moments divided by ship displacement. Calculating and displaying the curves as righting and heeling arms rather than moments is simply a matter of convenience.

APPLICATIONS OF HEELING MOMENT AND HEELING ARM CALCULATIONS

Off-Center Weights

Examples 4-3 and 4-4 have demonstrated how the angle of list of a ship can be determined when the ship's center of gravity is off center because of a lateral shift of cargo. Typical examples of actions that can produce relatively large angles of list are:

- Inadequately stowed cargo shifting within the holds while the ship is rolling violently in heavy weather.

- Cargo-loading operations, especially heavy-lift cargo handled by ship's gear.
- Passenger or personnel crowding—large numbers of people all moving to one side of a ship at the same time.
- Topside icing caused by operations in subfreezing temperatures with wind-driven spray freezing on the upper works of the ship mostly on one side.

All of these situations will cause the ship's center of gravity to shift laterally by a distance equal to GG′ = wd/Δ, as explained in Chapter 3. Except for the case of cargo loading, the practical application of this formula for GG′ in the examples described above requires a considerable amount of educated guesswork to determine reasonable estimates of the weight and the lateral distance of the weight shift or its distance off centerline if it is added to the ship. In addition, if the weight in question alters the ship's KG and/or displacement when it is loaded or taken aboard (topside icing, cargo loading) or when it shifts (passengers moving upward as well as laterally, for example), the new KG and displacement must be determined before calculating GG′, which must represent only the lateral shift of the center of gravity. In the case of cargo loading of a heavy item, both weight and distance will be known with some accuracy, but the suspended weight effect must be accounted for if it is lifted by a shipboard cargo boom by assuming that the weight is concentrated at the end of the cargo boom.

Beam Winds

Forces on the sail area of a ship (the projected area of the above-water profile, or silhouette) due to strong winds from abeam have the potential to produce large wind heel angles, especially on ships with large freeboards or large deck loads, such as containerships. Researchers have estimated wind forces and the heeling moments they produce based on measurements of wind velocities at sea, wind tunnel experiments on ship models, and theoretical studies in fluid dynamics. A number of formulas and calculation procedures have been published on the subject. One of the most frequently quoted formulas is given below and illustrated in Figure 4-16. It is a formula for wind heel "arm," which is wind heel moment divided by ship displacement. As mentioned previously, expressing heeling moments as arms by dividing by displacement is convenient because the resulting curve of heeling arm can be plotted on the ship's statical stability curve to determine the equilibrium angle of heel. The formula for wind heel arm (WHA) is:

$$\text{WHA} = \frac{0.0035 \, V_w^2 \, Al \cos^2 \varphi}{2{,}240 \, \Delta} \qquad (15)$$

126 APPLIED NAVAL ARCHITECTURE

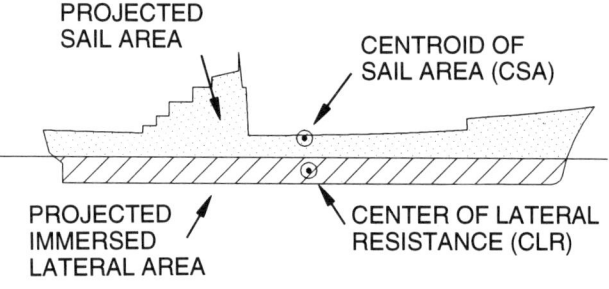

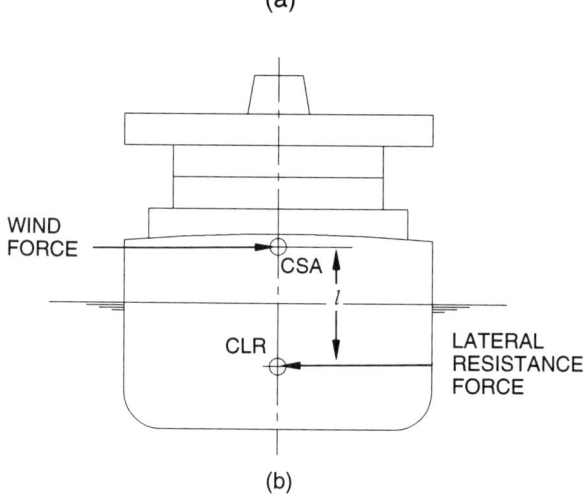

Figure 4-16. Wind heeling moment.

where: WHA = wind heel arm, feet
        V_w = wind velocity, knots
        A = projected "sail area," ft^2
        l = vertical distance between the centroid of the projected sail area and the centroid of the projected underwater hull profile (also called the center of lateral resistance)
        φ = angle of heel
        Δ = displacement, long tons

Converted into SI units, with WHA and l measured in meters, A in m^2, and Δ_m in metric tons, but retaining V_w in knots, the formula becomes

STABILITY AT LARGE ANGLES

$$\text{WHA} = \frac{0.0171 \, V_w^2 \, Al \cos^2 \varphi}{1{,}000 \, \Delta_m} \tag{15a}$$

These formulas are approximate because the coefficient (0.0035 or 0.0171) is an average value derived from wind tunnel tests, and because the wind velocity is assumed to be constant over the entire sail area. Actual wind velocities measured at sea show that there is considerable variation of the velocity, depending on the height of the anemometer above the sea surface. When these formulas are used with a constant wind velocity, it is intended that the velocity represent measurements taken at the standard anemometer height of 10 meters (33 feet) above the sea surface.

Example 4-5
A Mariner class ship is at a mean seawater draft of 22′3″ and with KG = 28.0 ft. At this draft, the sail area (projected area of the above-water profile) is 16,150 ft² and the centroid of the sail area is 18.9 ft above the waterline. Plot the statical stability curve of this ship and superimpose on it the wind heel arm curve for a 100-knot beam wind. What angle of heel will this wind produce?

The displacement at a draft of 22′3″ in seawater is 15,000 tons. The righting arms for this displacement and KG = 28 ft have been read directly from the cross curves of stability (Figure 4-6) without correction. They are listed in the table below and shown in Figure 4-17. The center of lateral resistance is at half the draft, or 11.1 ft below the waterline, hence the lever arm of the wind heel moment is

$$l = 18.9 + 11.1 = 30.0 \text{ ft}$$

The wind heel arm is, from equation (15),

$$\text{WHA} = \frac{0.0035 \, V_w^2 \, Al \cos^2 \varphi}{2{,}240 \Delta}$$

$$= \frac{0.0035 \times 100^2 \times 16{,}150 \times 30.0 \times \cos^2 \varphi}{2{,}240 \times 15{,}000}$$

$$= 0.504 \cos^2 \varphi$$

Values of the wind heel arm for various angles of heel are listed in the table below and plotted in Figure 4-17.

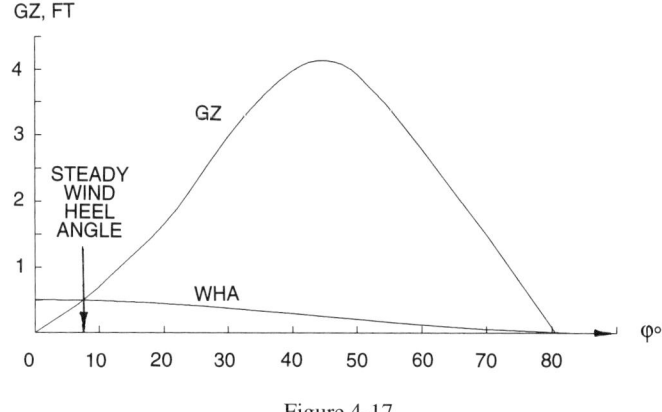

Figure 4-17.

	"1"	"2"	"3"
	$\varphi°$	GcZc (KG = 28) Δ = 15,000	WHA = $0.504 \cos^2 \varphi$
	0	0	0.50
	10	0.70	0.48
	20	1.65	0.44
	30	2.97	0.38
	40	3.98	0.29
	50	3.92	0.21
	60	2.78	0.13
	70	1.59	0.06
	80	0.08	0.01

The plotted curves in the figure intersect at the steady wind heel angle, which is 7.5 degrees.

The seven-degree heel angle of the ship in the above example is a rather modest angle of heel for a ship in hurricane-force winds of 100 knots. It should be noted, however, that this ship is a general dry cargo ship, which has a relatively small sail area. Wind heel can be much more severe on ships with higher profiles. For example, if the above ship were to be converted to a containership by removing the cargo-handling gear and stacking containers on the deck, the sail area would increase substantially, by 50 percent or more. Furthermore, the centroid of the added sail area would be well above the deck. The wind heel angle experienced by such ships in strong beam winds can be significant.

Heeling During a Steady Turn

Any mass moving in a circular path has components of acceleration tangential to the path in the direction of motion and normal to the path directed toward the center of the circle. The moving body subjected to these accelerations is held in dynamic equilibrium by inertial forces opposite in direction to the acceleration components. The equilibrating force in the normal direction is directed outwards, hence it is called the centrifugal inertia force, or more simply, the centrifugal force. The magnitude of the centrifugal force on a mass is

$$F_C = \frac{mv^2}{r} = \frac{wv^2}{gr} \tag{16}$$

where F_C = centrifugal force
  m = mass of moving body
  v = velocity of the body in the tangential direction
  r = radius of curvature of the circular path
  w = weight of the body
  g = acceleration of gravity

For a ship in a steady turn, the centrifugal force, acting at the center of gravity of the ship and directed outward with respect to the turning circle, is

$$F_C = \frac{(\Delta v^2)}{(gR)} \tag{17}$$

where Δ = ship displacement (weight)
  v = ship speed while turning
  R = radius of the turning circle

After the ship has reached a state of turning equilibrium and is in a steady turn, the centrifugal force is resisted by an equal and opposite force at the center of lateral resistance (at approximately the half-draft above the keel). The vertical separation of the two forces is equal to KG – T/2 when the ship is upright, and reduces to (KG – T/2) cos φ as the ship heels to angle φ. The heeling moment caused by this couple can be converted into a "turn heel arm" (THA) by dividing the moment by the ship's displacement:

$$\begin{aligned} THA &= \frac{F_C (KG - T/2)}{\Delta} \\ &= \left(\frac{\Delta v^2}{gR}\right) \frac{(KG - T/2) \cos \varphi}{\Delta} \\ &= \left(\frac{v^2}{gR}\right) (KG - T/2) \cos \varphi \end{aligned} \tag{18}$$

The ship speed in equation (18) must be in units consistent with the units of "g," the gravitational acceleration. The radius of a ship's turning circle is known from the results of turning circle tests, a standard part of the full-scale maneuvering trials that are prescribed for new ships. The turning diameter of most large merchant ships with the rudder at maximum rudder angle is between two and four ship lengths. Thus the typical limiting (minimum) value of R in equation (18) would be one to two times the ship length.

The heel angle that results during a steady turn can be estimated by plotting the curve of turn heel arm given by the above equation on top of the ship's statical stability curve. As noted previously, the steady heel angle is the angle determined by the intersection of the two curves.

The heeling angle at maximum possible turning speed and full rudder angle for most merchant ships is not large enough to be dangerous, mostly because the achievable ship speed in a turn is not excessive. For high-speed combatant ships, however, turning can produce angles of heel that are significant, possibly even dangerous. Officers and helmsmen of such ships should be aware of another, smaller, heeling moment that is also in effect during turning. The turning force on a ship's rudder acts at a height above the keel approximated by the centroid of the rudder, which is below the ship's center of lateral resistance. It creates a heeling moment in the opposite direction of that caused by the centrifugal force. This rudder force heeling moment actually reduces the angle of heel estimated by the above equation. It is ignored when using equation (18) because it is very small compared to the centrifugal moment. The small turning force moment is always in effect during a turn, but the large centrifugal moment takes some time to develop because the ship must be fully into the turn for it to cause its maximum heeling angle. In fact, however, when the outward heeling first develops, it reaches an angle much greater than (sometimes twice as large as) that predicted by the equation because of a dynamic overshooting effect. If a helmsman were frightened that the ship might capsize when it overshoots its steady heeling angle and were to reduce the rudder angle or reverse the rudder at that point, the dangerous heel would be instantly increased because the opposing rudder force moment would be reduced or reversed. The proper action to reduce heel while turning is to reduce speed immediately before (slowly) reducing rudder angle.

PROBLEMS

Note: After constructing each statical stability curve pertaining to the Mariner class ship in these problems, label the following items on the curve and determine their numerical values.

1. Range of stability.
2. Angle of maximum stability.

STABILITY AT LARGE ANGLES

3. Maximum righting arm.
4. If G is on centerline, GM (at 57.3°), and draw straight line initial slope.
5. If GM is negative, the angle of loll.
6. If G is off centerline, the angle of list.

1. A Mariner floats in seawater at a mean draft of 18′5″ with KG = 29.2 ft. Calculate, plot, and label the statical stability curve for this condition.

2. A Mariner displaces 17,500 tons of seawater and has KG = 26.8 ft. Calculate, plot, and label the statical stability curve for this condition.

3. A Mariner floats at a mean draft of 29′8″ with its center of gravity 27.2 ft above the keel and 1.3 ft to port of centerline. Calculate, plot, and label the statical stability curve for this condition. Compare the angle of list determined from the curve with that predicted by the inclining experiment formula.

4. A Mariner floats in seawater at drafts of 24 ft forward and 27 ft aft, and KG = 31.4 ft. Calculate, plot, and label the statical stability curve for this condition.

5. A Mariner at a 22-foot draft even keel and KG = 28.5 ft has yet to load a 320-ton weight on deck at 47 ft above the keel and 24 ft off centerline to starboard. Calculate, plot, and label the statical stability curve for this ship after loading is completed. Compare the angle of list determined from the curve with that predicted by the inclining experiment formula.

6. A 200-pound man has a raft 12 ft wide which weighs 5,800 pounds. He proposes to use it as a float for diving practice. When the man is standing on the centerline of the raft, the righting arms are:

φ, degrees	0	15	30	45	60
GZ, feet	0	0.23	0.67	0.75	0.10

and the freeboard of the raft is 1 ft. What angle of heel will he produce if he walks to the edge of the raft, 6 ft off centerline? Is this a good raft for the purpose? (To answer this question, determine the freeboard on the low side of the raft when he is standing at the edge.)

7. A Mariner floats with no list in seawater at drafts of 25′9″ forward and 26′3″ aft, with GM = 4.0 ft. Then an estimated 350 tons of cargo breaks

loose and shifts 20 ft to starboard. To see how this affects the ship's stability, calculate and plot the statical stability curves for both conditions, before and after the shift of cargo, on the same grid. Compare the angle of list taken from the curve with that predicted by the inclining experiment formula.

8. A Mariner floats at a mean draft of 20 ft, with no list and with KG = 29.0 ft. In hold #4, which is 80 ft long and 76 ft wide, grain in bulk (stowing at 48 ft^3 per long ton) shifts such that its surface (initially level) is lowered by 6 ft on one side and raised by 6 ft on the other side. Determine the angle of list caused by the shift of grain in two ways: by plotting the resulting statical stability curve, and by using the inclining experiment formula.

9. Calculate and plot statical stability curves for a Mariner in the following two conditions of loading:
   1. Full load draft = 29′9″, with GM = 2.0 ft
   2. After discharging from the full load condition 600 tons of cargo from 40 ft above the keel and 20 ft to starboard of centerline

   Compare the ranges of stability depicted by the two curves, and read out the angle of list from curve 2. To which side will the ship list?

10. A destroyer has the following righting arms at a mean draft of 13′0″, at which its displacement is 3,010 tons, with an assumed KG of 14.0 ft.

φ, degrees	0	10	20	30	40	50	60	70	80	90
GZ, ft	0	1.0	1.9	2.7	3.1	3.0	2.6	1.9	1.1	0.3

    If this ship is floating at 13′0″ draft with KG = 16.0 ft and KM = 19.5 ft, determine the steady heel angles that would result from the following three events:

    1. A beam wind of 100 knots blows on the destroyer's above-water profile, or sail area, of 9,000 ft^2. The centroid of the sail area is 13.5 ft above the waterline.
    2. The ship makes a turn at 25 knots. The radius of the turning circle is 625 ft.
    3. All 600 people on board (average weight of 165 pounds per person) crowd to one side of the ship, their centroids shifting, on average, 22 ft laterally.

CHAPTER FIVE

TRIM AND LONGITUDINAL STABILITY

The studies of stability in the previous chapters have been concerned with transverse stability, in which the external forces that disturbed the equilibrium of the ship created heeling moments that caused the ship to rotate about a longitudinal axis. Stability considerations apply to other modes of motion as well. In this chapter we are concerned with *longitudinal stability,* or the tendency of a ship to return to equilibrium after external forces have created moments that cause the ship to rotate about a transverse axis. Such moments are called *trimming moments,* and they result in *changes in trim,* which are observable as changes of the drafts at bow and stern.

THE GEOMETRY OF TRIM

Longitudinal stability is in many ways analogous to transverse stability, the principal difference being in their relative magnitudes. Similarly, trim is analogous to heel, but the measurement of the two is done differently. Before we examine the mechanics of longitudinal stability and trim, it is essential to understand how trim and changes of trim are defined and measured. The ship profile shown in Figure 5-1 illustrates the geometry of trim. Vertical distances in the figure are exaggerated to clarify the definitions.

Definitions
A ship floating at equal draft all along is said to be *on an even keel,* or to have *zero trim.* If the drafts are not the same from bow to stern, the ship is floating with a trim. Deeper drafts aft than forward result in *trim by the stern,* which is a typical operational condition. If the draft forward exceeds the draft aft, the condition is called *trim by the head* or *by the bow.* The measure of the amount of trim is simply the difference between the after and forward drafts.

134   APPLIED NAVAL ARCHITECTURE

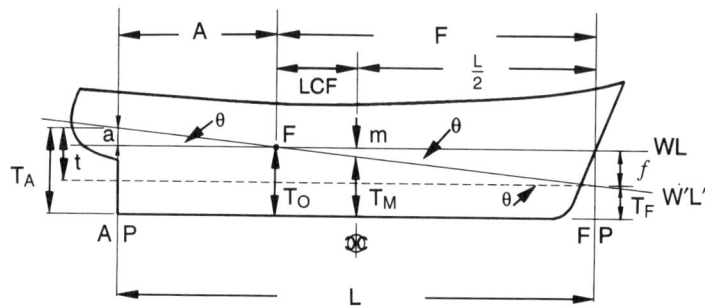

Figure 5-1. The geometry of trim.

Two waterlines (or waterplanes) are shown on the ship in Figure 5-1: a trimmed waterline and the even keel waterline corresponding to the same displacement. They are shown intersecting at the center of flotation of the even keel waterplane. This is the condition that the displacement be the same for both waterplanes, a fact that will be addressed later. The quantities shown in the figure are defined as follows:

L = length of ship between perpendiculars, assumed to be the same as between draft marks in this case
F = distance of center of flotation (point F on the figure) from the forward perpendicular or mark
A = distance of center of flotation from after perpendicular or mark
LCF = distance of center of flotation from amidships
T_F = draft forward
T_A = draft aft
T_M = mean draft or draft at amidships. It is the average of T_F and T_A.
T_O = draft at center of flotation, also called the corresponding even keel draft
t = trim
f = forward difference in drafts, even keel to trimmed waterlines
a = aft difference in drafts, even keel to trimmed waterlines
m = midships difference in drafts, even keel to trimmed waterlines
θ = angle of trim

Trim could be specified by stating the angle of trim, the way heel is characterized by its angle. But the practical consequences of trim are associated with the drafts at the ends of the ship rather than with the angle of trim, so trim is defined as the difference in the drafts aft and forward.

TRIM AND LONGITUDINAL STABILITY 135

$$t = T_A - T_F \tag{1}$$

or $\quad t = T_F - T_A$

The first version of equation (1) applies to trim by the stern and the second to trim by the head.

Angle of Trim
A number of similar triangles shown in Figure 5-1 are created by the two waterlines. Thus the tangent of the angle of trim may be expressed in any of the following forms:

$$\tan \theta = \frac{t}{L} = \frac{f}{F} = \frac{a}{A} = \frac{m}{LCF} \tag{2}$$

or, in general, $\tan \theta = x/X$ where X is the longitudinal distance from the center of flotation to any point on the ship, and x is the difference between the even keel and trimmed drafts at the point in question.

Conventional practice in making draft marks on ships and in reading the marks to record draft changes requires a conversion of units to be introduced to these equations. The conventions are:

	Drafts	Trim	Longitudinal measurements
U.S. units	feet and inches	inches or feet and inches	feet (decimal)
SI units	meters	centimeters	meters

Observing these conventions, the above equations may be rewritten to include the appropriate unit conversions. Note that, to avoid confusion of units, all quantities in Figure 5-1 that are measured in feet or meters are shown in uppercase, and those that are measured in inches or centimeters are shown in lowercase. Rewriting the equations in both systems of measurement:

U.S. units:

$$t = T_A - T_F \qquad \text{[drafts in feet and inches, t in inches]} \tag{1a}$$

$$\tan \theta = \frac{t}{12 L} = \frac{f}{12 F} = \frac{a}{12 A} = \frac{m}{12 LCF} \tag{2a}$$

SI units:

$$t = 100 (T_A - T_F) \qquad \text{[drafts in meters, t in centimeters]} \tag{1b}$$

$$\tan \theta = \frac{t}{100\,L} = \frac{f}{100\,F} = \frac{a}{100\,A} = \frac{m}{100\,\text{LCF}} \qquad (2b)$$

Change of Trim

When the trim of a ship changes as a result of loading, discharging, or shifting weights, the measure of the change of trim is independent of the drafts of the ship. Change of trim is defined as the final trim minus the initial trim, regardless of how small or how large the change of mean draft. Every trim and change of trim must be specified by both a magnitude and a direction or sense—by the head or by the stern. The calculation of a change of trim is therefore an algebraic subtraction. It can be done by assigning an algebraic sign to each direction, for example, plus for "by the stern" and minus for "by the head," before subtracting initial trim from final trim. The sign of the resulting change of trim will then correctly define the direction of the change. Alternatively, a simple rule may be applied: If the directions of the final and initial trims are the same as one another, the magnitude of the change of trim is the difference between final and initial trims. If the direction of trim changes from head to stern or from stern to head, the change of trim is determined by adding initial and final trims. The direction of the change is apparent from the magnitudes and directions of the initial and final trims.

THE MECHANICS OF TRIM

It is essential to understand the geometry of trim and the definitions associated with trim and change of trim before attempting to determine the changes in drafts that result from loading, discharging, and shifting weights aboard ship. We are prepared now to examine the mechanics of trim which will demonstrate the relationships between the physical causes of trim change and the measure of the consequent changes in drafts.

Longitudinal Stability

Suppose a ship is induced to incline longitudinally (i.e., change trim) such that its draft increases at one end and decreases at the other, as shown in Figure 5-2. The moment causing this inclination is assumed to be applied without changing the displacement, and the angle of inclination is assumed to be very small. As in the study of transverse stability (Chapter 3, Figure 3-8), the rotation causes a shift in buoyancy from an emerged wedge to an immersed wedge, and the center of buoyancy of the ship moves in a circular arc from B to B′. The lines of action of the buoyant force (equal to the ship's displacement) associated with B and B′ respectively intersect at a point defined as the *longitudinal metacenter* (M_L). The mechanics of this longitudinal transfer of buoyancy and its relationship to the properties of the hull form are entirely analogous to the transverse inclination ex-

TRIM AND LONGITUDINAL STABILITY

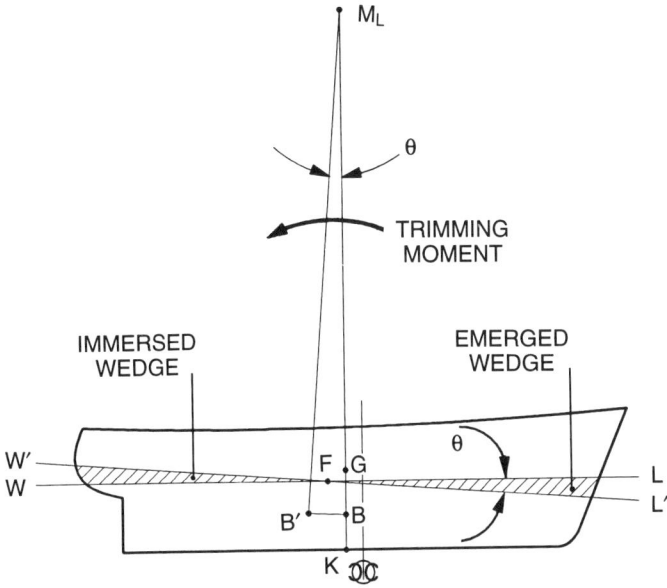

Figure 5-2. Longitudinal stability.

amined in Chapter 3. The resulting equation will be stated here by analogy to equation (18) of Chapter 3, thus:

$$BM_L = \frac{I_L}{\nabla} \quad (3)$$

where BM_L = longitudinal metacentric radius
I_L = longitudinal moment of inertia of the waterplane
∇ = volume of displacement of the ship

The derivation of the above formula for the longitudinal metacentric radius proceeds in the same way as that for the transverse metacentric radius, by integrating longitudinally to obtain the volumes and moments of volumes of the emerged and immersed wedges, respectively. Although the process is analogous to that shown in equations (11) through (18) of Chapter 3, there are some differences that arise from the lack of symmetry of a typical ship's forebody and afterbody and from the very different proportions of these longitudinal wedges compared to the transverse wedges caused by heeling.

Axis of Trim. It is evident, as shown in Figure 5-2, and from our knowledge of ship form (Chapter 2), that even for a very small trim angle, the emerged and immersed wedges will have totally different shapes from one another since the ship is not symmetrical longitudinally. For constant displacement, however, they must

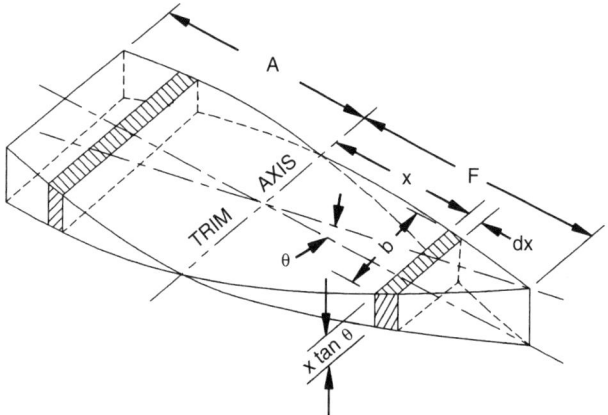

Figure 5-3. Geometry of the immersed and emerged wedges.

have equal volumes. The result of this requirement is that as the ship trims, the axis about which rotation takes place is not at amidships. In fact, the trimming axis will be at different longitudinal positions for each draft of the ship. Its location can be determined by equating the volumes of the emerged and immersed wedges.

The wedge geometry is shown in Figure 5-3, where x denotes the distance from the axis of trim to an elemental volume of one of the wedges, and b is the breadth of the waterplane at longitudinal location x. The height of the elemental wedge volume is $x \tan \theta$, where θ is the angle of trim. Thus the elemental volume (shaded in the figure) is equal to $b(x \tan \theta) dx$. The entire wedge volume is then:

$$v = \int bx \tan \theta \, dx = \tan \theta \int bx \, dx \qquad (4)$$

since θ is constant. The integration is performed longitudinally from the axis of rotation to the forward end of the waterplane. The same definitions apply to the other wedge, here shown from the axis of trim to the stern. Equating the two wedge volumes, we have

$$\tan \theta \int_{aft} bx \, dx = \tan \theta \int_{fwd} bx \, dx$$

or
$$\int_{aft} bx \, dx = \int_{fwd} bx \, dx \qquad (5)$$

in which the limits imply integrations from the axis of trim to the aft and forward ends of the waterplane, respectively. Note that since b dx is the elemental area of the waterplane at location x, the two sides of equation (5) are, respectively, the moments of area of the aft and forward portions of the waterplane about the axis of trim. Hence we conclude that the axis of trim coincides with the axis at which the waterplane achieves a longitudinal balance, namely its centroidal axis passing through the center of flotation. Thus we have the important conclusion that *a ship trims about a transverse axis passing through the center of flotation of its waterplane.* Ship's officers often refer to the axis of trim as the "tipping center."

Longitudinal vs. Transverse Metacenter. A second major difference between transverse and longitudinal stability should be noted. The two metacentric radii (transverse and longitudinal) are proportional respectively to the transverse and longitudinal moments of inertia of the waterplane. Because a ship's length is much greater than its beam, these two moments of inertia are vastly different in magnitude. Typical ships have length-to-beam ratios which vary from about 6 to 1 to 9 to 1. Consider a simple rectangular barge with the average proportions of a ship, that is, with a length 7.5 times its beam. This barge will have longitudinal and transverse waterplane moments of inertia as follows:

$$I_L = \frac{BL^3}{12} \tag{6}$$

$$I_T = \frac{LB^3}{12} \tag{7}$$

The ratio of its metacentric radii is

$$\frac{BM_L}{BM_T} = \frac{I_L}{I_T} = \frac{BL^3}{LB^3} = \left(\frac{L}{B}\right)^2 \tag{8}$$

so the longitudinal metacentric radius will be 7.5^2, or 56.25 times the transverse metacentric radius. This magnitude of the difference between longitudinal and transverse stability is typical of ships as well. A ship's longitudinal stability is so large that it is quite impossible to render it unstable longitudinally. The longitudinal metacenter of a typical surface ship is several hundred feet above its center of gravity.

Trimming Moment, Trim, and Drafts

Since longitudinal stability is so large that there is no concern about loss of stability or danger of capsizing in the longitudinal direction, we focus our attention only on the effect of longitudinal stability on static trim and change of trim caused by shifting, loading, or discharging weights aboard ship. The mechanics of these relationships can be shown by analogy to the effect of transverse stability on static heel and change of heel angle which were derived in Chapter 3, equation (23). The equation for angle of heel is:

140   APPLIED NAVAL ARCHITECTURE

$$\tan \varphi = \frac{GG'}{GM_T} = \frac{wd}{\Delta\, GM_T} \qquad (9)$$

in which GG' is the transverse shift of G in response to heeling moment wd, and GM_T is the transverse metacentric height. The equation is valid only if the angle of heel is small. The analogous equation for angle of trim change is:

$$\tan \theta = \frac{GG'}{GM_L} = \frac{wd}{\Delta\, GM_L} \qquad (10)$$

where θ   = angle of change of trim
  GG' = *longitudinal* shift of G
  wd = trimming moment, caused, for example, by shifting weight w *longitudinally* by distance d
  GM_L = *longitudinal* metacentric height

As in the transverse case, equation (10) is correct only for very small angles of trim. Unlike transverse inclinations, however, actual trim angles are never large enough to cause significant inaccuracies in the use of this equation because it is restricted to small angles. To illustrate this point, consider the angle of trim of a 600-foot-long ship with a trim of 30 feet (which is several times larger than any practical trim for this ship). The tangent of the angle of trim would be 30/600 = 0.05, and the corresponding angle of trim would be less than 3 degrees.

Combining equations (2) and (10) to eliminate the angle θ, which is not of practical interest, enables direct determination of the change in trim or the changes in drafts at the forward and after perpendiculars caused by a trimming moment. Thus, in U.S. units,

$$\frac{t}{12\,L} = \frac{wd}{\Delta\, GM_L}$$

or

$$t = \frac{wd}{\left(\dfrac{\Delta\, GM_L}{12\,L}\right)} \qquad (11)$$

will calculate directly the change in trim in inches produced by trimming moment wd (foot-tons) of a ship with displacement Δ (long tons weight), longitudinal metacentric height GM_L (feet), and length L (feet).

The equivalent expression in SI units with trim change expressed in centimeters, trimming moment in metric ton-meters, displacement in metric tons mass, and GM_L and L in meters is:

$$t = \frac{wd}{\left(\dfrac{\Delta_m\, GM_L}{100\,L}\right)} \qquad (11a)$$

TRIM AND LONGITUDINAL STABILITY

The resulting changes in draft at bow and stern are, from equation (2),

$$f = (\frac{F}{L})t \qquad (12)$$

$$a = (\frac{A}{L})t \qquad (13)$$

Draft changes f and a are opposite in sense, one representing an increase and the other a decrease in draft, dependent on the direction of the trimming moment, as illustrated in the following example.

Example 5-1
A tanker 1,045 feet long with a displacement of 305,000 tons floats in seawater at drafts of 62'0" forward and 65'0" aft. KB = 33.0 feet, KG = 45.0 feet, KM_L = 740 feet, and the center of flotation is 15.5 feet aft of amidships. What final drafts will result if 2,200 tons of cargo are shifted from cargo tanks #2 to #5, a distance of 285 feet?

The longitudinal metacentric height is:

$$GM_L = KM_L - KG = 740 - 45.0 = 695 \text{ feet}$$

Equation (11) calculates the change of trim:

$$t = \frac{wd}{(\frac{\Delta\, GM_L}{12\,L})} = \frac{2{,}200 \times 285}{(\frac{305{,}000 \times 695}{12 \times 1{,}045})}$$

$$= 37.1 \text{ inches change of trim}$$

Since the weight was shifted toward the stern, the trim change is *by the stern*, increasing the draft aft and decreasing the draft forward. The changes in draft at each end are calculated by determining the distances of the center of flotation from the forward and after perpendiculars.

$$F = \frac{L}{2} + LCF = \frac{1{,}045}{2} + 15.5 = 538 \text{ ft}$$

$$A = \frac{L}{2} - LCF = \frac{1{,}045}{2} - 15.5 = 507 \text{ ft}$$

(Or, we could write A = L − F = 1,045 − 538 = 507 ft)

$$f = (\frac{F}{L})t = (\frac{538}{1{,}045})\,37.1 = 0.515 \times 37.1 = 19.1 \text{ in}$$

$$= 1'7.1'' \text{ decrease of draft forward}$$

$$a = (\frac{A}{L})t = (\frac{507}{1,045})37.1 = 0.485 \times 37.1 = 18.0 \text{ in}$$
$$= 1'6.0'' \text{ increase of draft aft}$$
(Or, we could write $a = t - f = 37.1 - 19.1 = 18.0$ in)

The f and a calculations show that the change in draft at the bow amounts to 51.5 percent of the trim change, while that at the stern is 48.5 percent. This finding is consistent with the location of the center of flotation, which is aft of amidships. The smaller of the two draft changes will always be at the location closer to the center of flotation.

The final drafts are then determined:

$T_F = (62'0'') - (1'7.1'') = 60'4.9''$
$T_A = (65'0'') + (1'6.0'') = 66'6.0''$

Two simple checks can be made of the above calculations. The trim change is the *sum* of the changes of drafts forward and aft, since trim is by the stern both before and after the shift of weight. Thus $f + a = t$, or $19.1 + 18.0 = 37.1$ inches, verifying the change in trim. Also, the final trim should be equal to the initial trim plus the change of trim:

Initial trim = $(65'0'') - (62'0'') = 3'0'' = 36''$ (stern)
Change of trim = $37.1''$ (stern)
Final trim = $(66'6'') - (60'4.9'') = 6'1.1'' = 73.1''$ (stern)
Checking: $36.0 + 37.1 = 73.1''$ final trim by the stern

It will be apparent that some of the calculations in this example, introduced to clarify what happens at every step, could be shortened or eliminated. For example, it is not necessary to determine A, the distance of the center of flotation from the stern, if the change in draft aft is calculated using the alternate equation, $a = t - f$.

ASSUMPTIONS AND ACCURACY

When the methods of the above sample problem are used to determine the drafts of a ship of typical form, the results are not exact because some unexpressed assumptions have been made that are not precisely correct. All of the hydrostatic particulars that were used in the calculations (displacement, KM_L, and LCF, for example) were determined in the ordinary way, from plotted or tabulated results of the ship form calculations described in Chapter 2. But the hydrostatic particulars apply exactly only to the ship on an even keel, without trim. In applications involving a ship with trim, they are not exactly correct. Fortunately, for ships of normal form at ordinary operational amounts of trim, the assumption that even-

TRIM AND LONGITUDINAL STABILITY

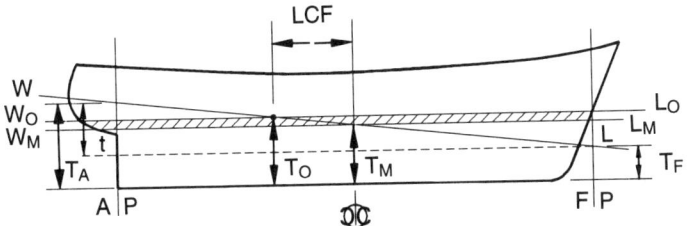

Figure 5-4. Displacement when trimmed—definitions.

keel hydrostatics may be applied to trimmed conditions is acceptable because the errors introduced are so small as to be negligible. An exception to this general observation can occur when determining the displacement of a trimmed ship, if the trim is large or the center of flotation is especially distant from amidships, or both. When necessary, a corrected displacement for a trimmed ship can be determined using the even-keel hydrostatic properties, as described below.

Displacement When Trimmed

Determining the displacement from its drafts forward and aft (as opposed to determining it by a summation of weights) for a ship with trim may be done in three ways, depending on the accuracy required. Figure 5-4 defines the drafts and other measurements referred to in the following descriptions. The methods, in order of increasing accuracy and increasing complexity, are:

- Determine the displacement from the hydrostatic curves at mean draft (T_M).
- Determine the displacement from the hydrostatic curves at the corresponding even-keel draft (T_O) which is the draft at the center of flotation.
- Integrate the immersed sectional areas corresponding to the trimmed waterline from bow to stern, to determine the actual immersed volume of the trimmed ship.

The first method is the least accurate. It is exact only if the trim is zero and is nearly correct for a trimmed ship only if the center of flotation (F) is at amidships. It is frequently assumed to be sufficiently accurate for operational calculations made by ship's officers, an assumption which is justified for slow to moderate-speed merchant ships with small or normal trim because their hull form is such that the center of flotation is relatively close to amidships. It should not be accepted for high-speed ships or ships with very broad sterns (for vehicle ramps, for example), which tend to have centers of flotation well aft of amidships.

The second method is acceptable for the high-speed and broad-stern ships for all applications except those requiring the utmost accuracy. It is based on the fact that as a ship changes trim at constant displacement, the draft at the center of flotation remains unchanged. Therefore in Figure 5-4, the displacement to trimmed waterline WL is the same as that to even-keel waterline W_oL_o. That displacement is also equal to the displacement to even-keel waterline W_ML_M, at mean draft T_M, plus the displacement of a correction layer of thickness $m = T_o - T_M$, shown shaded in the figure. The problem is that to determine m or T_o, the LCF of waterplane W_oL_o must be known, and that in turn requires that T_o is already known. The solution is to assume that the LCFs of the two even-keel waterplanes W_oL_o and W_ML_M are the same. This approximation is satisfactory because layer thickness is usually quite small. The procedure is as follows:

1. At mean draft $T_M = (T_F + T_A)/2$, the uncorrected displacement, the LCF, and the TPI or TPC are read from the curves of form.

2. From the slope of WL, or from equation (2), the layer thickness is m = (LCF/L)t. The units of m will be the same as those of t, preferably inches in U.S. units and centimeters in SI units.

3. The displacement of the correction layer is equal to the layer thickness times the TPI or TPC:

$$d\Delta = m \times TPI \quad \text{(m in inches, } d\Delta \text{ in tons)}$$

or $\quad d\Delta_m = m \times TPC \quad$ (m in cm, $d\Delta_m$ in metric tons)

4. Corrected displacement = uncorrected displacement + displacement of correction layer.

5. Even keel draft (T_o) corresponding to the trimmed waterline (WL) is given by:

$$T_O = T_M + m = T_M + \left(\frac{LCF}{L}\right)t \qquad (14)$$

The calculations described in steps 2 and 3 can be combined into a single equation for the *change of displacement with trim,* as follows:

$$d\Delta = \left(\frac{TPI \times LCF}{L}\right)t \qquad [t \text{ inches, } d\Delta \text{ long tons}] \qquad (15)$$

$$d\Delta_m = \left(\frac{TPC \times LCF}{L}\right)t \qquad [t \text{ cm, } d\Delta_m \text{ metric tons}] \qquad (15a)$$

The change in displacement can be positive or negative. For the situation illustrated in Figure 5-4, the correction is positive, the trimmed displacement being larger than the mean draft displacement. The sign of the correction changes if the center of flotation is forward of amidships or if the trim is in the opposite direction. Thus there are four possible cases:

TRIM AND LONGITUDINAL STABILITY 145

1. Trim by *stern*, LCF *aft*, correction must be *added*.
2. Trim by *stern*, LCF *fwd*, correction must be *subtracted*.
3. Trim by *head*, LCF *aft*, correction must be *subtracted*.
4. Trim by *head*, LCF *fwd*, correction must be *added*.

It is recommended that a simple sketch like the figure be made for each application of the calculations to determine if the correction should be added or subtracted.

To simplify the calculation of change of displacement with trim, some curves of form include precalculated values of the quantities TPI × LCF/L or TPC × LCF/L plotted against the draft of the ship. They are called, respectively, "increase in displacement per inch trim by the stern" and "increase in displacement per centimeter trim by the stern." The plotted functions are shown as negative if LCF is forward of amidships, and they must be used with reversed signs when trim is by the head. When these quantities are available, the calculation of the increase or decrease in displacement is simply a matter of multiplying the trim of the ship by the appropriate quantity from the curves of form.

It is essential that the words "increase in displacement" or "change in displacement" with trim are not misinterpreted. The trim in question takes place at *constant displacement,* which neither increases nor decreases because of trim. The increase or decrease of displacement is only in comparison to the displacement that would be determined from the curves of form *at the mean (midships) draft of the trimmed ship.*

Example 5-2

A ship of LBP = 150 m floats in seawater at drafts of 8.2 m forward and 9.8 m aft. At the mean draft of 9.0 m, the curves of form give Δ_m = 20,200 MT, TPC = 26.5, and LCF = 4.3 m aft of amidships. Determine the even-keel draft corresponding to the trimmed waterline and the displacement corrected for trim.

Calculation of trim:

$$t = T_A - T_F = 9.8 - 8.2 = 1.6 \text{ m} = 160 \text{ cm by the stern}$$

Corresponding even-keel draft:

$$T_o = T_M + \left(\frac{\text{LCF}}{L}\right)t = 9.00 + \left(\frac{4.3}{150}\right)1.6 = 9.04 \text{ m}$$

Note that the trim is expressed in meters, consistent with the mean draft value. The layer thickness is added to the mean draft because trim is by the stern and LCF is aft. Displacement of the correction layer:

$$d\Delta_m = \left(\frac{TPC \times LCF}{L}\right)t = \left(\frac{26.5 \times 4.3}{150}\right)160 = 122 \text{ MT}$$

In this expression, trim must be in centimeters to be consistent with the TPC. Finally, the displacement is:

$$\text{Trimmed } \Delta_m = 20{,}200 + 122 = 20{,}322 \text{ MT}$$

The third and most exact method of determining the displacement of a trimmed ship is to actually calculate by direct numerical integration the displaced volume of the ship floating at the trimmed waterline. Immersed areas of the stations, each to the draft at the given station, are picked off from the Bonjean's curves and put through a longitudinal integration from bow to stern exactly like the calculations that produced the even-keel displacement curve, as described in Chapter 2. This method is used only in cases requiring extreme accuracy, as in the calculation of displacement of a ship during an inclining experiment, or in special extreme trim situations like those involved in trimming a supertanker to bring the propeller or its shaft out of the water for inspection or repair.

Moment to Change Trim
Returning now to the relationship between trim change and trimming moment, equation (11):

$$t = \frac{wd}{\left(\frac{\Delta \, GM_L}{12 \, L}\right)} \tag{11}$$

The right side of this equation is expressed as a compound fraction which puts the moment that causes the trim change in the numerator, and the characteristics of the ship at the time of the trim change in the denominator. The denominator thus represents the *moment to change trim one inch,* because if trim change t is equal to one inch, the quantity in the denominator and the trimming moment wd must be equal to one another. Hence we define:

$$MT1 = \frac{\Delta \, GM_L}{12 \, L} \tag{16}$$

where MT1 is the moment to change trim one inch. The equivalent SI expression is:

$$MTcm = \frac{\Delta_m \, GM_L}{100 \, L} \tag{16a}$$

in which MTcm is the moment to change trim one centimeter.

TRIM AND LONGITUDINAL STABILITY

The convenience of defining these quantities is that, since they are independent of the trimming moment, they can be precalculated and treated as a property of the ship when making trim calculations. The equation for trim change can then be written:

$$t = \frac{wd}{MT1} \tag{17}$$

or

$$t = \frac{wd}{MTcm} \tag{17a}$$

Precalculating the moment to change trim for all possible conditions of a ship is, however, complicated by the fact that it is dependent on both ship form or hydrostatics and ship loading, since, to determine GM_L, KG must be known. Fortunately, its dependence on KG is minimal because the range of KG values for a typical ship is far smaller than the value of GM_L. Even considering extremes of loading, KG will almost certainly be between one-third and two-thirds of the depth of the ship, while GM_L is of the order of magnitude of the ship's length. Thus a close approximation to the MT1 or MTcm can be made using an assumed position of G. The most common assumption is that GM_L is approximated by BM_L. The potential error introduced by this assumption is in the order of several feet (or meters) out of several hundred. We therefore define the *approximate moment to change trim* as follows:

$$\text{approx. MT1} = \frac{\Delta BM_L}{12 L} \tag{18}$$

which can be simplified by noting that $\Delta = \rho g \nabla$, and $BM_L = \frac{I_L}{\nabla}$, therefore

$$\text{approx. MT1} = \frac{\rho g I_L}{12 L} \tag{19}$$

$$= \frac{I_L}{420 L} \quad \text{(seawater)}$$

$$= \frac{I_L}{432 L} \quad \text{(fresh water)}$$

The approximate moment to change trim is calculated routinely for drafts ranging from light ship to load draft and beyond and is included in the curves of form or hydrostatic curves described in Chapter 2. Since it is accurate enough for typical trim calculations, in the curves of form it is usually identified simply as the MT1. In most curves of form, only the seawater MT1 is plotted for seagoing ships and the fresh-water MT1 for ships intended to operate solely in fresh water, as in the Great Lakes. When a seagoing ship loads cargo in a fresh-water port, the ship's officer should correct the MT1 to its fresh-water equivalent when making trim calculations.

148   APPLIED NAVAL ARCHITECTURE

In the SI system, $\Delta_m = \rho \nabla$ and the expressions for approximate MTcm are:

$$\text{approx. MTcm} = \frac{\rho I_L}{100 \, L}$$

$$= \frac{1.025 \, I_L}{100 \, L} \quad \text{(seawater)} \quad (19a)$$

$$= \frac{I_L}{100 \, L} \quad \text{(fresh water)}$$

Example 5-3

Using the data given for the tanker of Example 5-1, calculate the exact and approximate MT1's. Use the approximate MT1 to determine the final drafts of the ship and compare them to the drafts calculated in Example 5-1.

Exact MT1:

$$\text{MT1} = \frac{\Delta \, GM_L}{12 \, L} = \frac{305{,}000 \times 695}{12 \times 1{,}045} = 16{,}904$$

Approximate MT1:

$$BM_L = KM_L - KB = 740 - 33 = 707 \text{ feet}$$

$$\text{MT1} = \frac{\Delta \, BM_L}{12 \, L} = \frac{305{,}000 \times 707}{12 \times 1{,}045} = 17{,}196$$

Trim calculation using approximate MT1:

$$t = \frac{wd}{MT1} = \frac{2{,}200 \times 285}{17{,}196} = 36.5 \text{ inches by stern}$$

$$f = \left(\frac{F}{L}\right) t = \left(\frac{538}{1{,}045}\right) 36.5 = 18.8 \text{ inches decrease forward}$$

$$a = t - f = 36.5 - 18.8 = 17.7 \text{ inches increase aft}$$

Final drafts, forward and aft:

$$T_F = (62'0'') - (1'6.8'') = 60'5.2'' \quad [\textit{cf. } 60'4.9'' \text{ exact}]$$
$$T_A = (65'0'') + (1'5.7'') = 66'5.7'' \quad [\textit{cf. } 66'6.0'' \text{ exact}]$$

This shows that the approximate MT1 differs from the exact value by only 1.5 percent in this case. The resulting drafts are 0.3" more than the exact drafts, a negligible difference that would not even be detectable on the ship.

TRIM AND LONGITUDINAL STABILITY

LOADING AND DISCHARGING SMALL WEIGHTS

When weights are loaded on or discharged from a ship, there is a change in displacement (and hence in mean draft) as well as a change in trim. If the change in displacement is relatively small, such that the waterplane properties (TPI, LCF, MT1) at the completion of loading are not very much changed from their values before loading began, the drafts after loading can be determined by adding just one step to the procedures described above for the case of a shifted weight. The additional calculation involves determining a *parallel sinkage* or *parallel rise* caused by loading or discharging the weights.

The procedure, called here the *small weights method,* is to model the loading of a weight by imagining it to be loaded at the longitudinal position of the center of flotation, then shifted to the actual location at which it is to be loaded. The final effect on the ship's drafts is the same as if the weight were loaded directly at its final destination, but the modeled procedure makes it possible to calculate separately the effects of parallel sinkage and change of trim that take place. If a weight is loaded at the center of flotation, there is no change of trim, but the drafts increase by a parallel sinkage equal to w/TPI inches all along the ship. The modeled shifting distance is the longitudinal distance from the center of flotation to the actual position of the loaded weight, and the change of trim and drafts from the parallel sinkage condition is determined as for a shifted weight exactly as shown in Example 5-3 above.

If a weight is discharged rather than loaded, the procedure is the same with all draft changes reversed in direction. The parallel effect is a rise (reduction on drafts all along) and the imagined shift is from the actual position of the weight to the center of flotation, rather than the reverse. If more than one weight is involved in the loading transaction, whether the weights be loaded or discharged or some of each, the best approach is to determine the weight and position of an equivalent single weight by making a "weights and moments" calculation on the group of weights before calculating sinkage and trim as described above. This procedure is much simpler than treating each weight individually, which would require repetitive calculations of sinkage and trim rather than just one. When dealing with multiple weights, it is important to remember that the procedure is limited to small weights in total. That is, the sum (algebraic) of the weights involved must be such that their combined loading does not very much change the area and shape of the waterplane.

The following three worked examples are applications of the small weights method for determining drafts after changes in loading.

Example 5-4
A Mariner class ship floats in seawater at drafts of 22'3" forward and 23'6" aft. Determine the drafts forward and aft after loading 450 tons of cargo in hold #5 on the second deck.

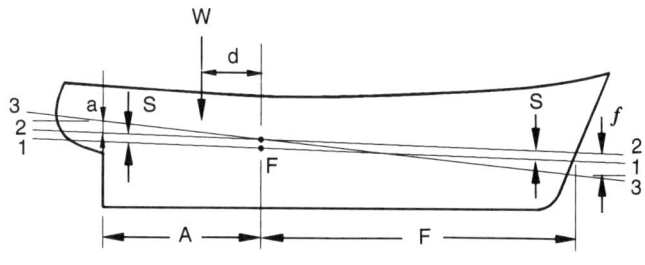

Figure 5-5.

The Mariner class hydrostatic data and loading tables are found in Appendix A. Figure 5-5 illustrates this worked example. The forward and aft draft marks are assumed to be at the forward and after perpendiculars (FP and AP) in this and in all examples.

Initial mean draft. Waterline #1 in Figure 5-5 represents the initial drafts of the ship before loading. The mean draft before loading is:

$$T_A = 23'6.0''$$
$$T_F = 22'3.0''$$
$$\text{Sum} = 45'9.0''$$

Therefore $T_M = \dfrac{\text{Sum}}{2} = \dfrac{(45'9.0'')}{2} = 22'10.5''$

Hydrostatic data. The following hydrostatic characteristics are read from the hydrostatic data at the mean draft:

$$\text{TPI} = 65.1 \text{ tons}$$
$$\text{MT1} = 1{,}580 \text{ ft-tons}$$
$$\text{LCF} = 273.9 \text{ feet from the FP} = F$$

At this point the distance of the center of flotation from the AP may be calculated. It is optional, but recommended.

$$A = 528.0 - 273.9 = 254.1 \text{ ft}$$

(In the Mariner hydrostatic data, the longitudinal positions of all hydrostatic centroids and centers of gravity of cargo spaces and tanks are recorded in feet from the FP. The LCF is therefore equal to the distance F shown in Figures 5-1 and 5-5. If centers are specified by their distances from amidships or from the after perpendicular, as they are in some trim and stability booklets, the corresponding F and A values must be calculated accordingly.)

TRIM AND LONGITUDINAL STABILITY 151

Parallel sinkage. Loading the weight at the center of flotation would cause parallel sinkage to an imagined intermediate waterline shown as waterline #2 in the figure. Waterline #2 is parallel to waterline #1. The parallel sinkage in inches is (see Chapter 3):

$$s = \frac{w}{TPI} = \frac{450}{65.1} = 6.9 \text{ in} \quad (add \text{ fwd and aft})$$

Change of trim. The loading table shows that the centroid of the cargo space in hold #5 on the second deck is 356.5 feet from the FP. Therefore the imagined shift of weight from the center of flotation (at 273.9 feet) is, in magnitude and direction:

$$d = 356.5 - 273.9 = 82.6 \text{ ft, toward the stern}$$

The shift is toward the stern, hence the resulting change of trim will be by the stern. From equation (17):

$$t = \frac{wd}{MT1} = \frac{450 \times 82.6}{1,580} = 23.5 \text{ in, by the stern}$$

Changes in drafts due to change of trim. A trim change by the stern causes the draft forward to decrease by f inches and the draft aft to increase by a inches in accordance with equations (12) and (13):

$$f = (\frac{F}{L})t = (\frac{273.9}{528})23.5 = 12.2 \text{ in} = 1'0.2'' \quad (subtract \text{ fwd})$$

$$a = (\frac{A}{L})t = (\frac{254.1}{528})23.5 = 11.3 \text{ in} \quad (add \text{ aft})$$

The quantities s, f, and a are measures of the changes in inches of the forward and aft drafts. The following table is a convenient way to arrange the calculation of the final drafts. It is especially helpful to use this format when doing "feet and inches" arithmetic that arises because draft marks in the hydrostatic charts are divided in inches and because draft changes are normally calculated in inches.

Drafts after loading

		Aft		Forward
Initial drafts		23' 6.0"		22' 3.0"
Parallel sinkage	+	6.9"	+	6.9"
		24' 0.9"		22' 9.9"
Change of trim	+	11.3"	−	1' 0.2"
Final drafts		25' 0.2"		21' 9.7"

152   APPLIED NAVAL ARCHITECTURE

Example 5-5
A containership with a length of 170.5 m floats at drafts of 8.80 m forward and 10.80 m aft in seawater. What drafts will result after 492 metric tons of containers are discharged from 52 m forward of amidships? Hydrostatic characteristics at the initial mean draft are: TPC = 32.8 t, MTcm = 268 t-m, LCF = 4.1 m aft of amidships.

For detailed explanations of each step, refer to Example 5-4 above. The procedures are similar.

Initial mean draft

$$T_M = \frac{(T_A + T_F)}{2} = \frac{10.80 + 8.80}{2} = 9.80 \text{ m}$$

Center of flotation. Since LCF is given as a distance aft of amidships, the distances to the forward and aft perpendiculars must be calculated:

$$F = \frac{L}{2} + LCF = \frac{170.5}{2} + 4.1 = 89.35 \text{ m}$$

$$A = \frac{L}{2} - LCF = \frac{170.5}{2} - 4.1 = 81.15 \text{ m}$$

Parallel rise. The cargo is *discharged*, so the drafts will *decrease* forward and aft.

$$s = \frac{w}{TPC} = \frac{492}{32.8} = 15 \text{ cm} = 0.15 \text{ m} \quad (subtract \text{ fwd and aft})$$

Change of trim

$$d = LCF \text{ aft of } \otimes + w \text{ fwd of } \otimes = 4.1 + 52.0 = 56.1 \text{ m}$$

Note that the trim change will be *by the stern*, because the weight is *discharged* from *forward* of the center of flotation.

$$t = \frac{wd}{MTcm} = \frac{492 \times 56.1}{268} = 103 \text{ cm} = 1.03 \text{ m by the stern}$$

Changes in drafts

$$f = (\frac{F}{L})t = (\frac{89.35}{170.5})1.03 = 0.54 \text{ m} \quad (subtract \text{ forward})$$

$$a = (\frac{A}{L})t = (\frac{81.15}{170.5})1.03 = 0.49 \text{ m} \quad (add \text{ aft})$$

TRIM AND LONGITUDINAL STABILITY

Drafts after Loading

		Aft		Foward
Initial drafts		10.80 m		8.80 m
Parallel rise	–	0.15	–	0.15
		10.65		8.65
Change of trim	+	0.49	–	0.54
Final drafts		11.14 m		8.11 m

Example 5-6

A Mariner class ship is floating in seawater at drafts of 25' 6" forward and 26'6" aft. The following loading then takes place:

Discharge 280 tons of cargo from hold #2, 2nd deck
Discharge 440 tons of cargo from hold #3, 2nd deck
Load 615 tons of cargo in hold #4, 3rd deck
Load 385 tons of cargo in hold #4, 2nd deck
Load 235 tons of cargo in hold #5, 2nd deck
Load 220 tons of fuel in double bottom tank #3, centerline

Determine the final drafts, using the small weights method.

The initial mean draft is, by inspection, 26'0". From the hydrostatic properties data in Appendix A, the following characteristics have been determined at this draft:

$$TPI = 67.3 \text{ tons}$$
$$MT1 = 1,720 \text{ tons}$$
$$LCF = 277.3 \text{ feet from FP} = F$$

Therefore $A = L - F = 528 - 277.3 = 250.7$ ft from the AP to the center of flotation.

Before calculating the parallel sinkage and the change in trim, it is expedient to reduce the list of weights loaded and discharged to a single equivalent weight. Then s, t, f, and a will be calculated only once each, as if only one weight were involved. Any other procedure that treats the items of loading separately would involve much more calculation without improving the accuracy of the result. The single equivalent weight is determined in the moment calculation below. The forward perpendicular (FP) is used as the axis of moments because the longitudinal centers of the spaces listed in the loading tables (see Appendix A) are measured from the FP.

154   APPLIED NAVAL ARCHITECTURE

Weight	LCG_{FP}	Moment
− 280	104.4	− 29,232.0
− 440	161.3	− 70,972.0
+ 615	221.9	+ 136,468.5
+ 385	221.5	+ 85,277.5
+ 235	356.5	+ 83,777.5
+ 220	161.6	+ 35,552.0
+ 735		+ 240,871.5

The loading activity is equivalent to loading a single weight of 735 tons at a point 240,871.5/735 = 327.7 ft from the FP. Its distance from the center of flotation is:

$$d = 327.7 - 277.3 = 50.4 \text{ ft aft}$$

Therefore it will cause a trim change by the stern. The parallel sinkage, change of trim, and changes in draft are calculated as in the previous examples for single weights. Thus:

$$s = \frac{w}{TPI} = \frac{735}{67.3} = 10.9 \text{ in}$$

$$t = \frac{wd}{MT1} = \frac{735 \times 50.4}{1,720} = 21.5 \text{ in}$$

$$f = (\frac{F}{L})t = (\frac{277.3}{528})21.5 = 11.3 \text{ in } (subtract)$$

$$a = (\frac{A}{L})t = (\frac{250.7}{528})21.5 = 10.2 \text{ in } (add)$$

Drafts after Loading

	Aft	Forward
Initial drafts	26′ 6.0″	25′ 6.0″
Parallel sinkage	+ 10.9″	+ 10.9″
	27′ 4.9″	26′ 4.9″
Change of trim	+ 10.2″	− 11.3″
Final drafts	28′ 3.1″	25′ 5.6″

Small Weights Method Generalized

Changes in drafts at any point in the ship caused by loading or discharging small weights can be calculated directly by combining the steps in the solutions above into a single formula. Thus the change in draft at location X in Figure 5-6 caused by loading weight w as shown may be determined as follows:

$$dT_x = s + x = \frac{w}{TPI} + (\frac{X}{L})t$$

TRIM AND LONGITUDINAL STABILITY

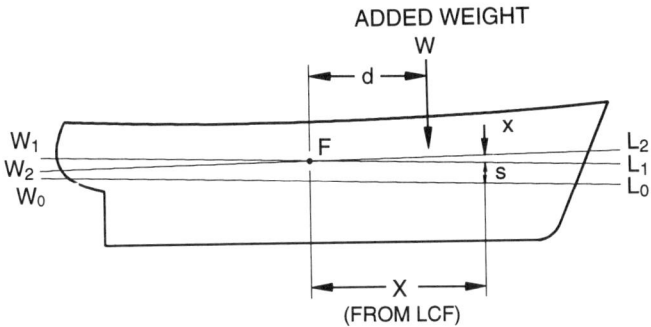

Figure 5-6. Draft change at any point.

$$dT_x = \frac{w}{TPI} + \left(\frac{X}{L}\right)\left(\frac{wd}{MT1}\right) \qquad (20)$$

where dT_x = change in draft (inches) at X feet from the center of flotation
s = parallel sinkage or rise (inches)
x = draft change at X due to trim change (inches)
w = weight added (or removed)
d = distance of weight w from LCF (feet)
W_0L_0 = waterline before loading
W_1L_1 = imagined intermediate waterline after parallel sinkage but before trim change
W_2L_2 = final waterline

The summation in equation (20) is algebraic, the sign of each term depending on whether the weight is added or removed and on the directions of both d and X from the center of flotation. Thus, for example, s is negative for a discharged weight, and x would be negative in Figure 5-6 if the point to be studied had been aft of the center of flotation rather than forward. As there are many combinations of positive and negative draft changes, a sketch like Figure 5-6 must always be made to avoid algebraic errors.

An important practical application of this single-step procedure arises when it is desired to change the draft at a given point by a specified amount by adding, discharging, or shifting weights. In such cases, dT_x is specified and equation (20) is to be solved for w (e.g., How much ballast must be taken into a given tank to increase the draft aft by a specified amount?) or for d (Where should 100 tons be loaded to achieve a required draft change?). In most cases, it is the draft at the bow or stern that must be changed by a specific amount, so that x and X are replaced by f and F or a and A, respectively, as defined in Figure 5-1. The following example illustrates a practical application of the procedure described by equation (20).

Example 5-7

A Mariner class ship approaches a seawater canal with drafts of 28'0" forward and 29'6" aft. The maximum permitted draft in the canal lock is 29'0". The master proposes to take on seawater ballast in deep tank #1 in order to enter the lock. Three questions must be answered:

1. How many tons of ballast will be needed?
2. Does deep tank #1 have enough capacity to do it?
3. Will the final forward draft be satisfactory?

Applying the general expression given by equation (20) to this specific case, we note that the requirement is to *decrease* the draft at the *stern* by six inches (from 29'6" to 29'0"). Since it is to be done by loading ballast forward, the parallel sinkage will *increase* the draft everywhere by s inches, while the change of trim will *decrease* the draft aft by a inches. The defining equation is, therefore:

$$s - a = -6.0$$

$$\frac{w}{TPI} - (\frac{A}{L})(\frac{wd}{MT1}) = -6.0$$

The required hydrostatic particulars are determined from the Mariner data in Appendix A at the mean draft of

$$T_M = \frac{T_F + T_A}{2} = \frac{(28'0'') + (29'6'')}{2} = \frac{57'6''}{2} = 28'9''$$

At 28'9" draft, we get: TPI = 69.4 tons
MT1 = 1,870 ft–tons
LCF = F = 280.5 ft
A = L − F = 528 − 280.5 = 247.5 ft

From the Mariner table of tank capacities (see Appendix A), deep tank #1 is seen to have a capacity of 137.4 tons of seawater, and its LCG is 40.3 feet aft of the forward perpendicular. Trimming arm d is the distance of the tank centroid from the center of flotation:

$$d = 280.5 - 40.3 = 240.2 \text{ ft.}$$

The change of trim will be by the head, as required to reduce the draft at the stern. Entering this data into the defining equation, we have:

$$\frac{w}{TPI} - (\frac{A}{L})(\frac{wd}{MT1}) = -6.0$$

TRIM AND LONGITUDINAL STABILITY

$$w\left(\frac{1}{TPI} - \frac{A\,d}{L\,MT1}\right) = -6.0$$

$$w\left(\frac{1}{69.4} - \frac{247.5 \times 240.2}{528 \times 1,870}\right) = -6.0$$

$$w(0.01441 - 0.06021) = -6.0$$

$$-0.04580\,w = -6.0$$

$$w = \frac{6.0}{0.04580} = 131.0 \text{ tons}$$

Sinkage and trim can thus be calculated:

$$s = \frac{w}{TPI} = \frac{131.0}{69.4} = 1.9 \text{ in } (add)$$

$$t = \frac{wd}{MT1} = \frac{131.0 \times 240.2}{1,870} = 16.8 \text{ in by the head}$$

$$f = \left(\frac{F}{L}\right)t = \left(\frac{280.5}{528}\right)16.8 = 8.9 \text{ in } (add)$$

$$a = \left(\frac{A}{L}\right)t = \left(\frac{247.5}{528}\right)16.8 = 7.9 \text{ in } (subtract)$$

Drafts after Ballasting:

	Aft	Forward
Initial drafts	29' 6.0"	28' 0.0"
Parallel sinkage	+ 1.9"	+ 1.9"
	29' 7.9"	28' 1.9"
Change of trim	− 7.9"	+ 8.9"
Final drafts	29' 0.0"	28' 10.8"

The answers to the three questions are:

1. Weight of ballast to be loaded is 131 tons.
2. The tank capacity of 137.4 tons is sufficient.
3. The forward draft, 28'10.8", is satisfactory because it is less than the 29'0" limit.

Trim Table

Trim and stability booklets prepared by the design office for use by ship's officers to calculate the stability and trim of the ship for each voyage usually include a *trim table* or *trim diagram.* As an example, the trim table for the Mariner class ship is shown in Figure 5-7. The table shows the increase or decrease of draft in inches at the forward and after draft marks caused by loading a specific weight (usually 100 tons) at specific longitudinal locations on the ship. For break-bulk ships, values are tabulated for points of loading at equal increments of length, say each 10 feet. The tabulated values are displayed directly below a scale drawing of the ship profile showing the locations of holds and tanks, so that an item to be loaded can be located graphically. The draft change values tabulated in a trim table have been calculated using equation (20), with weight w equal to 100 tons, arm d corresponding to the distance of the weight from the center of flotation for each assumed position of the weight, and two values of distance X, namely F and A, the distances from the center of flotation to the forward and aft draft marks, respectively. Since the tabulated draft changes will depend on the initial mean draft (LCF, TPI, and MT1 vary with draft), tables are calculated and displayed for two or more assumed mean drafts. Depending on the accuracy desired, the tabulated values for the draft closest to the actual mean draft may be used to estimate the final drafts, or values determined from tables at two drafts may be interpolated. A sample calculation is shown in the figure. Many officers prefer to use trim tables rather than the procedures described above for solving small weights trim problems. When only one or two weights are involved and interpolation is not necessary, the tables are very convenient. When several weights are loaded and unloaded and interpolation is required, the calculation using the tables can be as involved as the direct solution.

LOADING AND DISCHARGING LARGE WEIGHTS

The small weights method just described is not satisfactory for calculating the drafts that will result after large amounts of weight are loaded or discharged, because with large changes in displacement and mean draft the hull form characteristics (LCF, MT1, TPI) cannot be treated as constants during the draft changes that occur. Furthermore, tons per inch immersion (TPI) is not a satisfactory means of determining parallel sinkage because the assumption that a parallel sinkage layer is wall-sided is not valid throughout a large change of draft.

To circumvent the inaccuracies that would result by using small weights techniques when large changes in displacement are involved, we must use an approach that determines the displacement and the longitudinal position of the whole ship's center of gravity before and after changing the loading, and relate that information to the drafts in both conditions. The *large weights method* is also referred to as the *LCG method* or *whole ship method* of calculating trim.

The procedure is best understood by dividing it into three parts, as follows:

1. Given drafts forward and aft before loading, determine displacement and longitudinal center of gravity (LCG).

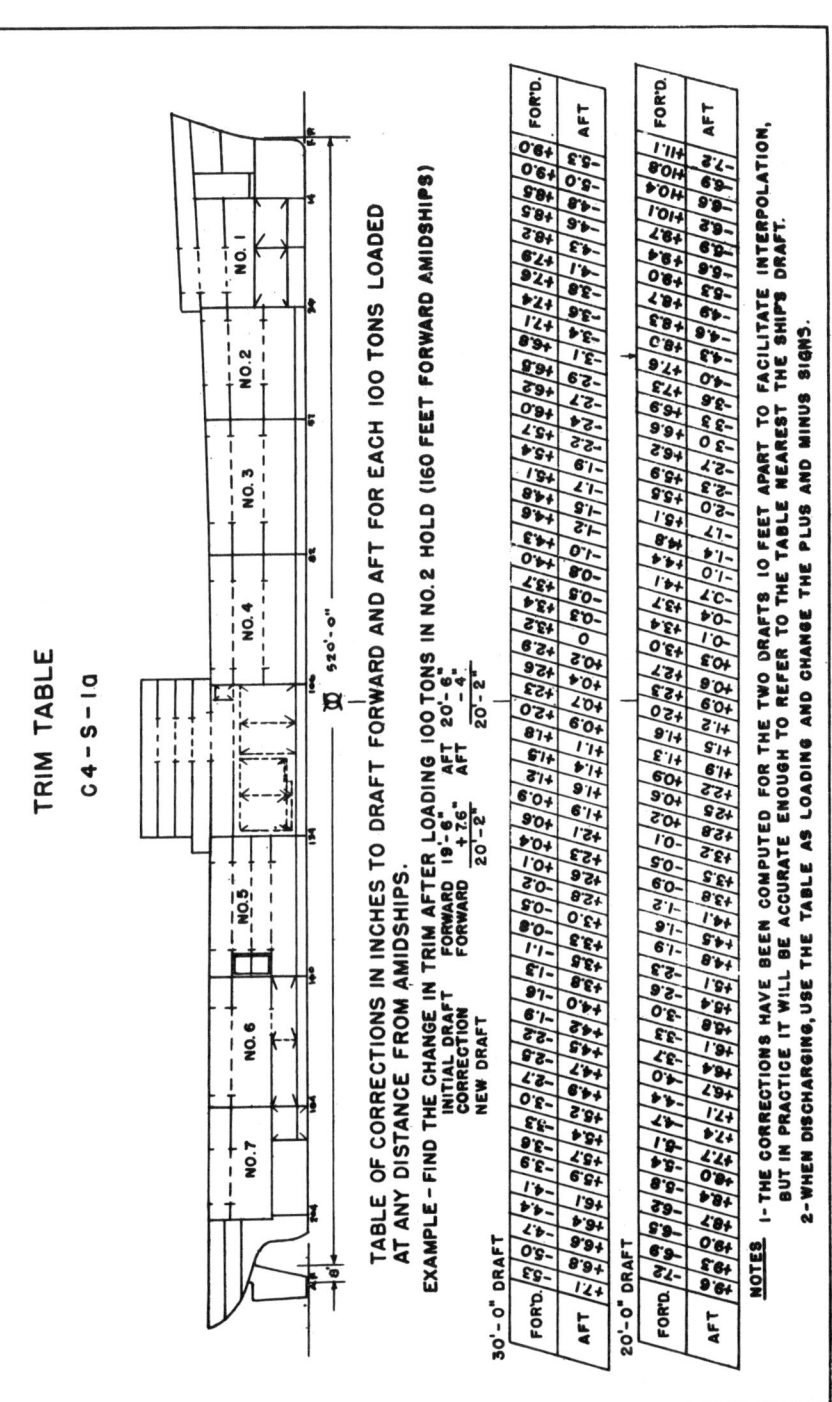

Figure 5-7. Mariner class trim table.

2. To the initial displacement, Δ, and its longitudinal moment, $\Delta \times$ LCG, add algebraically the weights and moments of the load changes and hence determine the final displacement and LCG after loading is complete.
3. Given the final displacement and LCG, determine the corresponding drafts forward and aft.

It will be immediately apparent that parts 1 and 3 are similar to each other, the procedure for part 3 being the inverse of that for part 1. Part 2 is an ordinary weights and moments calculation which includes the ship as well as the new loading.

Part 1: Determining Displacement and LCG from Drafts
The displacement before loading is determined from the drafts as in Example 5-2. If the trim is small and the center of flotation is near amidships, the displacement determined from the curves of form at the mean draft will be sufficiently accurate without making the correction to displacement for trim described in Example 5-2. To determine LCG, imagine that the ship at its initial displacement had previously been on an even keel as shown by even-keel waterline W_0L_0 in Figure 5-8. Its center of gravity in this reference condition would be at G_0, vertically in line with even-keel center of buoyancy B_0. The longitudinal center of buoyancy (LCB_0) for this waterline can be read from the curves of form. The actual trimmed condition, waterline WL in the figure, is supposed then to have been produced by a trimming moment, say wd, that caused G to shift from G_0 to G, measured by

$$GG_0 = \frac{wd}{\Delta} = \frac{\text{trimming moment}}{\text{displacement}}$$

and causing the actual observed trim,

$$t = T_A - T_F = \text{difference in drafts, forward and aft}$$

Since the trim is known, we can calculate trimming moment "wd" another way, from equation (17):

$$t = \frac{wd}{MT1}, \text{ hence trimming moment } wd = t \times MT1$$

Combining the two expressions for trimming moment above, we have:

$$\Delta \times GG_0 = t \times MT1$$

or
$$GG_0 = \frac{t \; MT1}{\Delta} \qquad (21)$$

TRIM AND LONGITUDINAL STABILITY

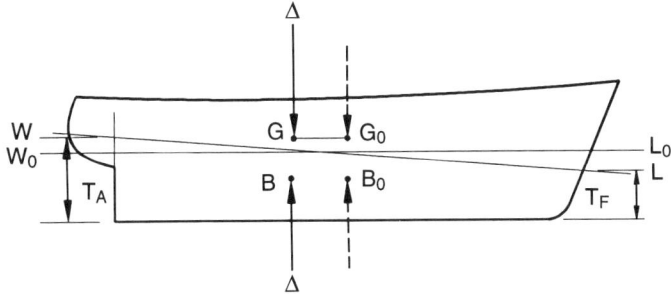

Figure 5-8. LCG of trimmed ship.

where GG_0 is the longitudinal distance of the actual center of gravity of the trimmed ship from the even-keel center of buoyancy. The MT1 is read from the curves of form at the draft at which displacement Δ was determined, that is, at the LCF draft if the displacement was corrected for trim, or at the mean draft if the correction was ignored. The LCG of the trimmed ship is then determined:

$$LCG = LCB_0 + GG_0 \qquad (22)$$

This is an algebraic sum, in which either term may be positive or negative if LCB_0 is given from amidships, depending on the direction of trim. Clearly G is aft of B_0 if trim is by the stern and forward of B_0 if trim is by the head.

Example 5-8
A Mariner class ship floats in seawater at drafts of 19'2" forward and 23'8" aft. Determine the ship's LCG in feet from the forward perpendicular.

First calculate the mean draft and the trim:

$$T_A = 23'8''$$
$$T_F = 19'2''$$
$$\text{Sum} = 42'10'' \qquad T_M = \frac{\text{Sum}}{2} = 21'5''$$
$$\text{Diff} = 4'6'' \qquad t = 4'6'' = 54 \text{ in by the stern}$$

At T = 21'5", the hydrostatic data show:

$$\Delta = 14{,}330 \text{ tons}$$
$$LCB_0 = 265.45 \text{ ft from FP (for zero trim)}$$
$$MT1 = 1{,}530 \text{ ft-tons}$$

162   APPLIED NAVAL ARCHITECTURE

The longitudinal distance between G and B_0 is determined as in equation (21):

$$GG_0 = \frac{t \times MT1}{\Delta} = \frac{54 \times 1,530}{14,330}$$

$$= 5.77 \text{ feet, G } \textit{aft} \text{ of } B_0$$

Since the trim is by the stern, G must be *aft* of even-keel center of buoyancy B_0. Therefore, as in equation (22):

$$LCG = LCB_0 + GG_0 = 265.45 + 5.77 = 271.22 \text{ ft from FP}$$

Part 2: Determining Final Displacement and LCG
The displacement and LCG just determined are entered in a calculation of weights and longitudinal moments into which all weights to be loaded or discharged are also entered along with their LCGs and longitudinal moments. The final displacement is simply the sum of the weights, and the final LCG is the total longitudinal moment divided by the final displacement. Care must be exercised here to make sure that the axis from which all LCGs are determined is the same.

Example 5-9
The Mariner ship in example 5-8 is to be loaded as follows:

   Load 320 tons in hold #2 on the tanktop.
   Load 355 tons in hold #2 on the 3rd deck.
   Load 640 tons in hold #3 on the 3rd deck.
   Load 386 tons in hold #3 on the 2nd deck.
   Load 652 tons in hold #4 on the 3rd deck.
   Load 448 tons in hold #5 on the 2nd deck.
   Load 685 tons in hold #6 on the 3rd deck.
   Load 422 tons in hold #6 on the 2nd deck.
   Load 355 tons in hold #7 on the 3rd deck.

Determine the displacement and LCG of this ship when loading is completed.
   From Example 5-8, the ship before loading displaced 14,330 tons and had LCG = 271.22 ft from the forward perpendicular. The items above are then loaded as specified, at LCGs determined from the Mariner loading table in Appendix A. The calculations are done in the table on page 163.
   After loading, the ship has a displacement of 18,593 tons and LCG = 271.88 ft from the forward perpendicular.

Weight	LCG_{FP}	Moment
14,330	271.22	3,886,582.6
320	106.2	33,984.0
355	105.3	37,381.5
640	161.6	103,424.0
386	161.3	62,261.8
652	221.9	144,678.8
448	356.5	159,712.0
685	415.5	284,617.5
422	416.5	175,763.0
355	469.4	166,637.0
18,593	271.88	5,055,042.2

Part 3: Determining Drafts from Displacement and LCG

This part reverses the procedures of Part 1. All values are different from those in Part 1, because a large change in mean draft has occurred. The curves of form are entered at the displacement determined in Part 2, and the following hydrostatic properties are read out:

Mean draft = T_0
LCB_0 corresponding to even-keel condition at T_0
MT1
LCF

Unless the LCG from Part 2 matches exactly the LCB_0 just determined, the ship will trim to bring the trimmed LCB in line with the actual LCG. Thus, compared with a reference even-keel condition, there is a trimming arm determined as in Part 1, equation (22), solved for GG_0, the longitudinal separation of the actual center of gravity (G) and the reference even-keel center of gravity (G_0) in line with B_0:

$$GG_0 = LCG - LCB_0 \tag{23}$$

Again, the signs of all terms must be determined in each case by making a sketch to show directions from amidships (if that is the axis of moments). The sketch will also indicate the direction of trim; if G is aft of B_0, trim is by the stern, and if G is forward of B_0, trim is by the head. Next, equation (21) is solved for trim in inches:

$$t = \frac{\Delta\, GG_0}{MT1} \tag{24}$$

The changes in draft forward and aft from the reference even-keel draft T_0 are determined as in the small weights method, equations (12) and (13):

$$f = (\frac{F}{L})t \tag{12}$$

$$a = (\frac{A}{L})t = t - f \tag{13}$$

where F and A are the distances of the forward and aft draft marks from the center of flotation. Finally, the drafts are:

$$T_F = T_0 \pm f \tag{25}$$

$$T_A = T_0 \pm a \tag{26}$$

in which the signs are chosen according to whether trim is by the head or by the stern.

Example 5-10
Determine the drafts forward and aft of the Mariner ship after loading as described in Example 5-9.

The Mariner graph of hydrostatic data is entered at the final displacement from Example 5-9, that is, at 18,593 tons. The following particulars are determined:

$T_M = 26'9''$
$LCB_0 = 267.70$ ft from FP (at zero trim)
$MT1 = 1,758$ ft-tons
$LCF = F = 278.15$ ft from FP

Therefore $A = 528 - 278.15 = 249.85$ ft

Trimming arm GG_0 is the longitudinal distance from the actual center of gravity (LCG) to the even-keel center of buoyancy (LCB_0), as given by equation (23):

$$GG_0 = LCG - LCB_0 = 271.88 - 267.70 = 4.18 \text{ ft (G aft of } B_0)$$

The trim in inches from the even-keel draft of 26'9'' is, from equation (24):

$$t = \frac{\Delta GG_0}{MT1} = \frac{18,593 \times 4.18}{1,758} = 44.2 \text{ in.}$$

The trim is by the stern, because G is *aft* of B_0. Changes in draft forward and aft from the even-keel draft are:

$$f = (\frac{F}{L})t = (\frac{278.15}{528})44.2 = 23.3 \text{ in.} = 1'11.3'' \text{ (subtract)}$$

TRIM AND LONGITUDINAL STABILITY

$$a = \left(\frac{A}{L}\right)t = \left(\frac{249.85}{528}\right)44.2 = 20.9 \text{ in} = 1'8.9'' \text{ (add)}$$

Drafts after Loading:

	Aft	Forward
Mean draft after loading	26' 9.0"	26' 9.0"
Change of trim	+ 1' 8.9"	− 1' 11.3"
Final drafts	28' 5.9"	24' 9.7"

The large weights method of determining trim and drafts is equally as correct for small or large weights, and can be applied to all ship-loading calculations. Forms included in standard trim and stability booklets placed aboard ship are designed to make calculations equivalent to Parts 2 and 3 of the process described above. Calculations like those in Part 1 are not necessary in filling out loading forms because the forms require that all weights be listed, including light ship and all deadweight items in full, not just those that are added or discharged in a particular loading operation.

PROBLEMS

Note: Problems 1 through 13 are to be done using the *small weights* method.

1. A Mariner at drafts of 27'0" forward and 29'0" aft pumps 115 tons of fuel from No. 5 center double bottom tank (DBT) to No. 3 center DBT. Determine the new drafts forward and aft.

2. A Mariner enters port at drafts of 28'0" forward and 27'0" aft. The following cargo is discharged:

    1,900 tons from No. 2 hold (106.2 ft from FP)
    850 tons from No. 4 hold (223.1 ft from FP)
    1,200 tons from No. 7 hold (469.4 ft from FP)

    The following cargo is loaded:

    100 tons in No. 1 hold (56.6 ft from FP)
    300 tons in No. 2 hold (106.2 ft from FP)
    800 tons in No. 5 hold (353.6 ft from FP)

    Determine the new drafts forward and aft.

3. A rectangular barge with the following characteristics is floating at even keel in seawater.

L = 150 ft   Displacement = 1,500 long tons
B = 35 ft   KG = 9.0 ft
D = 17 ft

(a) Determine the transverse and longitudinal metacentric heights of this barge after loading 200 tons on deck, the center of gravity of the added weight being at amidships and 3 feet above the deck.

(b) If the 200-ton weight is then shifted 10 ft forward and 5 ft to port, determine the resulting drafts forward and aft and the angle of heel.

4. A Mariner floats at drafts of 24'3" forward and 25'9" aft. Then 340 tons are discharged from the deck at 24 ft forward of amidships. Calculate the new drafts.

5. A railcar barge (rectangular) is 52 m long, 9 m wide, and it floats in fresh water at 2.4 m draft even keel, with KG = 2.3m. What will be the change in draft at each end of this barge if a 25,000 kg railcar (already on board) is rolled from one end to the other, a distance of 41 m traveled by the railcar's center of mass?

6. A Mariner arrives with drafts of 21'0" forward and 22'0" aft in seawater. It then discharges 600 tons from No. 3 hold, third deck, loads 1,200 tons in No. 4 hold, Tanktop, and shifts 320 tons from No. 2 hold, second deck to No. 5 hold, second deck. What will be the final drafts forward and aft?

7. A ship 300 ft long arrives in port with drafts of 10'6" forward and 14'3" aft. Then 160 tons of cargo is loaded at 80 ft forward of amidships and 80 tons of cargo is discharged from 40 ft aft of amidships. Calculate the new drafts, given that TPI = 20, MT1 = 320, and the center of flotation is 10 ft aft of amidships.

8. A Mariner floats at drafts of 23'3" forward and 25'9" aft. At what distance from the forward perpendicular must a weight of 180 tons be loaded so as to increase the draft aft to 26'0"? What will be the resulting draft forward?

9. What weight of seawater ballast must be taken into double bottom tanks No. 2 in order to bring a Mariner to even keel if its present drafts are 18'10" forward and 20'3" aft? Are these tanks of sufficient capacity to do this? What will be the final even-keel draft?

TRIM AND LONGITUDINAL STABILITY

10. A self-propelled rectangular barge 120 ft long and 40 ft wide is to be fitted for ice breaking in the Great Lakes (fresh water) by adding trim tanks at the bow and stern so that it can raise its bow above the ice by pumping water to the stern tank. When the water is pumped back to the bow tank, the bow comes down and breaks the ice. Drafts forward and aft are 6 ft when the water is in the forward tank. The length between the centers of volumes of the tanks is 90 ft, and KG = 5 ft. What must the minimum volume of each trim tank be so that the bow just clears the ice when water is pumped aft?

11. A Mariner is floating at drafts of 19'0" forward and 19'10" aft. The chief engineer requests that the draft at the propeller be increased to 23'0" to immerse the propeller blades. The propeller is located 12 ft forward of the after perpendicular. How much cargo must be loaded in hold No. 6 (on the second and third decks combined) to do this?

12. A Mariner is floating on even keel at a displacement of 15,600 tons, and there are 800 tons of cargo still to be loaded. Space is available in hold No. 2 (second deck) and hold No. 6 (third deck). Determine how much of the 800 tons should be loaded in each of these spaces to complete loading trimmed 2 ft by the stern.

13. A ship has the following characteristics when floating on even keel in seawater at a displacement of 37,100 MT:

   L = 230 m C_B = 0.580   MTcm = 663
   B = 30 m C_W = 0.750   LCF = 125 m from FP

   How far from the bow (FP) must a mass of 600 MT be loaded in order to increase the draft at the stern by 30 cm? What will be the resulting drafts forward and aft?

Note: Problems 14 through 19 are to be done using the *large weights* method.

14. A Mariner floats at drafts of 17'6" forward and 23'0" aft in seawater. Determine its displacement (long tons) and LCG (from the FP).

15. Determine the displacement and location of the center of gravity (KG and LCG) of a Mariner vessel in seawater whose drafts are 20'0" forward and 25'0" aft, and whose transverse metacentric height is 3.20 ft.

16. A ship of length 155 m displaces 18,290 MT when floating in seawater at drafts of 7.5 m forward and 8.96 m aft. The hydrostatic data at the corresponding even-keel draft gives MTcm = 210 MT-m and LCB = 0.12 m forward of amidships. Determine the LCG of this ship with respect to amidships.

17. Do problem 2 (above) by the large weights method and compare the drafts with those determined by the small weights method used in problem 2.

18. A rectangular barge of L = 200 ft, B = 50 ft, D = 25 ft is floating in seawater at a draft forward of 7 ft and a draft aft of 9 ft. A 100-ton weight already on board is shifted 10 ft transversely, causing an angle of heel of 1.5 degrees. Determine the barge's displacement, KG, and LCG from amidships. Do the LCG calculation two ways: exactly (determine GM_L) and approximately (determine approximate MT1 to calculate trim), and compare the two answers.

19. A Mariner arrives in a seawater port with drafts of 20'0" forward and 24'0" aft, with KG = 28.5 ft. The following loading then takes place:

	Tons	VCG*	LCG(FP)*
Discharge	1,060	32.6	351.5
Discharge	1,100	28.4	161.9
Load	850	26.6	105.6
Load	1,530	27.6	161.4
Load	900	21.5	353.9
Load	1,070	34.1	416.3

* VCG and LCG(FP) are, respectively, the height above keel and the distance from the forward perpendicular of each of the weights.

Determine the final GM and drafts forward and aft.

CHAPTER SIX

FLOODING AND SUBDIVISION

Every ship that puts to sea is at risk of taking flooding water aboard. Every year, some ships of the world's merchant fleet are lost because collision or grounding damage admits enough flooding water that they sink. Although ship loss statistics indicate that only a fraction of 1 percent of the world's active merchant ships are lost annually, concern of the maritime community for the lives, ships, and cargoes at risk has resulted in many studies, international conferences, and regulations that are intended to help ship designers and operators increase the probability of survival of their ships when damage is sustained. It is the purpose of this chapter to examine the causes and effects of flooding and how proper subdivision of ship hulls into separate watertight compartments can minimize the risk of total loss, should flooding occur.

CAUSES OF FLOODING

The most common causes of flooding are collision and grounding. A moving ship has a large momentum (mass times velocity) even when moving slowly, because of its large mass. Thus, when it strikes another ship or a stationary structure like a pier or bridge, or when it runs aground on a rocky bottom, it does not come to rest easily or immediately. Invariably, hull plating is ruptured and spaces are flooded. Other events that can breach the watertight integrity of a ship's hull are internal explosion, wastage of hull steel causing a leak, and enemy action using underwater attack. Ship officers should take note that all of the above events (except for enemy attack) can be attributed to human error, either of navigation, operation of machinery, or inadequate maintenance.

Flooding of spaces can sometimes take place without rupturing the hull below the waterline. Taking water aboard through openings in the deck in heavy weather, called *downflooding,* or water used to fight a shipboard fire, or erroneous opening of valves in piping systems connected to openings in the hull are examples. The behavior of the ship as a result of this kind of flooding differs from the cases in which there is uncontrolled free flow of seawater through a hull rupture. Since the hull remains intact, the flotation, stability, and trim of the ship can be evaluated using the principles of intact stability with free surface, as discussed in Chapters 3 through 5.

EFFECTS OF FLOODING

What happens to a ship when its hull is damaged (or holed) below the waterline and flooding water enters (or bilges) a compartment? In a typical case, as flooding progresses, many changes take place simultaneously in the draft, freeboard, trim, and transverse and longitudinal stability. In order to define terms and examine these events individually, see Figure 6-1, which illustrates the simplest possible case of a hypothetical vessel with symmetrical bow and stern and watertight transverse bulkheads extending to the deck. Such bulkheads are known as *subdivision bulkheads* because they subdivide the vessel into separate watertight compartments. The uppermost deck to which the subdivision bulkheads extend is called the *bulkhead deck*. The intact vessel, floating on even keel at waterline W_oL_o as in (a), is in a stable equilibrium condition. Buoyant volume ∇ is labeled here as the *active buoyancy*. The internal watertight volume of the compartments above the waterline and up to the bulkhead deck is denoted *reserve buoyancy*, because when flooding causes the vessel to settle deeper into the water, these spaces, except for the flooded compartment itself, become part of the active buoyant volume. If the central compartment is holed, so that water enters it as in (b), the flooded part of the compartment is no longer buoyant, and the draft increases to compensate for this loss in buoyancy. Flooding stops as shown in (c) when the sum of the active buoyant volumes forward and aft of the flooded compartment becomes equal to the volume of displacement (∇) of the ship before it was damaged, and equilibrium is restored at waterline $W'L'$. The flooding water is imagined to be part of the sea (as opposed to considering it as if its weight were added to the ship's weight), so that the buoyancy required to float the ship does not change.

It is also correct to say that equilibrium in the flooded condition is restored when the buoyancy lost as a result of flooding is regained from the reserve buoyancy. Part (d) of the figure illustrates this concept. The volume of the bilged compartment up to the original intact waterline (W_oL_o) is called *lost buoyancy*. In general, any part of the previously active buoyant volume that becomes flooded is considered lost buoyancy. The buoyant volume between intact waterline WL and flooded waterline $W'L'$ and outside of the flooded compartment is called *regained buoyancy*. In general, any part of the previous reserve buoyant volume that is active buoyant volume when flooding has ended is considered re-

FLOODING AND SUBDIVISION   171

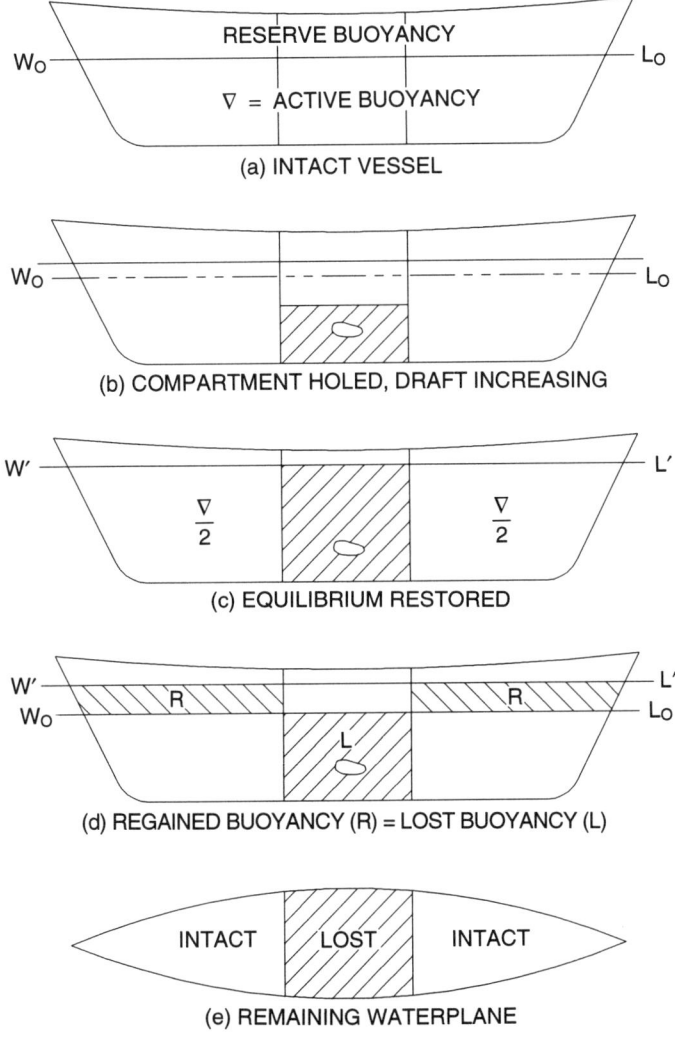

Figure 6-1. Effects of flooding.

gained buoyancy. Thus equilibrium can be said to be restored when the regained buoyancy equals the lost buoyancy, as shown in (d). It should be noted that the upper part of the bilged compartment does not contribute to lost buoyancy since it was not part of the original active buoyancy, nor is it regained buoyancy since it is flooded when equilibrium is restored. Thus it is clear that a vessel that has no internal subdivision must sink if it is holed anywhere below the waterline, because it can generate no regained buoyancy.

Part (e) of Figure 6-1 illustrates the condition of the waterplane after flooding. Freely flooding seawater has rendered that part of the waterplane within the bilged compartment ineffective, reducing the area (A_W) and the transverse and longitudinal moments of inertia (I_T and I_L) of the waterplane. Thus the vessel's TPI, BM_T, and BM_L are all reduced.

The effects of the flooding illustrated by this example are as follows.

Effect of Flooding on Draft and Freeboard

Flooding of a portion of a ship results, in general, in an increase in draft and a consequent decrease in freeboard. The only exception occurs if the bilged compartment is already filled with a liquid to a level above the final equilibrium waterline. In that case there will be a net decrease in draft as the liquid cargo is replaced by flooding water. In all other cases, the draft increases until equilibrium is restored or until flooding water progresses into other compartments. If the freeboard of the undamaged ship is so small that the flooding waterline reaches the bulkhead deck, the reserve buoyancy is used up and the vessel will sink. Loss of a ship due to complete loss of reserve buoyancy without heeling and with positive stability is called *foundering*. International regulations specifying the minimum permissible freeboard are based in part on providing adequate reserve buoyancy to prevent foundering.

Effect of Flooding on Stability

Ships of normal form suffer a decrease in both transverse and longitudinal stability when a major compartment is flooded. The predominant change in stability is caused by the loss of waterplane inertia, and hence of BM, as shown in Figure 6-1(e). On the other hand, as draft increases, so does the height of the center of buoyancy. The increase in KB offsets to some extent the decrease in BM, and in vessels of unusual form, a slight increase in stability can result. In the more usual case, and especially when the flooded compartment is at the widest part of the ship, the loss in transverse stability can be substantial and dangerous. The end result of complete loss of transverse stability is *capsizing*. This can take place even when the flooding water does not cause a heeling moment, if GM_T becomes negative and the freeboard has been so reduced that even a very small angle of loll will immerse the bulkhead deck. It is not feasible for ship's officers to predict the amount of remaining stability or the point at which capsizing will occur with any certainty; thus the seriousness of this situation cannot be overemphasized.

Two additional effects of flooding are not illustrated in the simple case described above because the flooded space was described as being symmetrical both transversely and longitudinally. In the more typical flooding situation there is a change of trim, since it is rare that a flooded compartment would be so located that sinkage is entirely parallel. If, in addition, the flooded compartment is

FLOODING AND SUBDIVISION 173

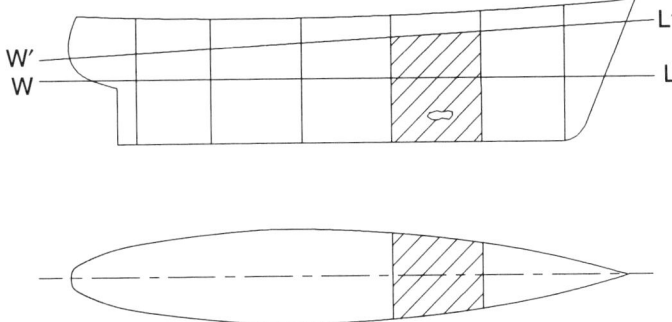

Figure 6-2. Flooding with trim change.

not symmetrical about the ship's centerline, the flooding water will cause the ship to heel as well.

Effect of Flooding on Trim
In addition to causing an increase in draft, or sinkage, flooding water usually also produces a change of trim. Flooding water entering a forward compartment as shown in Figure 6-2 will cause the ship to trim by the head until the center of buoyancy of the intact part of the underwater hull is in vertical alignment with the ship's center of gravity. If equilibrium is not restored before the trimmed waterline immerses any part of the bulkhead deck, downflooding through openings in the deck can take place (deck openings like cargo hatches are weathertight, but not watertight), and the ship will eventually be lost due to progressive flooding. As more compartments flood, the loss in longitudinal moment of inertia of the waterplane severely reduces longitudinal stability and the moment to change trim until the ship, upended, dives beneath the waves. This event is called *plunge*. The loss of a ship by plunging brought on by progressive flooding usually takes a considerable length of time, as opposed to the relatively rapid loss that accompanies capsizing.

Effect of Flooding on Heel
Flooding of any compartment puts a ship in a perilous condition, but if the flooded compartment is off center so as to cause heel as well as sinkage and trim, the resulting situation is most dangerous, as shown in Figure 6-3. The damaged ship will heel until the center of buoyancy of the intact underwater hull is vertically aligned with the ship's center of gravity. The freeboard on the low side decreases rapidly, and, if the flooded space is large, equilibrium in a heeled condition may not be achieved before the deck immerses. While a ship might survive for a time in such a precarious state because of the watertightness of the

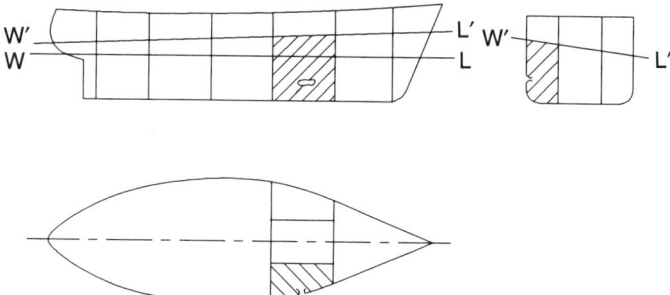

Figure 6-3. Off-center flooding with trim and heel.

deck plating, waves washing over the deck and the rolling motion of the ship are likely to be sufficient to cause water to enter nontight deck openings, flooding additional spaces and causing loss of the ship. Most hazardous of all is the fact that, as flooding progresses, the ship's transverse stability is also changing, usually becoming smaller. Negative metacentric height combined with heeling caused by asymmetrical flooding can cause the ship to capsize with little or no warning.

LIMITING FLOODING BY SUBDIVISION

In the examples of flooding described above, it was assumed that the interior of the hull was subdivided into watertight compartments by transverse subdivision bulkheads, so constructed that they are completely watertight up to the bulkhead deck. They must be of sufficient structural strength that they will remain watertight when subjected to the water pressure that would prevail if an adjacent compartment were flooded up to the bulkhead deck. It is readily apparent that a ship with no subdivision at all is doomed to sink if hull damage were to admit flooding water anywhere in the hull, because there would be no intact spaces available to preserve or recover any buoyant volume. Thus some degree of watertight subdivision is absolutely essential in any ship for it to survive damage that causes flooding water to enter the hull.

The need for some degree of subdivision to increase a ship's likelihood of surviving damage has long been recognized by naval architects, shipowners, ship and cargo insurers, and regulatory agencies. But recognition of the need is not enough. The critical question that must be addressed is: "How many bulkheads, how closely spaced?" The simplistic answer, "As many and as closely spaced as is required to assure survival of any conceivable extent of damage," imposes a requirement that is impossible to satisfy. The unsinkable ship is not achievable (unless it is entirely constructed of buoyant material) because the closer together bulkheads are spaced, the more likely it is that more than one of them will be damaged and that several adjacent compartments will be flooded. Given any arrangement of subdivision bulkheads, a location and length of damage can be found that will flood enough spaces to sink the ship. Furthermore, it is apparent that a very large number of very small compartments in a ship makes cargo stowage and loading inefficient or impossible and access to the spaces inconvenient. Thus the task of subdivision is one of devising a solution that provides an acceptable degree of safety without unreasonable economic or operational penalty.

Required Subdivision Bulkheads

Classification societies were among the first regulatory agencies to adopt standards of subdivision and to require the fitting of certain transverse watertight bulkheads. Their requirements are not based on flooding considerations alone, but on a combination of subdivision and transverse structural strength requirements. The rules of the American Bureau of Shipping (ABS) are typical of those of most of the world's classification societies as regards required transverse bulkheads.

A *collision bulkhead* is required to be constructed near the bow of all ships. Since in nearly every collision the bow of at least one of the ships (the striking ship) is damaged, the extreme bow is statistically the part of a ship most prone to damage. The collision bulkhead should be located close enough to the stem to limit the amount of flooding water so that the resulting sinkage and trim are not dangerously large, but far enough from the stem so that the bulkhead itself is not likely to be damaged in a collision. The rules require that it be not less than 0.05L (5 percent of waterline length as defined in the rules) abaft the forward perpendicular, and not more than 0.05L plus 10 feet (3.05 meters) for passenger ships, or not more than 0.08L for all other types of ships. The collision bulkhead or forepeak bulkhead as it is also known, is required to be constructed with extra heavy plating and stiffeners to withstand the dynamic forces of the sea if it is acting as a secondary bow as a result of collision damage to the bow of the ship.

An *afterpeak bulkhead* is required to enclose the propeller shaft tube or tubes in a watertight compartment at the stern. Thus flooding through the stern tube due to damage to a propeller or shafting will be limited to a relatively small amount. The only requirement for its location is that it be forward of the point at which the shafting penetrates the hull.

Machinery space bulkheads are required to enclose the machinery space in a separate watertight compartment. They serve not only as subdivision bulkheads, but also as fire screens.

Cargo hold bulkheads. In addition to the watertight bulkheads described above to set off the forepeak, afterpeak, and machinery spaces, additional subdivision bulkheads are arranged in the cargo spaces between the collision bulkhead and forward machinery space bulkhead and, when the machinery space is not fully aft, between the after machinery space bulkhead and the afterpeak bulk-

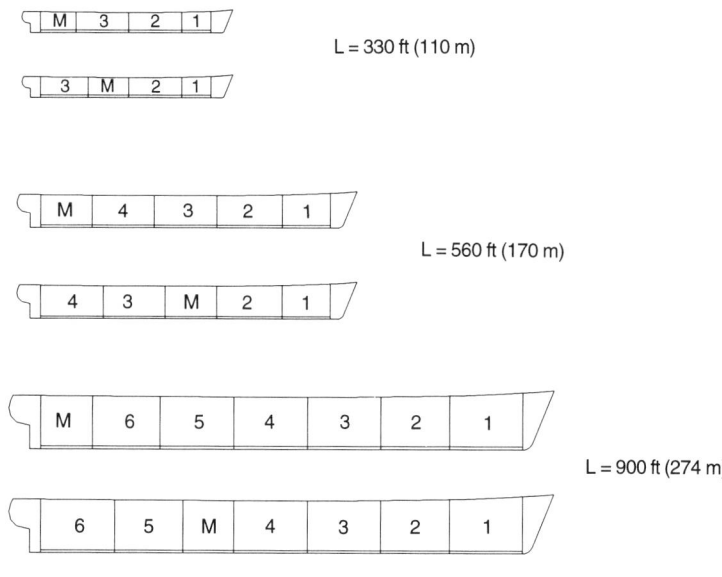

Figure 6-4. Typical watertight bulkhead arrangements for cargo ships.

head. The number and spacing of these additional cargo space bulkheads may be dictated for some types of ship by standards set forth in various international conventions, national regulations, or recommendations of the International Maritime Organization (IMO). In the case of dry cargo ships, most such regulations do not apply, so the ABS rules provide guidelines for the number and spacing of hold bulkheads. The rules require an increasing number of bulkheads as a ship gets longer, with a special provision that the first bulkhead aft of the collision bulkhead be not less than 0.2L from the stem. Figure 6-4 illustrates typical watertight bulkhead arrangements based on the ABS guidelines for cargo ships that are not subject to any other national or international subdivision standards. Examples of ships of three lengths are shown, each with two alternative positions of the machinery space.

CALCULATION OF CONDITION AFTER DAMAGE

Deciding the locations of subdivision bulkheads is only the beginning of the analysis a designer must make in order to determine the extent to which a ship can survive damage that causes flooding. Given a tentative arrangement of compartments, tanks, and other spaces separated by watertight bulkheads, calculations must be made to determine the drafts, trim, heel, and stability of the ship upon regaining equilibrium after the flooding of each of a number of defined portions of the ship. The extent of damage to be assumed in these calculations, the location on the ship of that damage, and the margin of safety required after equilibrium is reached are conditions that are defined in the various international or national regulations that may apply to the ship.

Permeability of Flooded Spaces

Flooding calculations cannot be expected to predict exact drafts and stability after flooding, because there are many uncertainties in actual flooding situations which can only be estimated when analyzing potential damage cases. Aside from the obvious uncertainties regarding the actual extent (length, height, and penetration) and location of damage, there is the question of the actual amount of flooding water that will enter a flooded compartment that already contains cargo, machinery, liquids, accommodations, or any other equipment or material. The ratio of the volume of flooding water (v_w) that enters a compartment to the molded volume of the compartment (v_c) is called the *volume permeability* (μ) of the space.

$$\mu = \frac{v_w}{v_c} \tag{1}$$

When used in calculations, μ is expressed in decimal form (0.85, for example), but values quoted in regulations and listed in tables are usually stated in percent (85 percent, or simply 85 permeability).

Approximate permeabilities have been determined for typical compartments, depending on their contents. Some typical values that have been determined by experiments or studies or that are recommended in various subdivision and damage stability criteria are listed below. Note that empty or void spaces do not have 100 percent permeability because the structure itself occupies part of the molded volume.

Space and Contents	Permeability, Percent
Empty or void compartment	95
Dry cargo or stores	60
Accommodation spaces	95
Machinery spaces	85
Container holds (containers flooded)	70
RO/RO holds (wheeled vehicles)	90
Barge ship holds (barges flooded)	76
Barge ship holds (barges intact)	30

One of the consequences of flooding is the loss of waterplane area and moments of inertia within a flooded space. The extent of these losses depends on how much structure, machinery, or cargo occupies the area of the waterplane at different levels of flooding of a space. The ratio of the area occupied by flooding water within a space (a) to the molded surface area of the space (a_s) is called the *surface permeability* (μ_s) of the space. Actual measurements and data regarding surface permeabilities are very sparse. For calculation purposes, μ_s is often assumed to be the same as the volume permeability of the same compartment, although it can be much higher, especially if the water level is near the top of the compartment, because the machinery or cargo in the space does not always extend all the way up to the deck above. A conservative choice so that stability loss will not be underestimated would be to assume a surface permeability of 95 percent, which would allow only for the area of structural elements that break the surface of the flooding water.

Two different approaches may be used to determine the waterline and stability after damage. They are the lost buoyancy method and the added weight method. Both methods involve a considerable amount of calculation, including iterative procedures necessitated by the changing hydrostatic properties as the draft increases. Calculations done by hand thus tend to be tedious and repetitious as successive approximations approach convergence on equilibrium. The sample calculations that follow are done for a simple hull geometry—the rectangular barge—so that the method can be shown with a minimum of repetitious iteration. Comprehensive calculation forms and descriptions of the calculations by hand for actual ship forms can be found in Volume 1 of *Principles of Naval Architecture* (see Bibliography). Since calculations of this kind must be done for numerous cases of assumed damage and flooded compartments, routine calculations are done almost exclusively by computer programs using as input information the ship's offsets and hydrostatic properties.

Lost Buoyancy Method

The lost buoyancy method treats the flooding water as part of the sea, hence the ship's displacement and volume of displacement and KG remain unchanged, but the shape and distribution of the buoyant volume changes as the lost buoyant volume below the original waterline is regained from the reserve buoyant volume above the original waterline. The following example is solved using the lost buoyancy method.

Example 6-1

A rectangular barge of L = 320 ft, B = 50 ft, D = 30 ft floats in seawater at an even keel draft of 10 ft, and KG = 12 ft. It is subdivided by five transverse bulkheads, located as shown in Figure 6-5. The second compartment from the bow is damaged and open to the sea. The permeability of the compartment is 75 percent, and the surface permeability is 95 percent. Determine the final equilibrium drafts and stability.

FLOODING AND SUBDIVISION 179

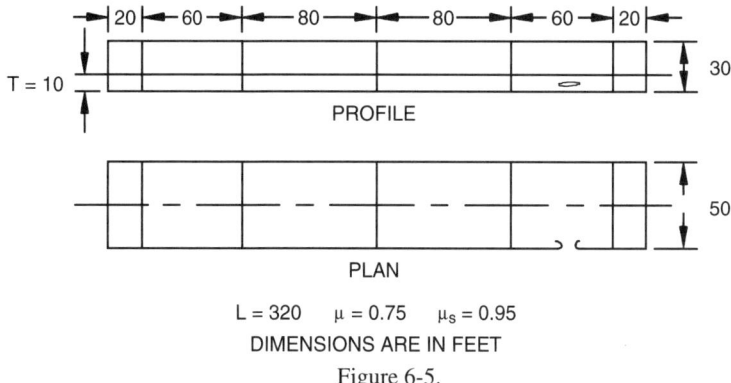

L = 320   μ = 0.75   μ$_s$ = 0.95
DIMENSIONS ARE IN FEET
Figure 6-5.

Displacement.

$$\nabla = LBT = 320 \times 50 \times 10 = 160{,}000 \text{ ft}^3$$

$$\Delta = \frac{\nabla}{35} = \frac{160{,}000}{35} = 4{,}571 \text{ tons}$$

Parallel Sinkage. As in ordinary trim problems (Chapter 5), the effect of flooding on the vessel's drafts is determined in two steps, parallel sinkage (trim restrained) and trim change (restraint removed). Quantities involved in determining parallel sinkage are:

v_c = molded volume of flooded compartment up to WL in Figure 6-6
v = lost buoyancy (volume) = volume of flooding water that enters the bilged compartment up to WL
μ = volume permeability of flooded compartment (%/100)
A = area of original undamaged waterplane WL
a_s = molded surface area of flooded compartment
a = net loss of waterplane area
A' = remaining waterplane area after damage
μ_s = surface permeability of flooded part of waterplane
s = parallel sinkage

From the dimensions given, the compartment volume is

$$v_c = 60 \times 50 \times 10 = 30{,}000 \text{ ft}^3$$

and the volume of lost buoyancy is

$$v = \mu v_c = 0.75 \times 30{,}000 = 22{,}500 \text{ ft}^3$$

Equilibrium will be restored when the lost buoyancy is regained. With the trim restrained, the regained buoyancy occupies the parallel sinkage layer shown in

180   APPLIED NAVAL ARCHITECTURE

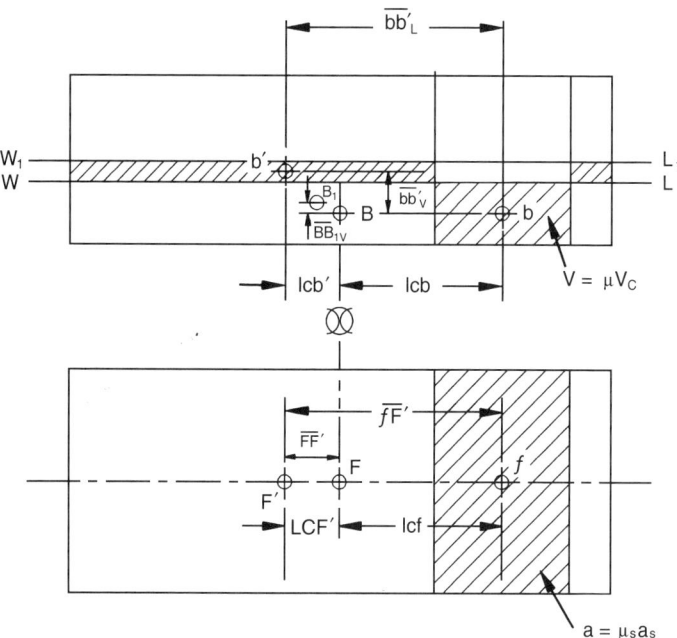

Figure 6-6.

Figure 6-6 between WL and W_1L_1, except for the part occupied by flooding water within the flooded compartment. That is, regained buoyancy equals remaining waterplane area times parallel sinkage.

The original waterplane area is

$$A = L \times B = 320 \times 50 = 16{,}000 \text{ ft}^2$$

The lost waterplane area is

$$a = \mu_s \times a_s = 0.95 \times 60 \times 50 = 2{,}850 \text{ ft}^2$$

Therefore the remaining waterplane area is

$$A' = A - a = 16{,}000 - 2{,}850 = 13{,}150 \text{ ft}^2$$

Thus regained buoyancy (A's) and parallel sinkage (s) will increase until the regained buoyancy equals the lost buoyancy (v).

$$s = \frac{v}{A'} = \frac{22{,}500}{13{,}150} = 1.711 \text{ ft}$$

FLOODING AND SUBDIVISION

Change of Trim. When the trim restraint is removed, the longitudinal moment of the shift in buoyancy from the centroid of the lost buoyant volume to that of the regained buoyancy will cause the barge to trim by the head. The quantities involved in the trim calculation are defined in Figure 6-6, which shows the barge profile (upper figure) and plan view of the flooded waterplane (lower figure). Longitudinal moment arms are measured from amidships, positive forward.

B = center of buoyancy, intact barge to WL
B_1 = center of buoyancy, damaged barge to W_1L_1
$\overline{BB_{1V}}$ = rise of B due to parallel sinkage
b = centroid of lost buoyancy
b' = centroid of regained buoyancy
$\overline{bb'}_L$ = longitudinal shift of buoyant volume (v)
$\overline{bb'}_V$ = vertical shift of buoyant volume (v)
lcb, lcb' = distances of b and b' from amidships
Kb, Kb' = heights of b and b' above keel
f = center of flotation of lost waterplane area (a)
F = center of flotation of original waterplane (A)
F' = center of flotation of remaining waterplane (A')
lcf, LCF, LCF' = distances of f, F, and F' from amidships

From the geometry of Figure 6-6 and the dimensions of the barge,

$$LCF = 0 \text{ (rectangular waterplane)}$$
$$lcb = lcf = 80 + 30 = 110 \text{ ft}$$

The centroids of the remaining waterplane (F') and the regained buoyancy (b') are determined by calculating the moments about amidships of the waterplane areas:

remaining = original − lost
$A' \times LCF' = A \times LCF - a \times lcf$
$13{,}150 \times LCF' = 16{,}000 \times 0 - 2{,}850 \times 110$
$$LCF' = -\frac{2{,}850 \times 110}{13{,}150} = -23.84 \text{ ft (aft of amidships)}$$

Since the barge is wall-sided, b' is vertically below F', and

$$lcb' = -23.84 \text{ ft}$$

The longitudinal shift of the buoyancy that was lost due to flooding is

$$\overline{bb'}_L = 110 + 23.84 = 133.84 \text{ ft aft}$$

The trimming moment (tm) caused by v/35 tons of lost buoyancy shifting 133.84 ft aft is

$$tm = \frac{v \times \overline{bb'_L}}{35} = \frac{22{,}500 \times 133.84}{35} = 86{,}040 \text{ ft-tons}$$

which causes a trim change of (t is in ft):

$$\frac{t}{L} = \frac{tm}{\Delta\, GM'_L} \approx \frac{tm}{\Delta\, BM'_L} = \frac{tm}{(I'_L/35)}$$

$$t = \frac{35 \times L \times tm}{I'_L}$$

The primes refer to the parallel sinkage condition. The substitution of BM'_L for GM'_L causes a negligible change in the result. We have to calculate I'_L, the longitudinal moment of inertia of the remaining waterplane about the transverse axis through its centroid, F'. The surface permeability (μ_s) is assumed to reduce the amount of lost waterplane inertia in proportion to the loss in waterplane area. The longitudinal moment of inertia of the remaining waterplane about the F' axis is equal to the longitudinal moment of inertia of the total waterplane (as if undamaged) about the F' axis minus that of the lost waterplane about the same F' axis:

$$I'_L = (I_{LO} + A\, \overline{FF'}^2) - \mu_s\, (i_{LO} + a_s\, \overline{fF'}^2)$$

where subscript O means "about its own centroidal axis."

$$I_{LO} = \frac{BL^3}{12} = \frac{50 \times 320^3}{12} = 136.533 \times 10^6 \text{ ft}^4$$

$$A \times \overline{FF'}^2 = 16{,}000\,(23.84)^2 = 9.094 \times 10^6 \text{ ft}^4$$

$$\mu_s\, i_{LO} = \frac{0.95 \times 50 \times 60^3}{12} = 0.855 \times 10^6 \text{ ft}^4$$

$$\mu_s\, a_s\, \overline{fF'}^2 = a \times \overline{fF'}^2 = 2{,}850\,(133.84)^2 = 51.052 \times 10^6 \text{ ft}^4$$

Therefore

$$I'_L = 10^6\,(136.533 + 9.094 - 0.855 - 51.052) = 93.720 \times 10^6 \text{ ft}^4$$

The trim change is

$$t = \frac{35 \times 320 \times 86{,}040}{93.720 \times 10^6} = 10.28 \text{ ft by the head}$$

Trimming the barge about center of flotation F', we have

$$F' \text{ from bow} = \frac{L}{2} + LCF' = \frac{320}{2} + 23.84 = 183.84 \text{ ft} = F$$

$$F' \text{ from stern} = \frac{L}{2} - LCF' = \frac{320}{2} - 23.84 = 136.16 \text{ ft} = A$$

and the drafts forward and aft are

$$T_F = T + s + \left(\frac{F}{L}\right)t = 10.00 + 1.71 + \left(\frac{183.84}{320}\right)10.28$$
$$= 17.62 \text{ ft}$$

$$T_A = T + s - \left(\frac{A}{L}\right)t = 10.00 + 1.71 - \left(\frac{136.16}{320}\right)10.28$$
$$= 7.34 \text{ ft}$$

Damaged Stability. The metacentric height in the damaged condition is given by

$$GM_{T2} = KB_2 + BM_{T2} - KG$$

where the subscript 2 refers to the final trimmed condition after flooding.

The KB increases in two stages when flooding occurs—an increase due to parallel sinkage, illustrated in Figure 6-6, and a further increase due to the change in trim, illustrated in Figure 6-7. The first of these increases is determined by equating the vertical shifting moment of the total buoyant volume to that of the lost buoyant volume. Referring to Figure 6-6, it is clear that

$$Kb = \frac{10.0}{2} = 5.0 \text{ ft}$$

$$Kb' = T + \frac{s}{2} = 10.0 + \frac{1.711}{2} = 10.86 \text{ ft}$$

$$\overline{bb'_V} = 10.86 - 5.00 = 5.86 \text{ ft up}$$

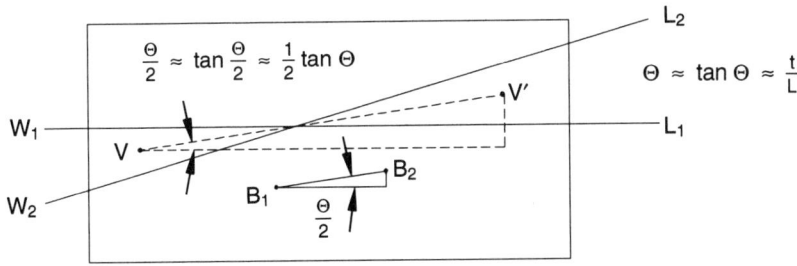

Figure 6-7.

Equating the moments of buoyant volumes,

$$\nabla \times \overline{BB_{1V}} = v \times \overline{bb'_V}$$

$$\overline{BB_{1V}} = \frac{22{,}500 \times 5.86}{160{,}000} = 0.824 \text{ ft}$$

The change in trim will cause a further increase in KB. This is made evident by referring to Figure 6-7, which shows a profile view of the barge drawn with a greatly exaggerated vertical scale for clarity, because the actual trim angles are very small. The figure shows the transfer of a wedge-shaped buoyant volume from point v to v', as the barge trims about the parallel sinkage waterplane W_1L_1. The vertical component of this transfer of buoyancy will be upward, since point v is always below the waterline and point v' is always above it, regardless of the direction of the trim change. Because all of the angles indicated in the figure are very small, the following approximations are valid:

1. The angle between vv' and W_1L_1 is approximately half of the angle of trim change (θ).
2. The small angles are approximately equal to their tangents.
3. The tangent function is approximately linear for very small angles.

Thus we have

$$\theta = \tan \theta$$

$$\frac{\theta}{2} = \tan \frac{\theta}{2} = \frac{1}{2} \tan \theta$$

Since θ is the angle of trim,

$$\tan \theta = \frac{t}{L} \quad \text{(t and L in feet)}$$

$$\tan \frac{\theta}{2} = \frac{t}{2L}$$

In the figure, the B triangle is geometrically similar to the v triangle. Let $\overline{B_1B_{2V}}$ and $\overline{B_1B_{2L}}$ be the vertical and longitudinal shifts of the center of buoyancy, respectively. Then

$$\frac{\overline{B_1B_{2V}}}{\overline{B_1B_{2L}}} = \tan \frac{\theta}{2} = \frac{t}{2L}$$

$$\overline{B_1B_{2V}} = \frac{t \times \overline{B_1B_{2L}}}{2L}$$

$\overline{B_1B_{2L}}$ is determined from the previously calculated trimming moment as follows:

$$tm = \Delta \times \overline{B_1B_{2L}}$$
$$\overline{B_1B_{2L}} = \frac{tm}{\Delta} = \frac{86{,}040}{4{,}571} = 18.82 \text{ ft}$$

Therefore

$$\overline{B_1B_{2V}} = \frac{t \times \overline{B_1B_{2L}}}{2L}$$
$$= \frac{10.28 \times 18.82}{2 \times 320}$$
$$= 0.302 \text{ ft}$$

The final KB_2 can now be determined:

$$\text{KB original was } \frac{T}{2} = \frac{10}{2} = 5.000 \text{ ft}$$
$$\text{Rise due to parallel sinkage} = 0.824$$
$$\text{Rise due to change of trim} = 0.302$$
$$KB_2 = 6.126 \text{ ft}$$

The metacentric radius in the damaged condition is

$$BM_{T2} = \frac{I_{T2}}{V}$$

where I_{T2} is the transverse moment of inertia of the waterplane remaining after damage. A close approximation to the final trimmed waterplane is the parallel sinkage waterplane. Since there is no off-center flooding, all parts of the waterplane (intact and flooded) have a common longitudinal axis of symmetry, and therefore transverse moments of inertia may be added or subtracted directly.

$$I_{T2} = I_{T0} - \mu_s\, i_{T0}$$
$$I_{T0} = \frac{LB^3}{12} = \frac{320 \times 50^3}{12} = 3{,}333{,}333 \text{ ft}^4$$
$$\mu_s\, i_{T0} = \frac{0.95 \times 60 \times 50^3}{12} = 593{,}750 \text{ ft}^4$$

Thus

$$I_{T2} = 3{,}333{,}333 - 593{,}750 = 2{,}739{,}583 \text{ ft}^4$$

$$BM_{T2} = \frac{2{,}739{,}583}{160{,}000} = 17.122 \text{ ft}$$

Finally, the damaged metacentric height is

$$GM_T = 6.126 + 17.122 - 12.000 = 11.248 \text{ ft}$$

The extent of the loss of stability caused by flooding can be seen by comparing GM_{T2} to the original undamaged GM with the intact barge floating at a draft of 10 ft:

$$GM_T = KB + BM_T - KG = \frac{T}{2} + \frac{B^2}{12T} - KG$$

$$= \frac{10}{2} + \frac{50^2}{12 \times 10} - 12$$

$$= 5.000 + 20.833 - 12.000 = 13.833 \text{ ft}$$

The loss in GM is therefore $13.833 - 11.248 = 2.585$ ft, which is about a 19 percent reduction.

Added Weight Method
The worked example illustrates the procedures involved in determining final drafts and stability after flooding, using a procedure known as the lost buoyancy method, in which the displacement and KG of the vessel are assumed to remain unchanged since the flooding water is modeled as if it were part of the sea. A different approach can also be used, called the *added weight method,* in which the flooding water that enters a bilged compartment is treated as a weight to be added to the ship's displacement, thus changing the displacement and KG as well as all of the other ship characteristics as flooding progresses. In general, the added weight method requires more calculation than the lost buoyancy method, because one has to guess at a final trim line after damage in order to calculate the weight and centroid of the flooding water that enters the ship. It thus involves iterative procedures even for simple barge calculations. The added weight method is not as convenient as the lost buoyancy method for flooding calculations done by hand, and it is not demonstrated in this text. It does have some advantages over the lost buoyancy method in some cases—especially if the flooded space or spaces are of very irregular geometry (as might often be the case in naval vessels, for example) or if it is desired to determine the stability at intermediate stages of flooding before the ship reaches flooded equilibrium. Detailed instructions and calculation forms for this method to be done using hand calculations are given in Volume 1 of *Principles of Naval Architecture* (see Bibliography and p. 178).

When applied to a particular flooding situation, both of these methods will, if done correctly, yield the same answers as to the trim line and stability when flooded equilibrium is reached. It should be noted, however, that the final metacentric heights determined by the two methods will not be equal to each other. This does not mean that the stability is different. The proper measure of stability is the righting moment at a given angle of heel or, for the upright condition, the product of the displacement and GM. Because the displacement is imagined to have increased during flooding in the added weight approach, its final GM will be proportionately smaller than that calculated using the lost buoyancy approach. The product of displacement and GM will be the same, however, regardless of the method of calculation.

Computer Methods
Although the sample calculation above is for a very simple hull form and simple geometry of the flooded space, it might be noted that flooding calculations can be somewhat involved and tedious. A complete set of calculations for a ship would require a similar amount of calculation for each of many flooded spaces, and additional complications if asymmetrical flooding causing the ship to list is involved. As a result, flooding calculations are today done almost exclusively by computer, and there are many commercially available programs for the purpose. Using such programs still involves a considerable amount of work, because their input requires detailed offsets; hydrostatics; dimensions that locate all bulkheads, tanks, and decks; specifications of the compartments to be flooded; and the permeabilities to be assumed.

Most computer programs for flooding calculations are not based on either of the methods mentioned above for hand calculations. They are done by calculations that converge on a final volume of displacement and the three coordinates of the center of buoyancy of the flooded ship by iteration, trying different drafts, trims, and heel angles until equilibrium is achieved. The basic calculations are done by direct integration of the immersed shapes of the hull as described by a table of offsets. Computer calculations have the obvious advantage of being able to compute many more flooding situations than hand calculations would allow, and they also determine the full set of righting arms and statical stability curves for the ship in each damaged condition—a task that would be prohibitive to do by hand.

FLOODABLE LENGTH

The calculations described above determine final drafts and stability of a ship when the space to be flooded is completely specified by its size, location, and permeability. They are not suitable for determining in the first instance the locations of the main subdivision bulkheads of a ship in the early stages of its design in order to provide a desired measure of survivability after flooding. Guidance for that task is provided, in part, by calculating the *floodable length* for any point in the ship.

To understand the notion of floodable length, it is helpful to conceptualize a hypothetical ship and a clear definition of sinking. The ship is imagined to have watertight bottom plating, side shell plating, and bulkheads. The side shell and subdivision bulkheads extend upward to the *bulkhead deck.* For calculation purposes, however, the bulkhead deck is imagined to be totally nonwatertight. In fact, the deck is imagined not to exist, because the shell and bulkheads are terminated at an imaginary line on the ship's sides below the bulkhead deck called the *margin line,* defined below. Above the margin line, the hypothetical ship ends. It has no deck. The ship is, conceptually, like an open-top canoe, but subdivided into separate watertight compartments by bulkheads. The sea in which it floats is supposed to be perfectly smooth and without waves. Only centerline symmetrical flooding is considered. That is, the behavior after flooding is limited to sinkage and trim, but without heel. The ship survives if it floats at any waterline that is below the margin line, and sinks if the waterline rises above the margin line. A waterline after flooding that is exactly tangent to the margin line at any point is thus a limiting trim line, putting the ship on the verge of sinking. The definitions that follow formalize the above concept of the ship for the purpose of determining floodable length.

- *Subdivision load line.* The deepest waterline permitted by the applicable subdivision regulations for the ship prior to flooding.
- *Bulkhead deck.* Uppermost continuous weathertight deck to which the transverse subdivision bulkheads and the side shell plating are carried.
- *Margin line.* A line defined variously, depending on the ship configuration. In no case may it be less than 3 in (76 mm) below the upper surface of the bulkhead deck at side. If the bulkheads extend to different decks in different portions of the ship, the margin line may be measured from the different decks in a manner prescribed by regulation.
- *Floodable length.* The floodable length at any point in the length of the ship is the maximum portion of the ship's length, having its center at the point in question, that can be symmetrically flooded at a given permeability, without immersing the margin line.

It will be made clear, by visualizing the ship defined above, that the floodable length varies with position in the ship and with the permeability of the space into which the flooding water is admitted. An example of a typical set of floodable length curves for a range of permeabilities is shown in Figure 6-8 superimposed on a sketch of the profile of the ship to which it pertains. An ordinate such as FL in the figure, drawn from a point on the baseline of the ship, such as point F, is a measure of the floodable length associated with that point in the ship. The scale for floodable length is the same as that for the horizontal ship length scale. Therefore, if the permeability is 60 percent, floodable length FL can be laid out

FLOODING AND SUBDIVISION 189

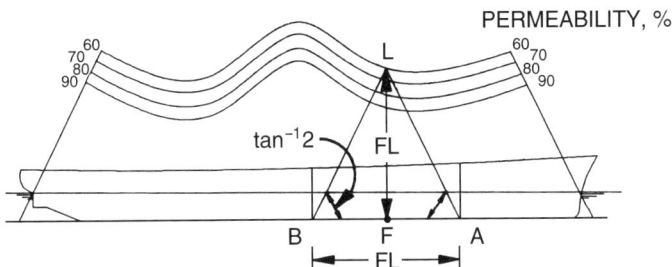

Figure 6-8. Floodable length definitions.

horizontally with its center at point F, to define the limits of an imagined compartment that has admitted flooding water. If subdivision bulkheads were to be located at A and B, flooding the compartment between them at 60 percent permeability would cause the ship to sink to a trimmed waterline that touches the margin line at only one point. It will be apparent that, with the given geometry, the ends of the floodable length could also be located by drawing the two sloping lines LA and LB at the angle of arctan 2.0 to the baseline, because that will produce AF and BF, each equal to half of FL. Similar tests of the locations of bulkheads that will admit just enough water to bring the ship to the brink of sinking could be made for any point in the ship and for any permeability.

Floodable length curves may be determined either manually, as described in *Principles of Naval Architecture,* or by computer. Specific calculation procedures vary, but all of them require some guesswork and iteration. Manual methods start by defining a final trimmed waterline tangent to the margin line, and use the Bonjean curves and special sectional area curves to the trimmed waterlines. A trial floodable length and trial location of the centroid of the flooding water that would enter the ship are used to calculate by numerical integration of the trim line sectional area curve the corresponding volume of flooding water and position of the centroid and midpoint of the flooded compartment that would result. A second try usually results in a defined compartment that actually satisfies the condition that the final trim line match the chosen limiting trim line. Computer methods may differ in procedure, but the results will be the same.

The manner in which the ship designer uses floodable length curves to choose appropriate locations of the subdivision bulkheads makes use of the isosceles triangle geometry shown in Figure 6-8. The collision and afterpeak bulkheads are located first, according to the classification rules. Engine space and cargo hold bulkheads are then positioned tentatively, and the triangles are drawn from their intersections with the baseline. If the vertex of the triangle corresponding to each compartment falls below the floodable length curve for the permeability appropriate to the compartment, the ship will survive flooding of that compartment without sinking due to plunge. (The floodable length curves described here are

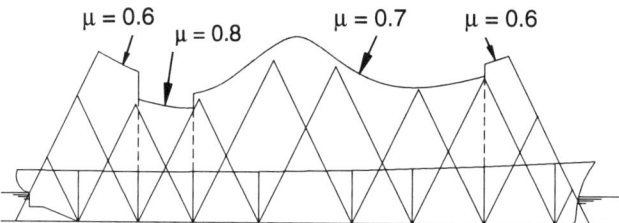

Figure 6-9. Testing bulkhead locations with a floodable length curve.

not tests for loss of transverse stability. That must be determined separately.) The altitudes of the set of triangles can be considered the "flooded lengths" associated with the flooding of every main compartment, one at a time. If the flooded length is much less than the floodable length, the final trim line will be well below the margin line, while flooding of a space whose flooded length (vertex) just touches the appropriate floodable length curve will bring the final trim line exactly to the margin line, so long as the actual permeability of the space matches exactly the permeability assumed in the calculations. Testing compartment lengths in this way can be done very rapidly, and changing the locations of the bulkheads is a very simple process. Furthermore, if the sides of the triangles are extended upward to a second tier of vertices, the ship can be tested for survivability in the event that two adjacent compartments are flooded by a hull penetration that damages a bulkhead as well as the side shell. If the full set of first-tier vertices test safely, the ship is said to be a *one-compartment ship* for sinkage and trim. If the second-tier vertices also lie below the floodable length curve, the ship is a *two-compartment ship,* and it will be appreciated that it will be much safer than a *one-compartment ship* because it can survive damage anywhere in its length, including on a subdivision bulkhead, unless the opening in the hull is so long that it damages two bulkheads and causes three compartments to be flooded. Figure 6-9 illustrates how a floodable length curve has been used to test the locations of the subdivision bulkheads in a ship. The curve is discontinuous because the permeabilities of the spaces vary as shown. The ship in this figure is seen to be a two-compartment ship, since the second-tier vertices all lie below the floodable length curve.

SUBDIVISION STANDARDS AND CRITERIA

In the early years of the twentieth century, when the oceans were regularly traversed by the now-legendary great passenger liners, it became clear to the authorities of the principal maritime nations of the world that there was a need for regulations for adequate subdivision beyond the classification society–required bulkhead rules. The first international conference on the subject was convened in 1913 as a direct result of the sinking of the *Titanic* in 1912. It was called the International Conference on Safety of Life at Sea (SOLAS). There have been, to date, five SOLAS conferences (1913, 1929, 1948, 1960, and 1974) and the essential work of the conferences today takes place continuously under the auspices of the International Maritime Organization (IMO) of the United Nations. Most of the provisions of these international conferences apply only to passenger ships, but some maritime nations, including the United States, have national standards of subdivision that apply to cargo ships as well.

Typical standards require the ship designer to assume a prescribed extent of damage (length, penetration, vertical extent), location of damage (between bulkheads or on bulkheads), and acceptable condition of flotation and stability after damage (location of margin line, maximum angle of list, minimum GM, for example). Although specific regulations are not discussed here, three types of approach used in various regulations to define the standards will be described briefly. They are the *integer compartmentation* approach, the *factor of subdivision* approach, and the *probability of survival* approach.

Integer Compartmentation
Regulations specifying integer compartmentation standards require that the ship be capable of surviving the flooding of an integer number of compartments simultaneously. One-compartment and two-compartment ships have been defined previously, and ships with three or more compartments are similarly defined. Regulations sometimes permit a ship to have different compartmentation standards for different portions of the ship, based on the knowledge that certain parts of a ship are statistically more likely to be damaged than others. An important threshold of safety is crossed when a ship becomes a two-compartment ship, because it can then sustain damage to a subdivision bulkhead as well as to the side shell and survive.

Factor of Subdivision
Since the earliest SOLAS conferences, the most enduring approach to subdivision has been to define a required factor of subdivision for each ship subject to the regulations. A factor of subdivision is a number equal to or less than one that multiplies the floodable length to define the permissible length of a ship's compartments. The assumption implicit in this procedure is that survivability after flooding always improves as the subdivision bulkheads are spaced closer together. It is the "closer the better" philosophy, which assumes that if the floodable length at a particular point in a ship is, say, 100 feet, spacing the bulkheads at 90, 80, 70, or 60 feet (that is, with a factor of subdivision of 0.9, 0.8, 0.7, or 0.6) will provide increasing safety as the spacing gets smaller. The choice of the required factor of subdivision is based on the concept of the ship's "criterion of service" numeral, an empirical number which is intended to describe the degree to which a ship is a passenger vessel. A small numeral is associated with

few passengers, a large number with many passengers. The required factor of subdivision gets smaller as the criterion of service numeral increases. It will be observed that the ship will be a one-compartment ship if the factor of subdivision is between 1 and 1/2, a two-compartment ship if it is between 1/2 and 1/3, a three-compartment ship if it is between 1/3 and 1/4, etc. While it is certain that safety increases as the integer compartmentation increases, it has been shown that reducing the factor of subdivision does not necessarily increase survivability until compartmentation is increased. The fact is that, as bulkhead spacing decreases, the probability that damage might involve a bulkhead increases, thus increasing the chances of flooding two compartments. Furthermore, as bulkheads get closer together, the length of damage that can be contained between bulkheads gets smaller, again increasing the likelihood of flooding more than one space. For these reasons, the factorial system standards are becoming less popular; and, while they are still acceptable in international regulations as an alternative to the probability of survival approach, it is likely that they will be phased out over a period of time.

Probability of Survival

A more rational approach to survivability was discussed at the 1960 SOLAS Conference, and regulations were included in the 1974 Convention which describe the probability of survival approach. The new regulations apply only to passenger ships at the present time, and they are called "alternate equivalent passenger vessel regulations" because the 1974 regulations permit passenger ship subdivision to be based either on the new probability calculations or on the existing 1960 regulations, which use the traditional factorial system of calculations.

The new regulations are based on studies of casualty data of about 860 ship damage cases, including data on the location and extent of flooding, the loading condition before damage occurred (draft, permeabilities, operating GM, etc.), and the sea state and wind conditions. Model tests that simulated the behavior of damaged ships in various wave conditions were also conducted and analyzed. There are three probabilities that have an effect on subdivision and damage stability:

1. The probability that the ship may be damaged
2. The probability as to the location and extent of flooding if the ship is damaged
3. The probability that the ship will survive the flooding

Probability theory was applied to the collected casualty data to formulate the new regulations. They define an "attained subdivision index" based on the probabilities regarding damage length and position, and on the effects of freeboard, stability, and heel in the flooded condition. The attained index of the ship must

FLOODING AND SUBDIVISION 193

be shown to be equal to or greater than a "required subdivision index," which is a function of the length of the ship and the number of persons it carries. Ships with higher passenger density have higher required subdivision indexes and hence more strict subdivision requirements.

This probabilistic approach to determining required subdivision is generally agreed by the experts working in the IMO study groups on the subject to be superior to the earlier procedures. Establishing appropriate levels of the probabilities is a difficult task, however. As sufficient data on casualties to types of ships other than passenger ships are gathered and analyzed, it can be expected that the probability approach will be applied to subdivision regulations for them as well.

PROBLEMS

Note: In all flooding problems, treat the flooding water as lost buoyancy.

1. A rectangular barge 100 ft long, 25 ft wide, and 20 ft deep floats in seawater at a draft of 10 ft, and KG = 8 ft. It is subdivided by four equally spaced transverse bulkheads, which divide the barge into five equal holds each 20 ft long and 25 ft wide. Determine the final drafts and GM of this barge if hold No. 3, the midship hold, is damaged and open to the sea. Make three such calculations, using permeabilities as follows:

    (a) $\mu = \mu_S = 100$ percent
    (b) $\mu = \mu_S = 75$ percent
    (c) $\mu = \mu_S = 50$ percent

2. Given the barge of problem 1, determine the final drafts and GM if hold No. 1, at the forward end, is flooded. Assume $\mu = 60$ percent and $\mu_S = 100$ percent.

3. Given the barge of problem 1, determine the final drafts and GM if hold No. 2 is flooded. Assume $\mu = 70$ percent and $\mu_S = 90$ percent.

4. A wall-sided barge 120 ft long and 20 ft wide has a central rectangular compartment 80 ft long, and watertight compartments at each end which are isosceles triangles in plan view, each 20 ft wide and 20 ft long. It floats in seawater at 5 ft draft even keel, and KG = 7.0 ft. If both end compartments are flooded by being open to the sea, determine the resulting drafts and GM. Assume $\mu = 95$ percent, $\mu_S = 100$ percent.

5. A rectangular barge 350 ft long and 40 ft wide floats on an even keel at a draft of 15 ft. An empty (μ = 95 percent, μ_S = 100 percent) forward end compartment 30 ft long has been bilged. Determine the forward and aft drafts that result.

CHAPTER SEVEN

SHIP STRENGTH

In the preceding chapters various aspects of the safety of ships have been addressed: safety against capsizing by providing adequate transverse stability, and safety against loss due to flooding by providing adequate subdivision, for example. In this chapter the safety of a ship against structural failure is considered by examining the arrangement and strength of ship structural members. The subject in all of its details is comprehensive and complex enough to fill many volumes, since it involves predicting the loads imposed on the ship in service, analyzing the stresses those loads cause in thousands of structural members, specifying the materials to use based on strength, cost, weldability, and ease of maintenance, designing each structural member, and choosing the structural arrangements to support the loads most efficiently. Since this book is not a book on ship structural design or advanced structural analysis, the approach taken will focus on the fundamentals rather than the details, and on the primary structure only, that is, the ship hull girder.

LOADS ON THE SHIP STRUCTURE

A ship must have sufficient structural strength to withstand its anticipated loads without failure or permanent distortion. The same could be said for any structure, machine, or device designed by engineers to function safely as intended by the designer. As with any device, adequate design of a ship's structure depends at the outset on an accurate assessment of the loads, or forces, that will be imposed on the structure in service. For ships at sea, the loads in question result from such a wide variety of sources that determining them with any accuracy is a formidable task. In fact, since many loads are imposed by nature, their magnitudes cannot be determined with certainty and must ultimately be predicted by probabilistic, rather than deterministic, methods. The probabilistic techniques will be examined later. For now, it is sufficient to consider the sources of the loads on a ship structure to appreciate the magnitude of the problem.

Static Loads
The static loads are those that have already been considered in connection with ship flotation, stability, and trim. There are the *weights* of the ship itself (structure, machinery, equipment) and its loading (cargo, fuel, etc.), which produce the downward-acting gravitational forces (mg forces), the sum of which is the ship's displacement. Balancing the total weight forces of a floating ship are its upward-acting *buoyancy* forces ($\rho g \nabla$ forces), which are the vertical components of the forces of water pressure acting on the immersed part of the ship's hull. The total buoyant force is also equal to the ship's displacement. Internal and external *pressures* on the sides of the ship and on tanks carrying liquids also produce static forces that stress the structure. *Thermal effects* can cause stresses in the ship structure due to the expansion or contraction of structural members that are attached to other members not subject to the temperature extremes. For structural design purposes, the loads described above are considered to be static, although in fact they change from voyage to voyage since the distribution of cargo and fuel weights will not always be the same.

Dynamic Loads
Superimposed on the static loads are a number of dynamic loads that vary constantly while the ship is at sea. The most obvious of these are the variable loads caused by a combination of the irregular ocean waves encountered by the ship and the motions of the ship itself as it moves through the waves and responds to them. *Wave-induced forces* cause the ship girder to undergo continuously changing longitudinal bending in the vertical and horizontal planes, torsion (twisting), and transverse bending. Since ship motions are periodic and changing, the accelerations resulting from the motions cause *inertial forces* (ma forces) on the structure wherever heavy loads are supported by the ship structure. Waves and ship motions also cause local loads on parts of the ship structure such as those arising from *shipping water* on deck and *wave impact* or "slap" on the side shell. Severe *impact loading* of the bottom plating forward can occur when a ship "slams," that is, the forefoot emerges from the water while pitching bow upward and, upon reentering the water, a significant amount of flat bottom plating strikes the nearly parallel water surface, causing a brief but intense pressure pulse that can cause bottom plating damage and induce a high-frequency vibratory stress wave in the hull that increases the wave-induced longitudinal bending stresses throughout the hull. A similar impact phenomenon can also cause slamming pressures to become excessive in the bow flare plating as it enters the water during deep pitching. High-frequency periodic loading of parts of a ship structure that support rotating machinery such as the main engines and the propellers may be induced

by the *forced vibrations* transmitted to the structure by such machinery. Inside tanks that contain slack liquids which do not fill the tanks, ship motions cause *sloshing forces* as the liquids surge within the tanks.

Occasional and Unusual Operational Loads
In addition to the loads described above that are likely to occur on all ships in normal operation, some ships must be designed to withstand structural loading imposed by special operational environments. Ships that must navigate in ice are subject to additional structural loads while *breaking ice*. These loads can increase the wave-induced bending stresses and cause high local forces at the point of contact with the ice. Combatant ships are subject to severe impact loads caused by *gun and missile blast* and recoil, underwater and above-water *explosions, aircraft landings,* and pressure loadings on interior bulkheads due to *internal flooding* caused by military attack. Special structural loads are also imposed on all ships during *launching* and *drydocking,* and occasionally during *berthing* if a pier is struck. Finally, unintentional or accidental loads may be caused by *collision, grounding,* and the flooding that may result from such accidents.

The complete loading situation on a ship at any given time is highly complex, as the above list of sources of loads indicates.

STRUCTURAL ARRANGEMENTS

To serve its intended purpose, a ship must be a floating watertight container capable of withstanding the kinds of loads listed above without failing structurally by rupturing or deforming permanently. The entire structure may be likened to a girder, that is, to a long structural member supported from below by buoyant forces and bearing the weights of its cargo and other deadweight items as well as its own structural weight, while bending and twisting in a seaway. The ship hull girder must be designed to resist longitudinal bending, the primary loading on a ship; thus it must consist mostly of longitudinal or fore and aft structural material. While many other structures consist largely of girders subjected to bending, the ship structure is unique in that broad expanses of plating are required to maintain watertightness. Combining the requirements of longitudinal strength and watertight integrity into a single "hull girder," while trying to achieve minimum weight of structure, has long been the task of ship structural designers. The structural arrangements that have resulted from their efforts are examined in this section.

Stiffened Plating
The structural configuration that has evolved in the design of steel ships as the fundamental structural unit is stiffened plating. An example of a stiffened plating panel is shown in Figure 7-1. The stiffeners may be rolled sections (angles, tees,

198   APPLIED NAVAL ARCHITECTURE

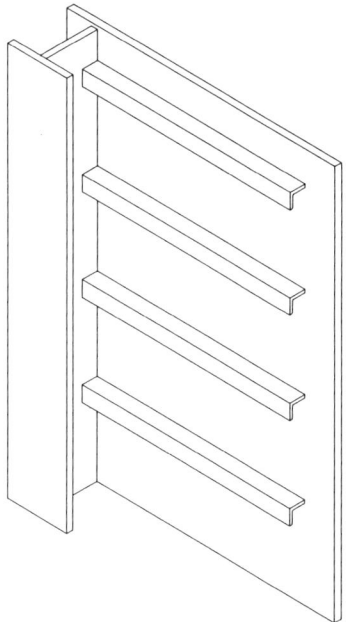

Figure 7-1. Stiffened plating.

etc.) welded to the plating; or large stiffeners may be built up of plates welded together and to the plating and smaller stiffeners that they support. For structural efficiency, stiffeners must run in two orthogonal directions as illustrated by the horizontal and vertical stiffeners shown in the figure. Small, closely spaced stiffeners support the plating directly, and the larger stiffeners support both plating and the small stiffeners that attach to them.

The structural designer must choose the orientation (longitudinal or transverse, vertical or horizontal) of each type of stiffener in each of the plated structures of the ship, such as bottom, sides, decks, and bulkheads. The choice is based on a number of considerations:

- *Structural efficiency.* This is determined by comparing the weights of alternative structures of equivalent strength. In general, the arrangement that has the least weight for a given strength is best. Exceptions to this general rule would occur when the least-weight solution would be much more costly than its alternatives.
- *Cost of materials and fabrication.* Alternative structural arrangements must be compared on the basis of cost as well as weight, and trade-off

studies made to decide how much additional cost may be justified to reduce the weight of the structure and thereby increase the payload or revenue-producing carrying capacity of the ship.

- *Structural continuity.* Structural members such as stiffeners must carry loads in the structure and transfer them to adjacent structure without creating abrupt changes in the stress level within the structural members. To assure that such "stress concentrations" are avoided, structural members must be designed to be continuous (or perfectly aligned if they are cut and then welded) through larger members or plated structures such as decks and bulkheads. Continuity of stiffeners is made effective "around corners" by attaching stiffeners on a panel directly to those on a perpendicular panel where the two meet.
- *Utilization of space.* In cross-stiffened panels, that is, large panels that are stiffened in both directions, the stiffeners in one direction are small and closely spaced and the orthogonal ones are much larger and more widely spaced, since they support the smaller ones. The structural designer's choice of stiffener orientation may sometimes be dictated by the need to avoid large structural members protruding into spaces and interfering with the effective utilization of the space.

Framing Systems

Although every ship has structural stiffeners running both longitudinally and transversely, the framing system of each ship can be characterized by the relative number, size, and spacing of its transverse stiffeners compared to the number, size, and spacing of its longitudinal stiffeners. Two very different framing systems—the transverse system and the longitudinal system—evolved in steel ship construction. They are described below, along with an example of a system that combines some features of both of the traditional systems.

Transverse Framing System. Figure 7-2 shows half of the midship section of a transversely framed ship. This framing system consists of many small, closely spaced transverse stiffeners and fewer larger, widely spaced longitudinal stiffeners. The transverse stiffeners are arranged so as to form structural rings that are closely spaced (from about 24 to 40 inches, or 600 to 1000 mm) throughout the length of the ship. Tracing the ring around the girth of the ship shown in Figure 7-2, we see that it consists of a deck beam under each deck, side frames supporting the side shell plating, and a deep floor plate supporting both the bottom plating and tank top (or inner bottom) plating. At each "corner" of the ring, brackets connect the members—beam knees at the deck beam to side frame connection, and margin brackets (also called hold frame brackets) connecting side frames to the floors. These frame rings with their brackets provide transverse strength to

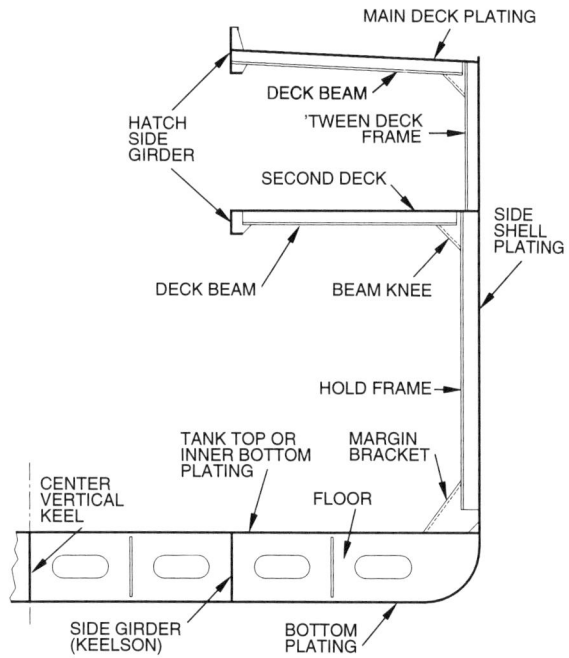

Figure 7-2. Transverse framing.

the structure, helping to maintain the cross-sectional shape of the hull, but they do not contribute anything to the ship's longitudinal strength.

Longitudinal strength in a transversely framed ship is provided by the shell plating and inner bottom plating, by the deck plating outboard of hatch and machinery casing openings, and by a number of large, widely spaced longitudinal members. Longitudinal deck girders support the deck beams. Deck girders adjacent to the hatches are called hatch side girders. In addition to providing longitudinal strength, deck girders reduce the span (length between supports) of deck beams so that the beams will not have to be very deep. The longitudinals in the double bottom are the center girder or center vertical keel, and the side girders, located so as to be directly beneath the deck girders to provide a rigid foundation for pillars as necessary to support the deck girders. Additional deck girders and side girders in the double bottom are provided in ships with a large beam, the typical spacing between girders being about 15 feet or 4.5 meters.

Longitudinal Framing System. The longitudinal framing system consists of many small, closely spaced longitudinals supporting the plating directly and being supported in turn by a few large, widely spaced longitudinals. A typical tanker midship section is depicted in Figure 7-3 to illustrate the longitudinal framing system because it is in tankers that this system in its purest form is most commonly employed. The tanker shown has no double bottom and no intermediate decks, and it is fitted with two longitudinal bulkheads, one of which is shown in the figure. Longitudinal stiffeners spaced about 24 to 36 inches apart (600 to 900 mm) give direct support to the plating of the deck, sides, bottom, and bulkheads. They also contribute to the longitudinal strength of the ship, making the longitudinal framing system more structurally efficient than the transverse system. The bottom longitudinal on centerline, or center girder, is extra large and heavy, principally to carry the loads imposed by keelblocks during drydocking. Very deep, heavy transverse structures called transverse webs or web frames are constructed at intervals of about 10 to 16 feet (3 to 5 meters) to provide transverse strength and to support the longitudinals. These heavy transverse webs consist of deck, side, and bottom transverses plus a vertical web on each longitudinal bulkhead. The side transverse and vertical web on the bulkhead are usually tied together with one to three horizontal struts or cross ties, as shown in the figure.

Two Systems Compared. During the early development of iron ships beginning in the 1830s, ship designers and builders, familiar with the structural arrangements of wooden ships, simply substituted equivalent iron members for their wooden counterparts. Wooden ship structure consisted of a framework or skeleton of transverse frame rings girding the ship, connected longitudinally by a massive timber keel structure, the planking of bottom and sides (often on both the outside and inside of the wooden framing members), and the deck planking. During the changeover from wood to iron, determining the required scantlings of iron structural members to provide the strength equivalent to the wooden members they replaced was a highly empirical procedure left for the most part to the individual designer or builder. (The word *scantlings* denotes the cross-sectional dimensions of structural members, as distinct from their lengths. Thus an angle bar whose scantlings are 8 × 6 × ½ has one 8-inch leg and one 6-inch leg and is ½ inch thick, regardless of its length. Steel plate scantlings normally refer only to the plate thickness regardless of its length or width.) As shipbuilding experience was gained with iron, and later with steel (in the 1870s), classification societies developed rules specifying the required scantlings of the structural members based largely on ship dimensions and past experience rather than on structural analysis. As a result, the transverse framing system predominated because it evolved naturally from the structural arrangement that had been success-

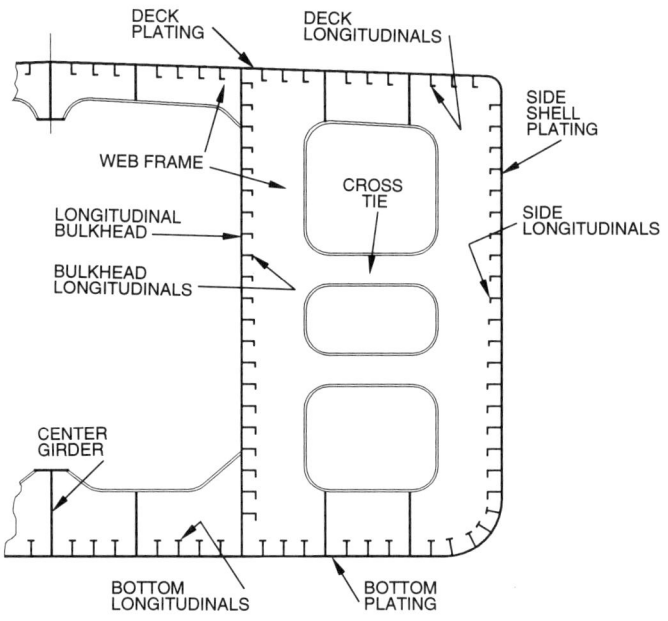

Figure 7-3. Longitudinally framed tanker.

ful in wooden ships. A notable early exception was the iron ship *Great Eastern*, launched in 1858, which had a longitudinally framed double hull across the bottom, halfway up the sides, and on the main deck.

Although the transversely framed ships were structurally sound and satisfactory in service, the fact that most of their stiffeners contributed nothing to the ship's resistance to longitudinal bending meant that the framing system was not optimal from the standpoint of structural efficiency, that is, of achieving the required strength for the least weight. The longitudinal system, in which most of the plating stiffeners are disposed in the fore and aft direction, has superior structural efficiency. This fact became clear as methods of structural analysis were increasingly applied to ship structures during the last quarter of the nineteenth century. In the early 1900s the advantages of longitudinal framing were clearly demonstrated, so it became increasingly common, especially in oil tankers and bulk carriers (grain, coal, ore, for example). It was not often employed in dry cargo ships because the deep web frames or transverses, which were spaced at every 12 to 15 feet in the holds, projected into the cargo space and interfered with the efficient stowage of packaged cargo.

The principal advantage of the longitudinal framing system is that the many longitudinal stiffeners serve dual purposes: they support the shell plating against local loading caused by water pressure and cargo loads (just as transverse frames in the transverse system do), and at the same time they contribute to the ship's resistance to longitudinal bending (which transverse frames do not). Therefore, a longitudinally framed ship has superior longitudinal strength to a transversely framed ship of equal size and structural weight. Put another way, for equal longitudinal strength, the longitudinally framed ship can be built with considerable weight saving compared to a transversely framed ship. The reduced structural weight permits the carriage of more cargo or payload. An additional advantage of stiffening the plating longitudinally in the deck and bottom plating is that longitudinally stiffened plating is more resistant to buckling between longitudinals when the deck or bottom is subjected to compressive stresses as it bends in a seaway than it would be if stiffened transversely. The buckling phenomenon is especially prevalent in a ship's upper deck. A final advantage that results in further weight savings arises out of the fact that longitudinal stiffeners supporting side shell and bulkhead plating are subjected to variable pressures from the sea or from liquid cargo. The hydrostatic pressure on each successive longitudinal and its associated plating increases with its depth below the waterline or below the surface of the liquid in the tank. Each longitudinal can be sized to withstand the maximum pressure associated with its depth in the ship, thus achieving an efficient use of the structural material. This "graduated size" configuration cannot be done effectively with transverse frames, which as a result are overly heavy at their upper ends.

In spite of its superior structural efficiency, a longitudinally framed ship is not without its drawbacks. The encroachment of deep webs into prime cargo spaces for ships carrying packaged cargo has already been mentioned. Another difficulty arises in the structural arrangements near the ends of the ship. The spacing of the longitudinals around the girth of the ship can be maintained constant so long as the girth does not change. As the hull narrows toward bow and stern, however, the girth necessarily reduces and the longitudinals become closer together. Difficulties in construction arise when they converge so closely that some longitudinals have to be eliminated. For this reason, transverse framing is usually resorted to at the bow and stern of longitudinally framed ships.

Combination Framing System. As a result of lessons learned by applying modern structural analysis techniques and studies of the strength-to-weight ratios of both transverse and longitudinal framing systems for ships of various types, very few ships today are built with the transverse framing system in its classic form. Tankers and dry bulk carriers are framed longitudinally, while most other ship types employ a combination framing system that exploits the best features of both traditional systems.

An example of a combined framing system is shown in Figure 7-4. Longitudinal framing is used in the bottom and decks, where the advantages of extra longitudinal strength and resistance to compressive plate buckling are most needed, and transverse framing is used in the sides, precluding the need for deep webs that might inhibit efficient cargo stowage. The transverse framing members shown in the figure are the deep, widely spaced transverses that support the longitudinals in the decks and bottom. More closely spaced, smaller transverse side frames (not shown) support the side shell plating between the deep transverses. In various forms and combinations of longitudinals and transverses suited to the needs of each particular design, combination systems are employed in many types of ships, including general dry cargo ships, containerships, and RO/RO ships, for example.

SHIP STRUCTURAL MATERIALS

The transition from wood to iron to steel in ship construction is a matter of historical fact. The reasons that steel has emerged as the predominant material for ship construction can be appreciated by considering what makes a material desirable for the job.

Requirements of Ship Hull Material

The material for any engineered product must be one that can be economically formed or processed into the product and that will perform its intended function in service for a reasonable, useful lifetime. Applying these general principles to a ship structure, we can identify a number of specific requirements or properties of a material that make it desirable for ship hull construction.

- *Availability and cost.* The material must be readily available and relatively inexpensive, as great quantities are required in the construction of large ships.
- *Uniformity.* The properties of the material must be uniform and dependable. This is possible only when the material is subjected to careful quality control during its manufacture, and when the processes used in its manufacture are controllable and repeatable. Lack of uniformity is one of the disadvantages of wood as a structural material, because natural variations in the structure of the wood itself (variations in grain, presence of knots, etc.) are unavoidable in a material not subject to quality control imposed by man.
- *Ease of fabrication.* The ideal structural material must be easily and cheaply formed into different shapes (plates, rolled sections, castings, etc.), and easily cut to size and joined together to build large, sometimes complex structures. The fabrication processes should not significantly al-

SHIP STRENGTH 205

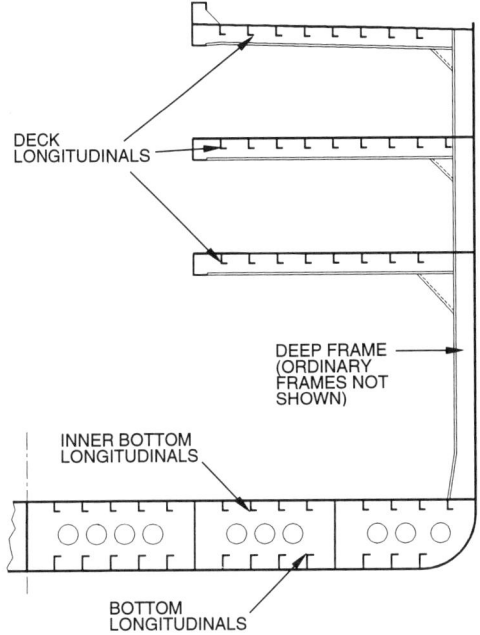

Figure 7-4. Combination framing system.

ter the properties of the material, so that its desirable properties are retained. Joints between structural members should be as strong as the materials being joined. This last requirement is one that could not be met by wooden ship structures or by riveted iron and steel structures. Since the development of welding processes, especially the electric arc welding that predominates in ship fabrication today, full-strength joints are possible and normal.

- *Ease of maintenance.* All materials are subject over a period of time to deterioration in service because of their exposure to liquids, gases, chemicals, radiation, or temperature changes. The choice of materials for particular engineering applications is frequently dictated by their resistance to oxidation, corrosion, dissolution, or thermal or radiation damage in service. For ship construction, there is no cheap, readily available material that will meet all of the demands of the ship's service life and that is not subject to corrosion or some other form of deterioration (rot and bio-

- *Strength vs. weight.* The strength of a material is, of course, an essential feature of any structural material. For any type of vehicle that must move, whether by land, sea, air, or space, a more important measure of a material's structural efficiency is its strength-to-weight ratio. A high strength-to-weight ratio is the most desirable. Metallurgists have devised many alloys and processes to increase this ratio, and the lighter metals such as aluminum, titanium, and magnesium have much higher strength-to-weight ratios than steel. Unfortunately, all of these materials are much more expensive than structural steel, and many have other undesirable properties as well.

- *Resistance to distortion under load.* An essential requirement of a structural material is that it submit to large loads without permanent distortion, while remaining elastic (deformations under load are recovered when load is removed) over a large range of loading. This characteristic is measured by the *modulus of elasticity* of a material, a property that will be described more fully later. Of all of the common, readily available structural materials, steel has the highest modulus of elasticity.

Considering all of the above requirements of a good structural material for ships, the various grades of steel that have been developed constitute the best available materials for hull construction. While some other materials are superior in one category or another, none has been found to be more cost effective while adequately meeting the varied service demands of ship structures.

Mechanical Properties of Steel

The properties of ship steels that relate to its strength are called its mechanical properties. Many of the mechanical properties of interest to ship structural designers are determined from a standard test known as the tensile test of the steel. A few basic definitions will promote a better understanding of the significance of the tensile test and the mechanical properties that it reveals.

Loads are the forces applied to a structural member or to a test specimen. External loads or forces on a member induce internal reactions called stresses, and they cause the member to deform. *Stress* is force per unit area, the area in question being the area that transmits the load, usually the cross-sectional area of a stressed member. Symbolically,

$$\sigma = \frac{P}{A} \tag{1}$$

where σ = stress
P = load on the stressed member
A = load-carrying area of the stressed member

There are three principal types of stresses into which all complex stress situations can be resolved. They are illustrated in Figure 7-5. *Tensile stress* or *tension* is produced when collinear forces on an element are directed away from each other, causing a deformation that lengthens the element. The stressed area is perpendicular to the direction of the forces. *Compressive stress* or *compression* is produced when collinear forces on an element are directed toward each other, tending to shorten the element. As in tension, the stressed area is perpendicular to the direction of the forces. *Shear stress* is produced when equal and opposite forces act on a member along parallel (not collinear) lines, the stressed area is parallel to the lines of action of the forces, and the element tends to slide or shear apart. The *deformation* that results from a stress may be expressed nondimensionally as a *strain*, which is defined as deformation per unit of original length. That is,

$$\varepsilon = \frac{\delta}{L} \quad (2)$$

where ε = strain
δ = deformation
L = original length

Thus a tensile test specimen 8 inches long that has stretched by 0.012 inches under load has a unit deformation or strain of $0.012/8.0 = 0.0015$ inches per inch.

In the standard tensile test, a test specimen is subjected to pure tensile loading increasing from zero to the load required to break the specimen, while frequent measurements of the load and the elongation of the specimen are made. A plot may be made of load against deformation, but the more common procedure is to plot stress ($\sigma = P/A$) against strain ($\varepsilon = \delta/L$), as shown in Figure 7-6. The curve

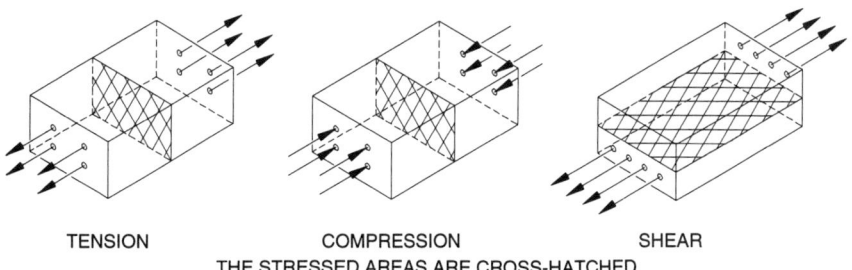

TENSION   COMPRESSION   SHEAR
THE STRESSED AREAS ARE CROSS-HATCHED

Figure 7-5. The three principal stresses.

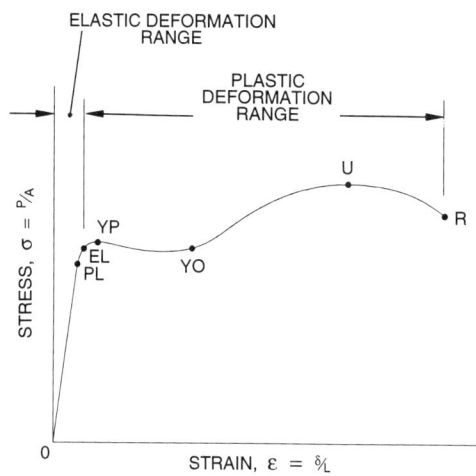

Figure 7-6. Stress-strain curve, mild steel.

shown in the figure is for typical mild steel or ordinary-strength hull structural steel, the most common steel used in hull construction.

Most of the mechanical properties of steel that are of importance to a structural engineer are shown in the stress-strain diagram. It can be seen from the figure that, initially, the relationship between stress and strain is a straight line; that is, stress is directly proportional to strain. This proportionality, known as Hooke's Law, pertains up to point PL on the diagram, the *proportional limit*. So long as the stress at the proportional limit is not exceeded, the stress-strain relationship can be defined with a single number, the slope of the stress-strain line. The proportional limit varies for different grades of steel depending on their carbon content and various alloying elements, but it is characteristic of all structural steels that the straight line relationship exists and that its slope is essentially the same regardless of the type of steel. At about the same stress as the proportional limit, or slightly above it, a point (EL in the figure) called the *elastic limit* is reached. When loaded to any point below the elastic limit, steel has a very remarkable property—the deformation or strain caused by the stress is completely recoverable when the load (or stress) is removed, and the piece of material returns to its original dimensions. This property is known as *elasticity,* and it is a most desirable quality of a structural material because a structure or machine part designed so that the elastic limit is never exceeded will never deform permanently. Deformations while under load will always be recovered when the load is removed.

Because the proportional limit and elastic limit are so close to one another, they are considered identical for practical purposes. The curve is presumed to be straight up to the elastic limit, and its slope is defined as the *modulus of elasticity* or *Young's modulus,* symbolically E.

$$E = \frac{\sigma}{\varepsilon} \qquad (3)$$

where E = modulus of elasticity (psi or MPa)
 σ = stress below the proportional limit (psi or MPa)
 ε = strain corresponding to the above stress (0-diml)

The modulus of elasticity is a measure of the *stiffness* of a material, that is, its capacity to resist deformation under load below the elastic limit. The modulus of elasticity of steel is 30×10^6 psi (207,000 MPa). It is three times as stiff as aluminum and its alloys ($E = 10 \times 10^6$ psi = 69,000 MPa) and nearly twice as stiff as copper and its alloys ($E = 16 \times 10^6$ psi = 110,000 MPa).

At a stress slightly higher than the elastic limit, a mild steel specimen begins to experience a rapid increase in strain without any increase in stress. This phenomenon is called *yield,* and the stress at which it occurs is the *yield stress* or *yield point,* marked YP on the figure. Deformations or strains associated with yielding are not recoverable, and the material is said to take a *permanent set.* Deformation beyond the elastic limit is called *plastic deformation.* Since the proportional limit has also been passed, the modulus of elasticity cannot be used to calculate plastic deformation. Knowledge of the yield point of a structural material is of utmost importance to the structural designer because permanent deformations of structural elements are undesirable at best and completely unacceptable in many applications. After mild steel has yielded with no increase in stress, a point is reached called *yield overcome* (YO) beyond which the steel again requires higher stress to deform any more. This remarkable characteristic of mild steel is beneficial as a warning sign that the member has been overstressed, since it may deform plastically and permanently in service and still not break.

Upon further loading, the steel will continue to deform plastically while the stress increases to its maximum value, called the *ultimate stress,* marked U on the diagram. The ultimate stress of a material is a measure of its strength. The apparent reduction in stress that takes place after the ultimate stress is reached, up to the point of *rupture,* marked R, does not represent a real reduction in stress. It arises because the stress that is plotted is defined as load divided by the original cross-sectional area of the piece, while the true stress would be determined by dividing the load by the actual cross-sectional area of the piece at each load. Since the area reduces materially by "necking down" at very high loading, the true stress at rupture is higher than the plotted stress.

During a tensile test, after the steel specimen has ruptured, it is fitted back together so that a measurement of its length at failure can be taken. It has by then stretched considerably. The amount of stretch, expressed as a percentage of its original length, is called its *elongation,* which is a measure of the *ductility* of the steel. Ductility is a very important property of structural materials, because in many applications they must be shaped and formed by bending, thus causing plastic deformation. Ductile steels can be deformed plastically without reducing their ultimate strength. Specifications for ship steels require that they have percentage elongations in the range of 20 to 25 percent as measured in various standard tensile tests.

The importance of ductility for ship steels became grimly apparent during World War II, when a rash of ship structural failures took place, many of them resulting in the ships breaking completely in two quite suddenly. It was discovered that the cracks in question started at sharp notches in the steel, that they took place only at low ambient temperatures, that the ship steel failed by brittle fracture rather than in the expected ductile mode, and that the energy required to cause the cracks was far below what would be needed to cause a ductile fracture. Research in fracture mechanics, metallurgy, and the microstructure of steels has resulted in the development of *notch tough steels* that do not fracture in a brittle mode at any temperatures experienced by a ship in service. Toughness is a property that implies high tensile strength combined with good ductility. The toughness of steel can be improved by keeping the carbon content of the steel low (below 0.25 percent), by alloying it with controlled amounts of manganese and sometimes nickel, by a deoxidation process known as killing, and by limiting grain size during the steel-making process with the addition of aluminum and with proper use of a heat treatment known as normalizing. Brittle fracture is no longer a problem in ships so long as the ship designer adheres to the steel specifications required by the classification societies, and so long as by careful design of ship structural details sharp notches are avoided in structural members.

Approved Ship Steels
The American Bureau of Shipping, like all classification societies, requires that the steel used in ship construction adhere to a rigid and detailed list of specifications, including the process of manufacture, method of deoxidation, chemical composition, heat treatment requirements, mechanical properties, and level of notch toughness. A single grade of steel cannot be used for all parts of the ship because certain parts of ship structures are more highly stressed than others, and because notch sensitivity is strongly dependent on steel thickness. Therefore classification societies specify a range of grades of acceptable structural steels with regulations pertaining to their applications aboard ship. They also specify higher-strength steels because ship designers sometimes choose extra high strength steels for critical, highly stressed parts of the structure, so long as the added expense can be justified by the weight savings.

- *Ordinary-strength steels.* Six grades of ordinary-strength hull structural steels are specified by the ABS. All must have a yield strength of at least 34,000 psi (235 MPa), an ultimate tensile strength between 58,000 and 71,000 psi (400 to 490 MPa), and a minimum of 24 percent elongation of a 2-inch tensile specimen. The various grades differ in their required level of notch toughness, the requirements increasing in severity as the steel plates get thicker.
- *Higher-strength steels.* Two strength levels of higher-strength steels are specified, with three "thickness grades" in each strength category. The

yield strength requirements are 45,500 psi (314 MPa) and 51,000 psi (353 MPa), respectively. Ultimate strengths range from 68,000 to 90,000 psi (471 to 618 MPa), and the minimum elongation is 22 percent. These steels are used when the premium cost of 25 to 50 percent over ordinary-strength steel and more difficult fabrication procedures can be justified. If even stronger steels are desired for special critical applications, steels specified by the American Society for Testing and Materials (ASTM) having yield strengths up to 100,000 psi (690 MPa) are permitted.

- *Special steels.* Other steels with special properties are sometimes needed for special applications in ship structures. For low service temperatures like those found in refrigerated ships and liquefied gas carriers, *low temperature steels* have been developed that have satisfactory notch toughness to service temperatures as low as −67°F (−55°C). For service in liquid cargo tanks that may be used to carry a wide variety of liquids, special *corrosion-resistant steels* may be used. Alternatively, ordinary steel clad with corrosion-resistant materials is also available. *Abrasion-resistant steels* that resist wear caused by abrasive materials dropped into holds, especially ores carried in bulk, are sometimes used in ore carrier holds.

Aluminum in Ship Structures

Although aluminum has not been used for the entire hull structure of large ships, its high strength-to-weight ratio makes it attractive for some special ship structural applications. Aluminum alloys are used as the principal hull material in a variety of small vessels, especially high-speed craft such as planing boats, hydrofoil craft, and surface effect craft. Besides the weight advantage, aluminum alloys have the advantage that they increase in strength at low temperatures, so they can be used for tanks in liquefied gas carriers. Aluminum is also very corrosion resistant, so long as care is taken in fabrication of aluminum structures to prevent them from being in direct contact with dissimilar metals by the use of gaskets or special coatings. The most prevalent use of aluminum in large ships has been in the superstructures, where the weight reduction results in improved stability.

Composites

Composite laminates such as glass-reinforced plastics have become commonplace as a hull structural material for a great variety of small craft such as recreational sailing and power yachts. These materials have the advantages of a high strength-to-weight ratio, low maintenance cost, and the ability to be fabricated into a virtually endless variety of types of laminates and shapes of hull. Their disadvantages, however, mitigate against their use as a hull material for large ships: the high initial cost of the material compared to steel, a very low modulus of elasticity (in the order of only 10 percent of that of steel), so that they deflect a great deal under load, and the fact that they are combustible, so that they cannot meet the fire-resistance regulations applicable to ships.

LONGITUDINAL STRENGTH

In the description of ship structural arrangements, the whole ship structure has been likened to a large girder, supported from below by buoyancy, and carrying its own weight plus the weights of machinery and equipment, cargo and consumables. The forces of weight (displacement) and buoyancy are the fundamental forces that have been dealt with throughout the study of ship theory. As our studies of floating ships progressed from flotation to equilibrium to stability, these same two forces had to be depicted with an increasing amount of detail at each step. Flotation studies require knowledge only of the magnitudes of the weight and buoyant forces. Floating equilibrium (trim, heel) studies require, in addition, the locations of their centroids, the centers of gravity and buoyancy. Stability can be determined only with the additional knowledge of how the center of buoyancy moves when floating equilibrium is disturbed temporarily. In the study of longitudinal strength, or hull girder strength, our knowledge of the weight and buoyant forces must again be expanded. The distribution of the weight and buoyant forces over the length of the ship must be known, in addition to their magnitudes and centroids. Also, in all of the earlier studies, the ship was treated as a rigid body. That is, it was imagined to maintain its shape and geometry even in the presence of the external forces of weight and buoyancy. In the study of the longitudinal strength of ships, the hull is treated as a deformable body, which can and does bend under the influence of the distributed forces of weight and buoyancy. The bending is caused by the internal stresses within the structural members of the hull, just as a steel specimen in a tensile test deforms in proportion to the stress imposed on it during the test.

Although the most realistic predictions of the forces, stresses, and deformations associated with the longitudinal bending of a ship in service require a statistical approach because the loads imposed by nature are not known with certainty, much can be learned by understanding the elements of beam theory of simple beams or girders subject to simple static loads and support forces. This simplified static approach will be used here, and the hull form will be restricted to simple geometries. Calculations of the longitudinal strength of realistic ship hulls in realistic seaways is today relegated to computer analysis because they require so much numerical detail.

Shear and Bending Moment in Beams
Bending of a beam occurs when there are loads and support forces on the beam that are perpendicular to the axis of the beam. The simplest case, used here to de-

fine what happens when a beam is subject to bending, is that of a hypothetically weightless beam propped up at both ends and supporting a single load (or force or weight) concentrated at a single point at midspan of the beam. This configuration is diagramed in Figure 7-7(a). The supports are called simple supports, which means that they can sustain no side force components, hence the reactions R_A and R_B are directed vertically upward.

First the magnitudes of the supporting reactions, which are externally applied forces, are determined, then the internal forces and moments in the beam. The external reactions are determined by solving the equations of static equilibrium for the whole beam, by expressing the fact that the sums of both the vertical forces (F_V) and the moments about any point (M) are zero. The chosen point for moments is the middle of the beam, where $x = L/2$.

$$\Sigma F_V = 0: \quad R_A + R_B = P$$

$$\Sigma M = 0: \quad \frac{R_A L}{2} = \frac{R_B L}{2}$$

$$\text{Therefore} \quad R_A = R_B = \frac{P}{2}$$

Clearly, in this simple case, conditions of symmetry would lead immediately to the conclusion that the reactions are each equal to half the load. The formal solutions of the force and moment balance equations are shown to indicate the method, which will also work in more complicated cases.

The internal forces in the beam are determined by writing the equilibrium equations for a portion of the beam from one end to an exploratory cut section such as section A-A in part (b) of the figure. In general, the required equilibrium of such a beam segment can be maintained only if there exists on the cut section both a force and a moment about the centroid of the cross section, designated as point C in the figure. The internal force and moment are defined as the *shear (V)* and *bending moment (M)* for the particular section of the beam in question. In part (c) of the figure, it is made clear that in this case the shear force must be directed downward to balance the upward support force, and the bending moment must be in the sense shown to counteract the rotation that would be produced by the forces P/2 and V. Convention has it that the both V and M in this figure are positive. The rules are that *if the left-hand end of a beam is analyzed, upward external forces produce positive shear and positive bending moment* at the cross section. Conversely, downward external forces would cause negative shear and bending moment at this section. It should be noted that the right-hand end of a beam could also be isolated for analysis. If it is, the above rule must be reversed for shear, but will remain the same for bending moment. Thus *if the right-hand end of a beam is analyzed, upward external forces produce negative shear, but positive bending moment* at the cross section.

214   APPLIED NAVAL ARCHITECTURE

What are the magnitudes of V and M in Figure 7-7(c)? They are determined by writing the usual equations of equilibrium.

$\Sigma F_V = 0$: $V = \dfrac{P}{2}$

$\Sigma M_C = 0$: $M = \dfrac{Px}{2}$

These expressions, which are applicable to the particular beam-loading situation shown in the figure, can be generalized by imagining that there are additional forces (positive or negative) acting on the left-hand portion of the beam. The equations for V and M would then be written as summations of the forces and moments of the forces on the beam segment:

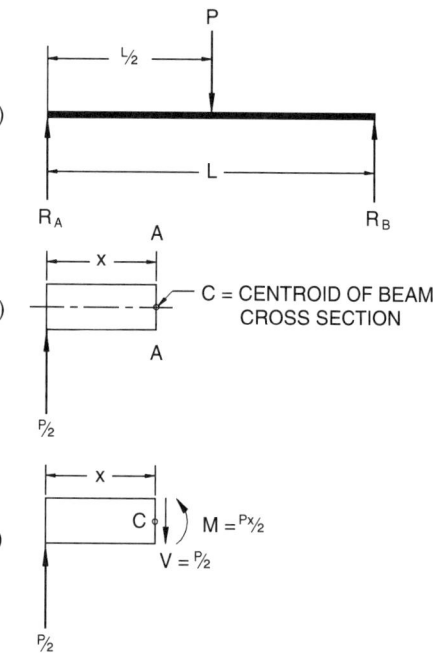

Figure 7-7. Shear and bending moment in a beam.

$$V = \Sigma_{\text{left}} F_V = -\Sigma_{\text{right}} F_V \qquad (4)$$

$$M = \Sigma_{\text{left}} M_C = \Sigma_{\text{right}} M_C \qquad (5)$$

where left and right refer to the portion of the beam that has been analyzed

F_V = vertical externally applied forces and reactions, taken as positive when directed upward

M_C = moments about centroid of cross section of the externally applied forces, taken as positive when the force is directed upward

Shear and Bending Moment Diagrams. The equations pertaining to the beam in Figure 7-7 show that for any position to the left of the central load ($0 < x < L/2$), the shear force V is constant and equal to the reaction P/2. The bending moment, however, is not constant, but increases linearly from zero at the left end (x = 0) to (P/2)(L/2) = PL/4 at midspan (x = L/2). These trends of shear and bending moment with position in a beam are best displayed by drawing shear and bending moment diagrams, which are plots of V and M against x, respectively. The diagrams for this beam are shown in Figure 7-8, where it can be seen that the shear of +P/2 and the bending moment increasing from zero to +PL/4 are plotted. Continuing the analysis to include the rest of the beam requires using the summations expressed by equations (4) and (5); or, alternatively, the right-hand half of the beam could be analyzed. If the analysis continues from the left, note that the shear from midspan to the right-hand half is equal to

$$V = \frac{P}{2} - P = -\frac{P}{2}$$

Note that the sign convention for left-end analysis is used, upward-directed loads being positive and downward-directed loads being negative. The bending moment for the right-hand half is

$$M = \left(\frac{P}{2}\right)x - P\left(x - \frac{L}{2}\right) = \frac{PL}{2} - \frac{Px}{2}$$

At $x = L/2$, this evaluates to $PL/2 - PL/4 = PL/4$, which verifies the midspan bending moment already determined. At the right-hand end, where $x = L$, it is $PL/2 - PL/2 = 0$, and the decrease from midspan is linear. The bending moment diagram is therefore triangular, the maximum moment occurring at midspan. As with the calculation of the reactions, it will be noted that the obvious simple symmetry of this particular beam could have been used to develop these diagrams with less formal analysis.

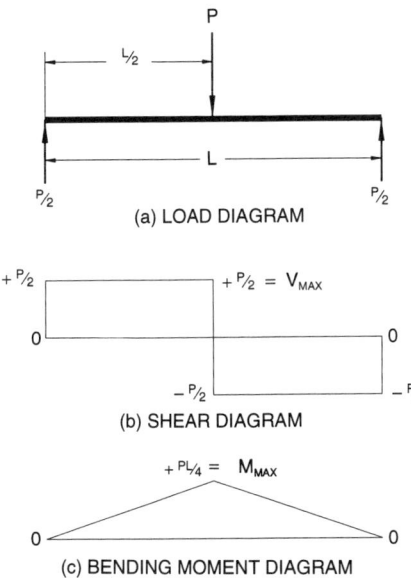

Figure 7-8. Shear and bending moment diagrams.

The load and reactions in the above example were all imagined to be focused at a single point on the beam. Many loads that beams support are spread over a discrete portion of the beam. They are called *distributed loads,* and the nature of their distribution must be specified in the load diagram. A common example of a distributed load is the weight of the beam itself. If the beam has a constant cross section from end to end, its weight is said to have a uniform distribution. Some loads may have nonuniform distributions, for example, the cargo loads carried in ship holds. Distributed loads are of particular significance in the bending analysis of ship hulls because all loads pertaining to the ship and its cargo are treated as distributed loads. Furthermore, the ship is not supported by concentrated simple reaction forces, but by hydrostatic support—its own buoyancy—which is distributed over the length of the ship with a distribution that is a function of the hull shape or form. In a load diagram, distributed forces, whether loads or support forces, are depicted graphically by plotting the weight per foot (pounds or tons per foot, or newtons or metric tons per meter, for example) against the length of the beam or structure. The total load is measured by the area enclosed by such a load diagram. Often arrows are shown within the distributed load diagram to indicate the direction of the loading. Examples are shown in Figure 7-9. The example shown in part (c) of the figure is a simplified plot of a typical ship loading diagram. The weight distribution diagram is plotted above the axis, and the buoyancy distribution diagram is plotted below the axis. When these two curves are plotted correctly for a ship floating in static equilibrium, their properties are:

- The area enclosed by the weight curve is equal to the displacement of the ship, expressed as the summation of the weights of the ship and its loading.
- The longitudinal position of the centroid of the weight curve is at the longitudinal center of gravity of the ship.
- The area enclosed by the buoyancy curve is equal to the displacement of the ship, expressed as the total buoyant force.
- The longitudinal position of the centroid of the buoyancy curve is at the longitudinal center of buoyancy of the ship.

Note that these four properties are really expressions of the equations of static equilibrium applied to a floating ship.

Load, Shear, and Bending Moment Relationships. When the loads on a beam are distributed, as is the case for a floating ship, it is helpful to plot a special load diagram in addition to the diagram showing separately the distributions of weight and buoyancy, as illustrated by Figure 7-9(c). The load diagram depicts the net vertical force per unit length acting at each point in the ship. That is,

$$p = b - w \qquad (6)$$

where b = buoyancy per unit length
 w = weight per unit length
 p = load per unit length

The load diagram can be determined graphically by superimposing the weight distribution and buoyancy distribution diagrams, plotted as if each of them were positive, then plotting the difference $(b - w)$ between the two curves against the length of the ship. The procedure is illustrated in the next worked example.

If all loads are distributed, the shear expressed by equation (4) is more appropriately written as an integral, rather than as a summation.

SHIP STRENGTH   217

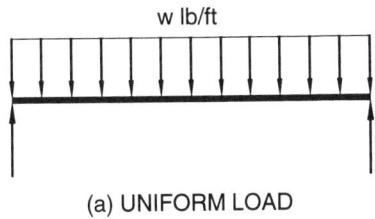

(a) UNIFORM LOAD

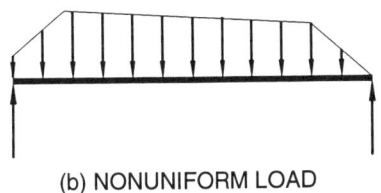

(b) NONUNIFORM LOAD

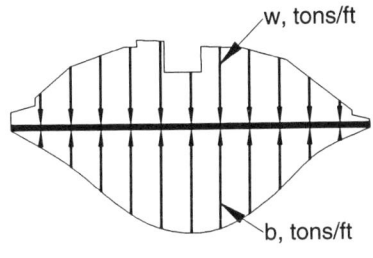

(c) NONUNIFORM WEIGHT, NONUNIFORM SUPPORT (BUOYANT SUPPORT)

Figure 7-9. Distributed loads.

$$V = \int (b-w)\, dx = \int p\, dx \quad (7)$$

in which the integral is taken from the left-hand end of the beam, that is, from the stern of the ship, to the point at which the shear force is to be calculated. It should be noted that equation (7) expresses the net area enclosed by the load diagram from the stern to the point in question. Thus we have the important observation that *the shear force at any point in a beam is equal to the area enclosed by the load diagram from the end of the beam to the point in question.*

Areas above the axis are positive (buoyancy exceeds weight) and areas below the axis are negative (weight exceeds buoyancy). Since the ship is floating in static equilibrium, the total weight is equal to the total buoyancy. Thus a condition of a properly constructed load diagram is that the total net area enclosed by the curve be zero, or that the positive area above the axis be equal to the negative area below the axis.

The relationship of the bending moment to the shear force is exactly the same as the relationship of the shear force to the load; that is, *the bending moment at any point in a beam is equal to the area enclosed by the shear diagram from the end of the beam to the point in question.*

This is expressed in equation form by

$$M = \int V\, dx \quad (8)$$

This statement can be verified for the simply supported beam of Figure 7-8 by noting the relationship between the maximum ordinate of the bending moment diagram and the area of the shear diagram between the left end and the midpoint of the beam.

$$\text{Shear diagram area} = \left(\frac{P}{2}\right)\left(\frac{L}{2}\right) = \frac{PL}{4} = M_{max}$$

Try some other points on this beam to verify that this "shear area equals bending moment" relationship works for any point. It is valid for any beam, regardless of the nature of the loading.

The load, shear, and bending moment diagram relationships defined by equations (7) and (8) may be used directly to construct shear and bending moment diagrams from the load diagram. The inverse relationships defining the derivatives of the shear and bending moment are helpful when checking the diagrams, because they describe the slopes of their respective curves. Thus, at any given point in a beam,

$$P = \frac{dV}{dx} \tag{9}$$

$$V = \frac{dM}{dx} \tag{10}$$

Put into words, we have two relationships that will help to determine the shapes of the diagrams:

1. The slope of the shear diagram at any point is equal to the load at that point.
2. The slope of the bending moment diagram at any point is equal to the shear at that point.

In a continuous curve, zero slope defines either a maximum or minimum (or maximum negative) value, therefore the above slope relationships have the following practical consequences:

1. The maximum value of shear occurs at that point in the length of the ship at which the load curve is zero, i.e., where the load curve crosses its axis.
2. The maximum value of bending moment occurs at that point in the length of the ship at which the shear curve is zero, i.e., where the shear curve crosses its axis.

The Ship as a Floating Beam

Most of the beams in stationary shoreside structures are subject to loads in service that may vary from time to time, but they are rarely reversed. That is to say, a beam such as the floor joist in a warehouse will bend because of its own weight, the weight of the flooring that it supports, and the variable weight of the packages or material that are stacked on the warehouse floor. Although the loading is variable, one would never expect loads on a floor joist to cause it to bend by arching upward, rather than by sagging downward at midspan. Considering the ship hull at sea as a "beam" supported by buoyancy and loaded with its own weight and the weights of the cargo and other material it carries, however, it

SHIP STRENGTH   219

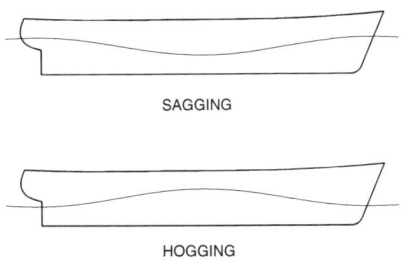

Figure 7-10. Hull girder bending in waves.

must be recognized that it sometimes has a tendency to bend downward at amidships like a floor joist, but at other times it is forced to arch upward as the buoyant forces are redistributed. This reversal of the direction of bending is not a rare occurrence. Indeed, it takes place continuously throughout the seagoing lifetime of any ship. It has been estimated that in a 20-year ship lifetime, a typical ship undergoes on the order of 100 million (that is, 10^8) such bending reversals.

The two directions of hull girder bending, illustrated in Figure 7-10, are called *sagging,* when the structure deflects downward at amidships, and *hogging,* when it arches upward. These opposing deformations reach their extremes when the ship moves directly into (head seas) or along with (following seas) waves whose crest-to-crest length is about equal to the length of the ship, as shown in the figure. When wave crests support the ends of the ship and a wave trough is at amidships, the hull tends to sag because of a lack of buoyant support amidships. The hogging condition occurs soon after, when a wave crest is at amidships and troughs are at the ends. Reversals of the direction of bending also reverse the type of direct stresses and strains that occur in the structures of the deck and bottom of the ship. Sagging causes compression in the deck and tension in the bottom structure, while hogging causes the deck to be in tension and the bottom structure to be in compression. Since a ship does not always move in head seas of ship-length waves, the sag/hog cycles illustrated here are not always as extreme. Nevertheless, such reversals do take place continuously in other sea conditions, at a lower stress level. As long as the weight distribution and the buoyancy distribution of a floating ship are not exactly equal to one another at every point in the length of the ship, the hull girder will be subject to shear forces, bending moments, and the stresses, strains, and deformations that result from them.

Sample Calculation for a Loaded Barge

The above comments apply even for a simple barge in still water, as the following worked example illustrates.

Example 7-1

A rectangular barge of 250-foot length, 35-foot beam and 20-foot depth floats in seawater at a draft of 2 feet when it is empty. The light ship weight may be assumed to be uniformly distributed over the length of the barge. It has 5 holds, each 50 feet long. The barge floats in seawater loaded with cargo as shown in Figure 7-11. Cargo weights within each hold are to be assumed uniformly dis-

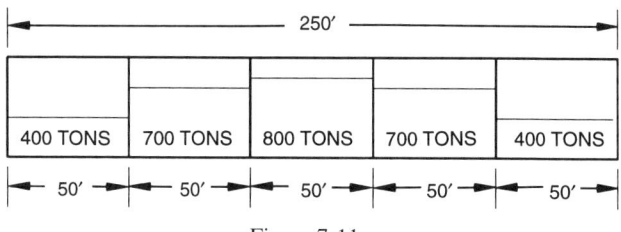

Figure 7-11.

tributed over the length of the hold. Calculate and plot diagrams of the distributions of weight, buoyancy, load, shear, and bending moment. Determine the values of the curves at each bulkhead and at their maximum points.

The weight per foot of the hull is Δ/L, where

$$\Delta = \frac{LBT}{35} = \frac{250 \times 35 \times 2}{35} = 500 \text{ tons}$$

Thus hull weight per foot = 500/250 = 2.0 tons per foot. Cargo weights per foot for the various holds are determined as follows, by dividing each cargo weight by the length of the hold:

Holds 1 and 5:   400/50 =   8 tons/ft
Holds 2 and 4:   700/50 = 14 tons/ft
Hold 3:   800/50 = 16 tons/ft

Adding the light ship weight of 2 tons/ft to each of the above cargo weights, we get loaded weights per foot for the five segments of the barge from end to end of 10, 16, 18, 16, and 10 tons/ft, respectively. These values constitute the weight curve as plotted in Figure 7-12(a).

The buoyancy per foot of a rectangular barge floating at an even keel is constant from bow to stern and is equal to the load displacement divided by the length of the barge. The load displacement (light ship plus cargo) is

$$\Delta = 500 + 400 + 700 + 800 + 700 + 400 = 3{,}500 \text{ tons}$$

Thus buoyancy per foot = b = 3,500/250 = 14 tons/ft. The buoyancy curve is plotted as a dashed line superimposed on the weight curve in part (a) of the figure. Note that the area enclosed by the weight curve equals that enclosed by the buoyancy curve.

The load curve is a plot of $b - w$, determined by calculating the differences between the buoyancy and weight curves shown in the figure. An excess of buoyancy is plotted as a positive value. The load curve is shown in part (b) of the figure. The ordinates are in tons per foot.

SHIP STRENGTH 221

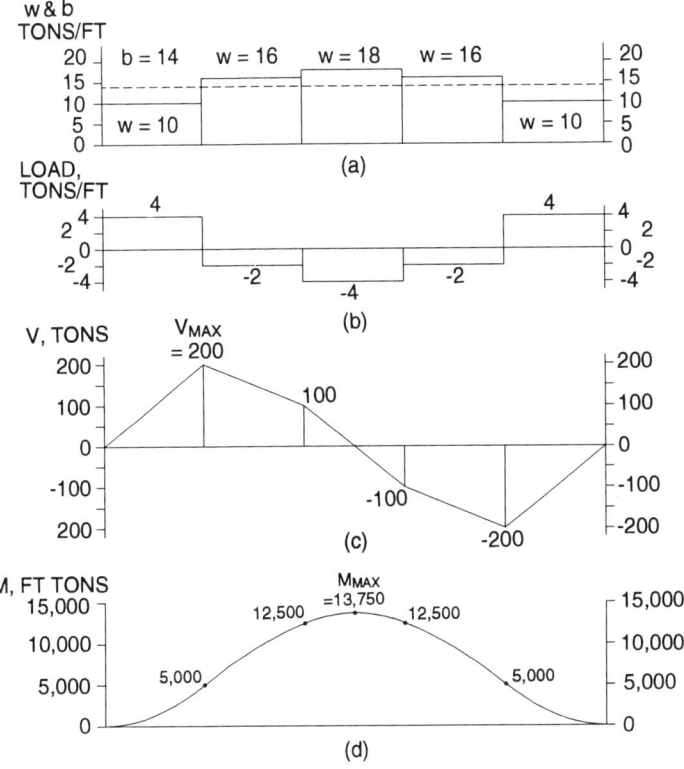

Figure 7-12.

The shear diagram is determined by calculating the areas enclosed by the load curve from the left end to each bulkhead. The cumulative values of these areas must be plotted. Areas of the load curve below the axis are negative; that is, they are subtracted from the cumulative total.

Bhd #1: $V = 4(50) = 200$ tons
Bhd #2: $V = 200 - 2(50) = 200 - 100 = 100$ tons
Bhd #3: $V = 100 - 4(50) = 100 - 200 = -100$ tons
Bhd #4: $V = -100 - 2(50) = -100 - 100 = -200$ tons
Fwd end: $V = -200 + 4(50) = -200 + 200 = 0$ tons

These values are shown on the shear diagram, part (c) of the figure. Note that the relationships mentioned above pertain; for example, the load curve crosses its axis at bulkheads #1 and #4, and the shear there reaches its maximum positive and negative values. It can be verified that each shear curve segment has a slope equal to the ordinate of the load curve segment above it. The shear curve has a positive slope only in the two end segments, where the load is positive. Note also that the shear is zero at both ends of the barge. This will be the case always for ships or for any loading in which there are no concentrated loads or supports.

Coming now to the bending moment diagram, it is determined from the cumulative areas enclosed by the shear diagram. Because the shear is zero at the ends of the barge, both *the bending moment and the slope of the bending moment curve will be zero at the ends*. Segments of the bending moment curve must be second-order polynomials, or parabolas, because the first-order straight lines of the shear curve segments integrate to second-order curves. In regions in which the shear is increasing (the two end segments in this problem), the bending moment curve will have an increasing slope. In segments where the shear curve has decreasing values, the slope of the bending moment curve will decrease. Where the shear curve crosses the axis (at amidships in the present case), the slope of the bending moment curve must be zero, hence it will reach its maximum value. The bending moment curve will be symmetrical about amidships in this case because of the symmetry about amidships of both the hull and the loading. Given these preliminary shape descriptions of the bending moment curve, we can proceed with the calculation of the areas of the various portions of the shear diagram.

$$\text{Bhd \#1: } M = \frac{1}{2} \times 200 \times 50 = 5{,}000 \text{ ft-tons}$$

$$\text{Bhd \#2: } M = 5{,}000 + \frac{1}{2}(200 + 100)\,50 = 5{,}000 + 7{,}500$$

$$= 12{,}500 \text{ ft-tons}$$

$$\text{Amidships: } M = 12{,}500 + \frac{1}{2} \times 100 \times 25 = 12{,}500 + 1{,}250$$

$$= 13{,}750 \text{ ft-tons}$$

Calculations for the rest of the curve may be omitted because of symmetry. The bending moment diagram is plotted in part (d) of the figure.

It is possible to determine the bending moment values directly from the load curve, if shear values are not needed. The bending moment at any point is equal to the moment about that point of the portion of the load diagram from the point to either end. As an example, we could check the bending moment calculation for the first bulkhead by determining the moment about that bulkhead of the load diagram to its left. The moment is equal to the area of the first segment of the load diagram times the moment arm from bulkhead #1 to the centroid of that area.

Load curve area to Bhd #1 = 4 (50) = 200 tons
Arm, centroid of area to Bhd #1 = 25 ft
M = 200 (25) = 5,000 ft-tons

A similar calculation could be done for any point on the bending moment curve, including points between bulkheads.

The above example demonstrates the procedures that can be used to determine the distributions of vertical shear forces and bending moments in a ship hull girder. The maximum values of shear force and bending moment and their locations in the ship structure are important to the ship structural designer. In order to simplify the calculations for this demonstration, a simplified hull form and a symmetrical loading condition have been chosen.

Complications Introduced by Ship Form and Loading
In the similar calculation for a ship, neither the hull nor the loading is symmetrical about amidships, so that a trim calculation must be done first to determine the waterline at which the ship floats in static equilibrium. If the calculation is to determine the static wave bending moments in sagging and hogging waves, the static balance is further complicated by the wave pattern. In any case, the areas of the ship stations up to the equilibrium waterline are determined and a sectional area curve is plotted. At any point in the ship, the buoyancy per unit length is equal to the sectional area times the weight density of the water (ρg). Thus the buoyancy curve has exactly the same shape as the sectional area curve. The buoyancy curve encloses an area equal to the ship's displacement and its centroid is located longitudinally at the position of the ship's center of buoyancy. The smooth buoyancy curve is often represented by a "stair-step" approximation in which the buoyancy per unit length values at 20 to 40 equally spaced segments of the length of the ship are assumed to be constant over the length of each segment, consistent with the weight curve segments described in the next paragraph. This approximation is especially adaptable to computer program and spreadsheet calculations.

Determining the weight distribution for a ship calculation is a tedious, exacting process done with the aid of a computer program. The ship is divided into a number of segments (usually 20 or 40) of equal length. From the detailed documentation of the light ship weight and the weights of all of the deadweight items (cargo, fuel, etc.) for a particular loading condition, the total weight of each of the segments is determined. The weights so deduced are assumed to be uniformly distributed over the length of the segment, so that a plot of weight per unit length against the ship length is approximated by 20 to 40 rectangles, a "stair-step" approximation, which has the properties that the enclosed area equals the ship's displacement and the longitudinal position of its centroid represents the ship's center of gravity.

After the weight and buoyancy curves are determined and the static balance achieved such that their respective areas and centroids match, the calculation procedure for the ship case is done exactly like that for the simple barge calculation demonstrated in the above worked example. If 20 ship segments have been defined in constructing the weight and buoyancy curves as described above, the load curve will consist of 20 rectangles, the shear curve of 20 straight line segments joined end to end, and the bending moment curve of 20 parabolic segments joined in a smooth curve.

STRUCTURAL RESPONSE OF THE SHIP

The ship hull that is undergoing bending responds, like any beam, by subjecting its structural members to both stress and deformation. Figure 7-13 shows a beam subjected to pure bending, with the amount of bending exaggerated to clarify what happens. Transverse cross sections of the beam remain plane as the beam bends, so that a segment of the beam (like the segment marked *abcd* in the figure) that was rectangular before bending took place, is distorted as shown when the beam bends. It will be clear that the upper portion of such a beam segment has been squeezed, and is thus in compression, while the lower portion has been stretched, and is thus in tension. It follows also from the shape of this segment that there is a neutral plane, shown by the *neutral axis (NA)*, along which there is neither tension nor compression. The amount of tensile or compressive strain (and hence stress as well) on any fiber of the beam is directly proportional to the distance y of the fiber below or above the neutral axis. Therefore the extreme fibers at the top and bottom surfaces of the beam are most highly stressed. (This is why structural members to be used in bending are usually shaped so that most of the material is concentrated at the top and bottom, as in an I beam. It also explains why the deck plating and bottom plating are the most highly stressed portions of a ship structure when the ship is hogging or sagging.) The neutral axis of any beam is located at the centroid of the cross section of the beam. It is therefore at mid-depth of a beam whose cross section is symmetrical top and bottom.

The beam shown in the figure is bending like a ship subjected to sagging, in which the bottom plating is in tension and the deck plating is in compression. For the hogging condition, the stresses are reversed, the deck being in tension and the bottom in compression.

Stresses Caused by Longitudinal Bending

The longitudinal tensile and compressive stresses caused by bending in the fibers of a beam are given by a formula known as the *flexure formula*. Its derivation, which will not be given here, can be found in any strength of materials textbook.

$$\sigma_y = \frac{My}{I} \qquad (11)$$

where σ_y = tensile or compressive stress at point y
M = bending moment at the cross section under consideration
y = vertical distance, NA to fiber under consideration

SHIP STRENGTH   225

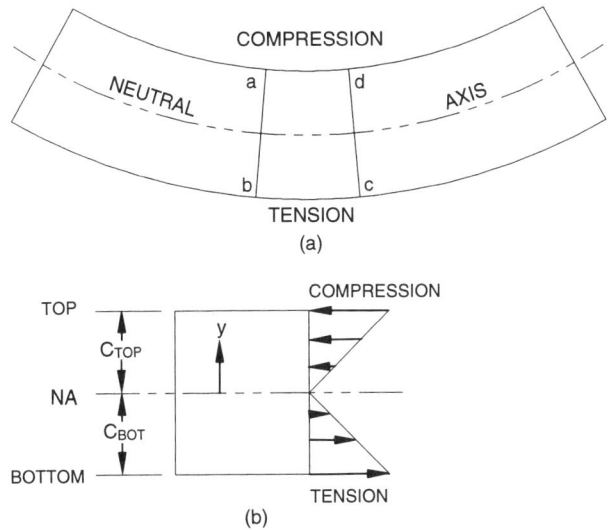

Figure 7-13. Pure bending of a beam.

I = moment of inertia of the cross section of the beam about its neutral (or centroidal) axis

The method of calculating the moment of inertia of the structural cross section is described below.

The formula is often applied only to determine the extreme stresses at the top and bottom of the beam, at which the distance from the neutral axis is designated c. For ship structural cross sections, which are not symmetrical about their neutral axes, there are two different c values, leading to two expressions for maximum bending stresses as follows:

$$\sigma_{top} = \frac{M\, c_{top}}{I} \quad (12)$$

$$\sigma_{bot} = \frac{M\, c_{bot}}{I} \quad (13)$$

These equations evaluate the stresses in the deck (top) and bottom plating of a ship subjected to a given bending moment.

Units for the Ship Calculations. Because of the large size of a ship hull girder cross section and the very large bending moments that are dealt with in ship calculations, the units that are often used in the calculations are unique to ship calculations. In fact, the traditional units that were developed to avoid very large numbers when calculations were all done by hand resulted in a kind of hybrid unit system in which inches and feet or centimeters and meters were used selectively. The table below describes this special unit system and compares it to the units that would normally be used in standard strength of materials textbooks. The areas (a) referred to are the cross-sectional areas of structural members.

	Strength of materials		Ship calculations	
	U.S.	SI	U.S.	SI
σ	lb/in^2	Pa	LTon/in^2	MN/m^2
M	in-lb	N-m	ft-tons	MN-m
c	in	cm	ft	m
I	in^4	cm^4	in^2-ft^2	cm^2-m^2
a	in^2	cm^2	in^2	cm^2

The Section Modulus
In the flexure formula, the bending moment is dependent on the length and loading on the beam, while both I and c are properties of the structural cross section of the beam. Since this is the case, the quantity I/c can be determined as a separate geometrical property of the beam cross section. It is called the *section modulus* of the beam, designated here by the symbol Z. Thus

$$Z = \frac{I}{c} \qquad (14)$$

The section modulus of a beam cross section is a measure of its ability to resist bending. The units of Z are length-cubed units. Using the special ship calculation units for I and c listed in the table above, the units of the section modulus would be in^2-ft or cm^2-m. For ship calculations, two section moduli, Z_{top} and Z_{bot}, are calculated by putting c_{top} and c_{bot} into equation (14). Flexure formulas (12) and (13) can be rewritten in terms of these section moduli as follows:

$$\sigma_{top} = \frac{M}{Z_{top}} \qquad (15)$$

$$\sigma_{bot} = \frac{M}{Z_{bot}} \qquad (16)$$

Section Modulus Calculation. In order to determine the maximum stresses associated with the longitudinal bending of a ship, the section moduli (top and bottom) of the ship structure must be determined for the ship section in the vicinity of the maximum bending moment. The maximum bending moment occurs not far from amidships for a wide range of ship forms and loading conditions. To make sure that adequate longitudinal strength is provided throughout the middle portion of the ship to withstand the maximum bending moment defined by classification society rules, the rules state that the required midship section modulus must be extended throughout at least four-tenths of the ship's length centered at amidships.

Only longitudinal structural members that are continuous through transverse bulkheads are included in the calculation of the section modulus, because transverse members and longitudinals of short extent do not act as part of the hull girder to withstand the tensile and compressive stresses caused by longitudinal hull bending. Therefore, to evaluate the minimum section modulus within the midship four-tenths length, the section considered must be chosen in way of the widest hatch opening in the deck. The structural members that may be included in the calculation of the section modulus, provided that they are continuous, are:

- Deck plating outboard of hatches
- Plating of side shell, bottom shell, inner bottom, and longitudinal bulkheads
- Center vertical keel, side girders (keelsons), and deck girders
- Longitudinals on deck, sides, bottom, inner bottom, and longitudinal bulkheads

Since the midship section of a ship is symmetrical about its centerline, the structural members on one side of centerline are included in the calculations pertaining to individual parts, and the resulting moment of inertia is multiplied by two. The calculation procedure follows. The choice of the position of the assumed axis of moments is entirely arbitrary. If the calculations are to be done by hand, taking the assumed axis at about mid-depth will minimize the magnitudes of the moment arms and the first and second moments of the individual parts, but positive and negative arms will be involved. For calculations done by computer (a spreadsheet procedure is ideal), large numbers are no less convenient than small ones, so the baseline may be taken as the assumed axis. All moment arms will then be positive.

After the longitudinally continuous structural members have been determined and a (half) midship section has been drawn, the steps in the calculation are:

1. Choose the axis of moments. The baseline is recommended.
2. Determine the following three quantities for each plate or structural shape. Nominal or molded width and height dimensions and moment arms may be used in the calculations. It is not necessary to adjust such dimensions to allow for the thickness of the plates.

    a = cross section area, in^2 or cm^2
    d_n = vertical distance of the center of gravity of the plate or shape from the assumed axis, ft or m

i_o = moment of inertia about the horizontal axis passing through the centroid of the plate or shape, in^2-ft^2 or cm^2-m^2

For standard steel shapes, i_o is listed in tables published by the manufacturers and by the American Institute of Steel Construction (AISC). The units may have to be changed to those mentioned above. For vertical plates, $i_o = (1/12)ah^2$, where a is the area (in^2 or cm^2), and h is the height of the plate in *feet* or *meters*. The i_o's of horizontal plates may be omitted from the calculations because they are negligibly small.

3. Calculate, for each member, its first and second moments about the assumed axis:

$$ad_n, \text{ in}^2\text{-ft or cm}^2\text{-m}$$
$$ad_n^2, \text{ in}^2\text{-ft}^2 \text{ or cm}^2\text{-m}^2$$

4. Calculate the following sums over all of the members, and multiply them by 2 to determine properties of the whole ship section:

A = total section area = $2\Sigma a$
Total moment of section about assumed axis = $2\Sigma \, ad_n$
I_n = moment of inertia of section about assumed axis = $2\Sigma \, (i_o + ad_n^2)$.

5. Calculate the vertical distance from the assumed axis to the true neutral axis:

$$Dg = \frac{\Sigma \, ad_n}{\Sigma a}$$

(Multiplication by two is not necessary here, since a ratio is being determined.)

6. Calculate, using the parallel axis theorem, the moment of inertia of the whole ship section about its true neutral axis:

$$I_o = I_n - AD_g^2$$

7. Calculate top and bottom values of c and Z:

c_{top} = distance of upper deck at side from neutral axis, in feet or meters
c_{bot} = distance of bottom plating from neutral axis, in feet or meters

$$Z_{top} = \frac{I_o}{c_{top}}$$

$$Z_{bot} = \frac{I_o}{c_{bot}}$$

The section moduli will be in units of in^2-ft or cm^2-m.

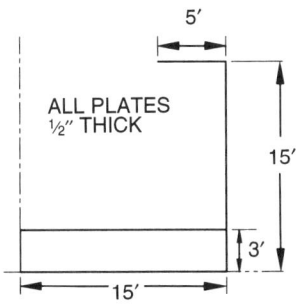

Figure 7-14.

A section modulus calculation is most conveniently arranged in tabular fashion, as demonstrated in the worked example that follows.

Example 7-2

The midship section including only the longitudinally continuous structure of a rectangular barge of L = 150 ft, B = 30 ft, D = 15 ft, is shown in Figure 7-14. All plating is 1/2 inch thick. Determine the location of the neutral axis, the total moment of inertia of the midship section (both sides), and the top and bottom section moduli. Determine also the stresses in the deck and bottom plating when the barge is subjected to a hogging bending moment of 10,000 ft-tons.

The calculations are shown in the table below. In the scantlings column, the dimensions of the items are shown in inches, so that the products shown will compute the areas in square inches. Thus, for example, the deck plating, 5 feet wide (i.e., 60 inches) and 1/2 inch thick, is listed as 60 × 0.50, and its area is then 30.0 square inches. Note that the center vertical keel is split by the centerline of the barge, hence only half of it is entered into the area calculation. All summations then represent the areas and moments of half of the structure.

The chosen axis of moments is the baseline of the barge. The calculations that follow are done as described in the above calculation procedure. Calculations of i_o are done only for the vertical plates of the side shell and the center vertical keel. Other i_o's are negligible.

Item	Scantlings	a in^2	d_n ft	ad_n $in^2 ft$	ad_n^2 $in^2 ft^2$	h ft	i_o $in^2 ft^2$
Main deck	60 × 0.5	30	15.0	450.0	6,750.0		
Side shell	180 × 0.5	90	7.5	675.0	5,062.5	15	1,688
Tank top	180 × 0.5	90	3.0	270.0	810.0		
Bottom shell	180 × 0.5	90	0.0	0.0	0.0		
Center keel	$\frac{1}{2}$(36 × 0.5)	9	1.5	13.5	20.3	3	7
Half-section totals		309		1,408.5	12,642.8		1,695

$$A = 2\Sigma a = 2 \times 309 = 618 \text{ in}^2$$

$$D_g = \frac{\Sigma \, ad_n}{\Sigma \, a} = \frac{1,408.5}{309} = 4.56 \text{ ft}$$

$$I_n = 2\Sigma (i_o + ad_n^2) = 2(1,695 + 12,642.8) = 28,676 \text{ in}^2 ft^2$$

$$I_o = I_n - AD_g^2 = 28,676 - 618(4.56)^2 = 15,826 \text{ in}^2 ft^2$$

$$c_{top} = \text{Depth} - D_g = 15.00 - 4.56 = 10.44 \text{ ft}$$

$$Z_{top} = \frac{I_o}{c_{top}} = \frac{15,826}{10.44} = 1,516 \text{ in}^2\text{ft}$$

$$c_{bot} = D_g = 4.56 \text{ ft}$$

$$Z_{bot} = \frac{I_o}{c_{bot}} = \frac{15,826}{4.56} = 3,471 \text{ in}^2\text{ft}$$

The stresses in the deck and bottom plating are given by equations (15) and (16). The bending moment of 10,000 ft-tons hogging will put the deck plating in tension and the bottom plating in compression.

Tensile stress in the deck =

$$\sigma_{top} = \frac{M}{Z_{top}} = \frac{10,000}{1,516} = 6.60 \text{ tons/in}^2 = 14,784 \text{ lb/in}^2$$

Compressive stress in the bottom plating =

$$\sigma_{bot} = \frac{M}{Z_{bot}} = \frac{10,000}{3,471} = 2.88 \text{ tons/in}^2 = 6,451 \text{ lb/in}^2$$

Permissible Bending Stress and Required Section Modulus
All of the world's classification societies have published standards that define adequate longitudinal strength of merchant ships. Checking the structural midship section design of a ship against classification society rules involves a considerable amount of calculation using a series of formulas and determining whether the size, type, and service of the ship lie within certain conditions of the applicability of the formulas. The rules of the American Bureau of Shipping are a typical example of the approach used. The following discussion describes briefly that approach, but formulas and calculations are not given in detail here. The ABS rules section on longitudinal strength should be consulted for details.

The ABS longitudinal strength standards are expressed by the following quantities:

- Maximum bending moment amidships, which equals still-water bending moment plus wave-induced bending moment
- Nominal permissible bending stress
- Required hull-girder section modulus amidships

Other parts of the longitudinal strength section of the rules (not discussed here) address shear force and stress, deck and hatch structure, and the use of higher-strength steels. The formulas quoted by the ABS rules for determining the above bending moments and section modulus are expressed as functions of ship length, beam, and block coefficient. Nominal permissible bending stress is expressed as a function only of ship length. In order to provide the reader with an idea of the magnitudes of the above quantities for ships of different sizes and services, the formulas have been applied to the four ships listed in the table below, and the results are given in the following paragraphs.

	Type	L, ft (m)	B, ft (m)	C_B
Ship 1	General cargo ship	528 (161)	76.0 (23.2)	0.612
Ship 2	Product tanker	630 (192)	90.0 (27.4)	0.772
Ship 3	Containership	810 (247)	105.7 (32.2)	0.579
Ship 4	Crude oil carrier	1,060 (323)	178.0 (54.3)	0.842

Bending Moment Amidships. The *total maximum bending moment (M_t)* that a given ship will have to be designed to withstand is given by

$$M_t = M_{sw} + M_w \tag{17}$$

The *still-water bending moment (M_{sw})* is calculated for a wide range of ship loading conditions as a matter of course during the ship design process, and the maximum one is to be used. In addition, a "standard" still-water bending moment is also defined by the ABS rules to be used instead of the detailed calculations when the ship designer wants an approximation of M_{sw} before the detailed calculations have been done. The *wave-induced bending moment (M_w)* is expressed by an equation that was derived on the basis of many years of research and statistical analysis of measurements of ship motions and stresses experienced by actual ships at sea. A ship designer is allowed to substitute statistical analyses based on approved computer programs of ship motions and stresses in realistic sea states for the M_w calculation. The bending moments below were determined by applying the standard formulas for both bending moments to each of the ships as described in the previous table. Values have been rounded to the nearest 1,000 ft-tons and to the nearest 100 MT-m.

	Ship 1 ft-tons	Ship 2 ft-tons	Ship 3 ft-tons	Ship 4 ft-tons
M_{sw}	158,000	318,000	619,000	2,222,000
M_w	229,000	466,000	912,000	3,247,000
M_t	387,000	784,000	1,531,000	5,469,000

	Ship 1 MT-m	Ship 2 MT-m	Ship 3 MT-m	Ship 4 MT-m
M_{sw}	48,900	98,500	191,700	688,100
M_w	70,900	144,300	282,400	1,005,600
M_t	119,800	242,800	474,100	1,693,700

Nominal Permissible Bending Stress. Building codes for any kind of machine or structure must be written so as to assure, as much as it is possible, that structures built in accordance with the code will not fail in service if they are properly operated and maintained. Only part of that assurance arises from specifying the maximum loads (i.e, bending moment) that the structure may have to withstand. The code must also specify the stress level that must not be exceeded within the structure, even when the maximum imaginable loads are exerted on it. The most obvious requirement is that neither the yield stress nor the ultimate stress must be exceeded in service. Permanent deformation of structural members can severely limit their usefulness, and rupture of main structure can be catastrophic.

On a structure like a ship, the most extreme loads are imposed by nature and are therefore beyond man's control. The best that engineers and scientists can do by undertaking research programs to study such loads is to estimate as closely as possible their maximum values. Seaway loads are statistically random, so that there is always some uncertainty associated with predicting lifetime maxima. That uncertainty, combined with some assumed reduction in the strength of the structure over the years because of natural wastage in spite of proper maintenance, and in light of possible human error by ship operators, however well trained, suggests that some margin of safety should be put into the specification of permissible stresses. This is done by prescribing a permissible bending stress that is somewhat lower than the yield strength of the steel.

The ABS *nominal permissible bending stress* (f_p), calculated for each of the four ships identified above, is as follows:

	Nominal permissible bending stress (f_p)	
	$tons/in^2$	MT/cm^2
Ship 1	10.25	1.614
Ship 2	10.37	1.633
Ship 3	10.57	1.665
Ship 4	10.69	1.684

The ABS specifies a minimum yield strength for ordinary strength hull structural steel of 45,500 psi, which is 20.31 tons/in^2, or 3.20 MT/cm^2. Since the above permissible stresses are approximately half of the required yield stress, the implied safety factor based on yield strength is about two. The fact that the nominal permissible stress increases as ship length increases should not be interpreted to mean that the longer ship may be built with a smaller safety factor. The permissible stress values have been worked out to include the effect of a predicted rate of corrosion of the load-bearing plates and shapes, which causes wastage of the plates after years of service. Plates are made thicker than necessary at the start so that they will be strong enough after, say 25 years, to withstand the maximum bending moments. Since corrosion rates are the same for thick or thin plates, the thicker plates required on the longer ships are made thicker by a smaller percentage than are the thinner plates of the shorter ships. Therefore the nominal stress at the beginning of the life of each ship is different so that at the ends of their useful lives the stresses will be about equal. Putting it another way, a new short ship has proportionately more excess steel than a new long ship, so that after the normal amount of lifetime wastage, each will have just the right steel thickness.

Required Section Modulus. The *required hull-girder section modulus amidships (SM)* is determined from the total bending moment and the nominal permissible stress according to the following equation, which is a rearrangement of equation (15) or (16).

$$SM = \frac{M_t}{f_p} \qquad (18)$$

That is, the required section modulus is determined by dividing the permissible stress into the maximum bending moment. For the example ships, this calculation results in the following table.

During the structural design stage of a ship, after the sizes of the longitudinal structural members have been determined based on their individual loading, the ship's two section moduli are calculated according to equations (15) and (16), using procedures similar to those shown in Example 7-2. Both section moduli must equal or exceed the minimum section modulus calculated according to the classification society formulas. If either of them does not, structural members must be increased in strength in the appropriate places (deck or bottom) until the minimum value is achieved.

Required minimum section modulus amidships (SM)

	in^2-ft	cm^2-m
Ship 1	37,800	74,200
Ship 2	75,600	148,700
Ship 3	144,800	284,700
Ship 4	511,600	1,005,800

PROBLEMS

1. A ship of 10,000 tons displacement and 350 feet long has a maximum hogging bending moment (foot-tons) of $\Delta L/30$. The depth of the midship section is 37 ft, and the neutral axis is 19.1 ft above the keel. The moment of inertia of the midship section is 4,000 ft^4. Calculate the maximum tensile and compressive stresses and state where each occurs (deck or bottom).

2. A vessel 400 ft long has a uniform cross section below water. The weights of hull, machinery, and cargo are 3,200; 800; and 6,400 tons, respectively. The hull weight is uniformly distributed over the entire ship length, the machinery weight extends uniformly over 1/5 of the length amidships, and the cargo weight extends uniformly over 2/5 of the length from each end. Draw curves of weight, buoyancy, load, shear, and bending moment, and determine their values at break points and at their maximum points.

3. A hypothetical vessel has a weight curve that varies linearly from zero at both ends to maximum at amidships, and a buoyancy curve that varies linearly from zero at amidships to maximum at both ends. Plot weight, buoyancy, load, shear, and bending moment curves and determine the maximum values of each in terms of displacement (Δ) and length L.

4. The mean values of weight per foot and buoyancy per foot of a 540-foot-long ship represented by six equal-length segments are:

Segment	w, tons/ft	b, tons/ft
1	26	11
2	51	42
3	29	49
4	25	47
5	21	28
6	31	6

    Construct curves of weight, buoyancy, load, shear, and bending moment, and determine their values at break points and at their maximum points.

5. A rectangular barge is 128 ft long, 30 ft wide, and 18 ft deep, and it weighs 544 tons when empty. The barge light weight may be assumed uniformly distributed over its length. It is divided into four holds of

equal lengths. Cargo is loaded into each hold as follows, stowed uniformly and level in each hold:

Hold #1	192 tons
Hold #2	224 tons
Hold #3	272 tons
Hold #4	176 tons

Construct weight, buoyancy, load, shear, and bending moment diagrams for the loaded barge, and calculate the values of each at the ends, at each bulkhead, and at their maximum points.

6. Calculate the minimum required section modulus (in^2-ft) of the barge in problem 5 such that in the loaded condition the bending stress does not exceed 15,000 psi.

7. A barge has a plan view as shown. All waterplanes are identical. Cargo is loaded evenly in the four rectangular holds as shown. Neglecting the weight of the barge itself, construct curves of weight, buoyancy, load, shear, and bending moment for the loaded barge in still seawater. Label the values of each curve at each bulkhead, and identify the maximum shear and bending moment.

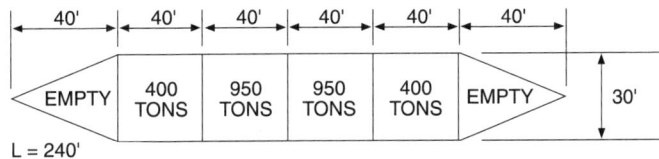

8. A vessel 600 ft long is assumed to have a uniform underwater cross section throughout its length. The hull weight is uniformly distributed over the entire length of the ship and is equal to 2,400 long tons. The ship has six holds, each of equal length, the contents of which are, starting from forward:

Hold	Tons cargo	Tons fuel	Tons mach'y
1	400	100	–
2	700	200	–
3	800	300	–
4	800	300	–
5	–	100	800
6	500	–	–

Each of these weights is uniformly distributed in its respective hold. Draw weight, buoyancy, load, shear, and bending moment curves, labeling the values of each at each change point and at their maximum points.

9. A rectangular barge 320 ft long, 45 ft beam, and 15 ft deep weighs 1,920 tons when empty, the light ship weight being evenly distributed over its length. It is then loaded at each end for a length of 100 ft with 1,200 tons of cargo, also evenly distributed (total cargo = 2,400 tons). Determine the maximum bending moment of this barge when loaded, and thus calculate the maximum stress in both the deck and bottom plating, if the structure has I = 65,600 in^2-ft^2 and the neutral axis is 6.80 ft above the keel.

10. Calculate the top and bottom section moduli (in^2-ft) of the structural section shown. All plating is 1/4-inch thick. What will be the stress (tons per square inch) in the deck plating if this vessel is subjected to a bending moment of 6,000 ft-tons?

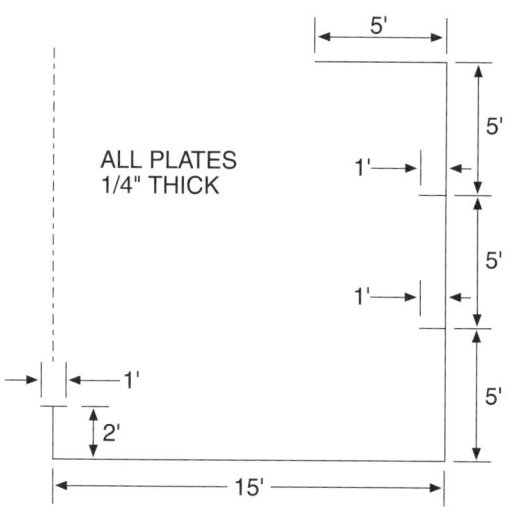

11. Calculate the top and bottom section moduli (in^2-ft) of the structural section shown below. Plating dimensions and thicknesses are as labeled. If the steel has a yield stress of 34,000 psi, what safety factor based on yield will this structure have when it is subjected to a bending moment of 12,000 ft-tons?

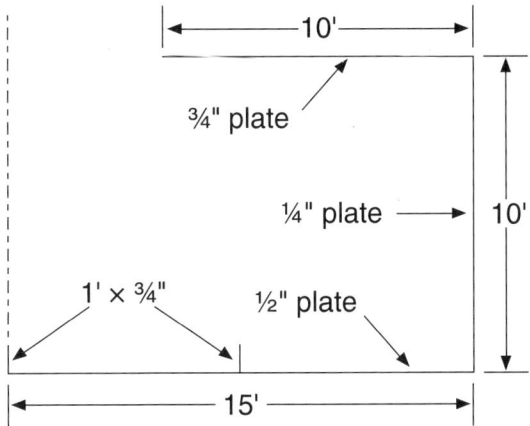

12. A ship 560 ft long and 42 ft deep (to main deck) has a maximum bending moment of 150,000 ft-tons. The neutral axis is 19 ft above the keel. For a maximum allowable bending stress of 9.0 tons per square inch, determine the required moment of inertia (in^2-ft^2) of the ship's midship section. Does the maximum stress occur in the deck plating or in the bottom plating? What stress occurs in the other flange (bottom or deck) at the same time?

CHAPTER EIGHT

SHIP RESISTANCE

Searching for a hull form of least resistance has occupied the attention of naval architects for a very long time. Many prominent and not-so-prominent scientists and experimenters from the early eighteenth century on attempted to discover the hydrodynamic laws and principles that would enable them to calculate the resistance of a ship moving through water and to relate it to hull form. Although many theories were claimed by their proponents to have solved the problem of the form of least resistance, none has withstood the test of time and of modern hydrodynamic knowledge. The evasive and sought-after "form of least resistance" was an impossible dream. Naval architects and hydrodynamicists have been led to the conclusion that the complex problem of determining ship resistance can only be solved by a combination of theoretical and experimental methods.

When ships were powered by sail and the propulsive forces that drove them were subject to the whims of nature, naval architects had to rely on their empirical knowledge of the long-term performances of previous designs to suggest changes in hull form or sail design and arrangements that would improve the performance of new designs. But the success of previous ships depended in large part on the seamanship skills of their officers and crews. Under the command of two different masters, a ship might perform very differently in similar weather conditions. Furthermore, a hull that had superior performance in light airs might be poorly suited to sailing well in heavy winds, and vice versa. Successful naval architects of necessity relied heavily on experience, empiricism, and trial and error to design a ship that would perform well on the average.

The advent of the application of mechanical power to ship propulsion in the early eighteenth century brought with it a new challenge to naval architects. Now they were free to specify the propulsive power to be installed, while their new partners in ship design, the marine engineers, set to work designing efficient and reliable power plants. In order to utilize that power effectively, naval architects needed reliable methods to determine ship resistance, to estimate what power must be installed to overcome the resistance at a required speed, and to design efficient propellers. In this chapter we find out how the challenge has been met, we examine what has been learned about the complex nature of fluid flow around a ship hull, and we describe modern methods of determining ship resistance, power, and efficiency of propulsion.

RESISTANCE, SPEED, AND POWER

The most accurate way to determine the resistance of a new ship design would be to build the ship and tow it through undisturbed smooth water while measuring the force exerted on the towing cable. If this test were repeated at a number of speeds, the towing force would increase as the speed increased, and the resulting relationship of towing force to speed would be a measure of the ship's resistance over the range of speeds at which it was tested. The resistance measured at design speed would be used to determine the power that must be produced to overcome the resistance, as follows:

$$\text{power} = \text{resistance} \times \text{speed}$$

It is immediately apparent that a full-scale towing test like the one described is not a practical way to determine the power required for a new ship. One cannot build the ship before deciding how much power to install. Moreover, if the designer determined as a result of the measurements that an alteration in the hull form would be beneficial, the ship would have to be rebuilt. And there are many other problems involving changes in resistance that take place when the propulsive force is produced by a propeller instead of a towline.

While the full-scale test is not a practical method for determining ship resistance, it does serve the purpose of defining what is meant by ship resistance. It is the force needed to tow a ship at a given speed in perfectly smooth water. The power associated with this resistance is called the *effective power* or *effective horsepower* of the ship at the speed in question. The calculation of effective horsepower will be described in a later section.

Components of Resistance

Because it is impractical to measure a ship's smooth-water resistance using full-scale tests, the naval architect must rely on a combination of model tests and analytical techniques to establish the relationship between speed and resistance of a new hull with sufficient accuracy to determine its power requirements: Designing and analyzing proper model tests requires an understanding of the physical nature of ship resistance and of the way model forces and speeds are expanded to full scale.

The resistance of a ship moving in smooth water is caused by a number of different fluid flow phenomena which interact and combine in very complex ways, so complex that hydrodynamicists have not found theoretical methods that can calculate ship resistance accurately enough to obviate the need for model tests. That is not to say that little is known of the nature of the problem. Indeed, hydrodynamic research has provided ship designers with much insight into the causes of ship resistance.

There are four main components of ship resistance in smooth water, each of which is associated with a particular flow phenomenon. They are:

1. *Frictional resistance,* which results from the fact that a solid surface (the ship's hull) moving through a viscous fluid carries with it some of the fluid immediately adjacent to the hull within a region called the boundary layer. The aftward acting component of the force necessary to set this fluid in motion is the frictional, or viscous, resistance.

2. *Wave-making resistance,* which is caused by the wave pattern produced by a ship moving through calm water. The energy expended to produce these waves is a measure of work done by the ship on the water, thus causing a resistive force on the ship. When model test results are expanded to full scale, the wave-making component is combined with the eddy-making component (described below) in what is called the *residuary resistance.*

3. *Eddy-making resistance,* which results from the inability of the water to flow in smooth streamlines around abrupt discontinuities at stern frames, bossings, and other appendages. The flow breaks clear and reverses, causing eddies to fill the void that would otherwise result. Energy put into eddy formation causes additional resistance to the ship's forward motion. The same phenomenon occurs where streamline flow separates from the hull near the stern if the curvature of the hull is too sharp for the flow lines to follow it. It is in that case called *separation resistance.*

4. *Air resistance,* which consists of both frictional and eddy-making resistances caused by the flow of air around the upper works of the ship.

In the following more-detailed analysis of these components, interactions among them will be described by defining additional components as necessary, but the phenomena of friction, wave-making, and eddy-making will be seen to be the fundamental fluid flow mechanisms that give rise to all components of ship resistance or drag, whatever they are called.

THE LAWS OF COMPARISON

Since the resistance of a ship cannot be calculated directly, model tests must be conducted to measure the resistance of the model at various speeds, and the measured resistances must be expanded to the corresponding resistances of the full-scale ship at ship speeds corresponding to the chosen model speeds. To conduct and analyze a proper model test, one must know how to "scale up" the dimensions, velocities, and forces measured on the model to the corresponding dimensions, velocities, and forces on the full-scale ship.

Complete Physical Equation

The appropriate scaling laws, or *laws of comparison*, between model and ship have been deduced by applying the techniques of dimensional analysis to a functional relationship that declares all of the variables on which the resistance of the model or ship depends. Experimentation and intuition are necessary to determine such a relationship, which is called a "complete physical equation." Experience has shown that the resistance of a ship depends on:

Length of ship (L)
Speed of ship (V)
Mass density of water (ρ)
Kinematic viscosity of water (ν)
Gravitational acceleration (g)
Hull form coefficients (C_B, C_P, etc.)

Instead of kinematic viscosity (ν), the absolute or dynamic viscosity (μ) may be substituted. The relationship between them is $\mu = \rho\nu$. In special cases in which extremely high speeds cause cavitation to occur at some points around the hull, the above list must also include the water pressure at the point of cavitation and the vapor pressure of the water. As this phenomenon is rarely a problem in ordinary resistance tests, it is not included in the present analysis. Cavitation will be considered later when dealing with propellers.

Since the purpose of this analysis is to determine scaling laws relating model measurements to the corresponding full-scale quantities, the hull form coefficients may be left out of the complete physical equation because geometrical similarity of the model to the ship requires that the coefficients of form be identical for model and ship.

The complete physical equation for ship model testing is, therefore:

$$R = f(L, V, \rho, \nu, g) \tag{1}$$

in which R is the resistance of the ship or model. This equation does not specify how the variables on the right side influence the resistance. It states only that the resistance is a function of the quantities shown on the right-hand side of the equation at the time of the test. Put another way, the complete physical equation asserts that a change in the value of any of the independent variables (right-hand side of the equation) will change the value of the dependent variable R.

Because of the number of variables involved, an experimental testing program intended to determine how the resistance is influenced by each of the independent variables would require a prohibitive number of tests to examine all possible combinations of the variables. The number of quantities to be examined can be reduced by applying the techniques of dimensional analysis to the complete physical equation. Dimensional analysis rests on the principle of dimensional homogeneity, which states that a complete physical equation can be rewritten by grouping the variables in such a way that:

1. Each group is independent of the others; that is, no group can be derived by combining the other groups.
2. The dimensions of each of the grouped variables are the same. The most common and useful way to assure this is to construct dimensionless groups of the given physical variables.

The procedures of dimensional analysis have been applied to equation (1), and thus it has been shown that the equation may be rewritten in the following form:

$$\frac{R}{\rho L^2 V^2} = f\left(\frac{VL}{\nu}, \frac{V^2}{gL}\right) \qquad (2)$$

There are several advantages to writing the complete physical equation in this form. Since each of the above groups are dimensionless, numerical values of each from a given test will be identical regardless of the measurement system used in the test (SI or U.S.). Furthermore, the number of variables has been reduced from six to three, so that fewer experiments are required to produce useful results.

The Resistance Coefficient

In the analysis of conventional model resistance tests, the three dimensionless numbers in equation (2) usually appear in slightly modified form. The quantity on the left side of equation (2), containing the resistance, is a resistance coefficient. The quantity $\rho L^2 V^2$ has the dimensions of a force, and since L^2 has the dimensions of an area, ρV^2 is a pressure. There is special hydrodynamic significance in potential flow to the stagnation pressure, which is $\frac{1}{2}\rho V^2$. It is the pressure at a stagnation point, or the "nose" of a blunt immersed body in potential flow. The stagnation pressure multiplied by an area which is characteristic and meaningful in the flow situation is used as a reference force in the definitions of many nondimensional coefficients, such as drag coefficients, in fluid flow studies. For ship resistance studies, the characteristic area is the wetted surface (S) of the ship hull. Thus the conventional form of ship resistance coefficient is

$$C = \frac{R}{\frac{\rho}{2} S V^2} \qquad (3)$$

The Reynolds Number
Of the two dimensionless numbers on the right side of equation (2), the first contains the viscosity of the water, thus it is associated with frictional resistance. It is the Reynolds number (R_n) which can be written in terms of either the kinematic viscosity or the absolute viscosity:

$$R_n = \frac{VL}{\nu} = \frac{\rho VL}{\mu} \qquad (4)$$

The Froude Number
The other dimensionless number contains the gravitational acceleration (g) which suggests that it is associated with wave-making resistance, since the crests of a ship's wave train are raised against the force of gravity. The square root of the quantity shown in equation (2) is called the Froude number (F_n) named for William Froude, who was the first person to devise a reliable procedure for model tests and the expansion of model measurements to full scale. The Froude number is defined as

$$F_n = \frac{V}{\sqrt{gL}} \qquad (5)$$

Equation (2) can be rewritten symbolically as follows:

$$C = f(R_n, F_n) \qquad (6)$$

Although we have succeeded in transforming the complete physical equation (1) into a simpler form as given by equations (6) or (2), the nature of the function f is still not known. Dimensional analysis cannot determine numerical answers, nor can it predict exactly how the resistance is dependent on the Froude number and the Reynolds number. But there is one way to apply this information to the problem of determining ship resistance from model tests. Equations (2) and (6) state that if the Froude and Reynolds numbers of the model are equal to those of the ship (or of any geometrically similar body of a different scale), their resistance coefficients will also be identical, regardless of the form of the function f.

Let us examine this approach to determine what restrictions it places on the proper conduct of a model test.

Corresponding Speeds
Geometrical similarity of the model to the ship is assumed, and subscripts M and S represent model and ship, respectively. The linear scale ratio (λ) is defined as the ratio of any ship dimension to the corresponding model dimension:

$$\lambda = \frac{L_S}{L_M} = \frac{B_S}{B_M} = \frac{T_S}{T_M}, \text{etc.} \tag{7}$$

Equating the Froude numbers of the ship and its model:

$$F_{nS} = F_{nM}$$

$$\left(\frac{V}{\sqrt{gL}}\right)_S = \left(\frac{V}{\sqrt{gL}}\right)_M$$

$$\frac{V_S}{V_M} = \frac{\sqrt{g_S L_S}}{\sqrt{g_M L_M}}$$

and, since the gravitational acceleration is the same for model and ship,

$$\frac{V_S}{V_M} = \sqrt{\frac{L_S}{L_M}} = \sqrt{\lambda} \tag{8}$$

Thus, testing a model at the same Froude number as that of the ship places a condition or restriction on the model speed. The speed at which a model must be tested to represent a particular ship speed is determined from equation (8) as follows:

$$V_M = \frac{V_S}{\sqrt{\lambda}} \tag{9}$$

Equations (8) and (9) define what Froude called the "corresponding speeds" of the ship and its model, although he deduced the corresponding speeds in a different way from the above analysis. He observed the wave patterns produced by a ship at a given speed, then he towed a geometrically similar model at various speeds until he observed a wave pattern that was geometrically similar to that produced by the ship, which occurred when "their speeds are in the ratio of the square roots of their linear dimensions." He was not concerned in his work in 1868 with expressing his findings in dimensionless form. In his work and that of many investigators since, the corresponding speed concept was expressed by equating the dimensional ratio $V_K/\sqrt{L}$ for model and ship. This quantity, called the *speed-length ratio,* is defined only with the speed expressed in knots and the length in feet. The dimensionless Froude number was named for Froude in recognition of the importance of his work.

Dynamical Similitude

Before examining the Reynolds number equivalency of ship and model, we show how the corresponding speeds defined above will determine the scaling factor for ship resistance. It has been previously stated that if the model Froude and Reynolds numbers are equal to those of the ship, their resistance coefficients will also be equal. Thus,

$$\left(\frac{R}{\frac{1}{2}\rho S V^2}\right)_S = \left(\frac{R}{\frac{1}{2}\rho S V^2}\right)_M \qquad (10)$$

Thus, the scaling factor for resistance is:

$$\frac{R_S}{R_M} = \left(\frac{\rho_S}{\rho_M}\right)\left(\frac{S_S}{S_M}\right)\left(\frac{V_S}{V_M}\right)^2 \qquad (11)$$

Because of geometrical similarity, the scaling factors for corresponding areas, volumes, and displacements are:

$$\frac{S_S}{S_M} = \left(\frac{L_S}{L_M}\right)^2 = \lambda^2 \qquad (12)$$

$$\frac{\nabla_S}{\nabla_M} = \left(\frac{L_S}{L_M}\right)^3 = \lambda^3 \qquad (13)$$

$$\frac{\Delta_S}{\Delta_M} = \frac{\rho_S g \nabla_S}{\rho_M g \nabla_M} = \left(\frac{\rho_S}{\rho_M}\right)\lambda^3 \qquad (14)$$

and, from the Froude speed scaling law, equation (8):

$$\left(\frac{V_S}{V_M}\right)^2 = \frac{L_S}{L_M} = \lambda \qquad (15)$$

Substitution of these scaling factors into equation (11) gives

$$\frac{R_S}{R_M} = \left(\frac{\rho_S}{\rho_M}\right)\left(\frac{S_S}{S_M}\right)\left(\frac{V_S}{V_M}\right)^2 = \left(\frac{\rho_S}{\rho_M}\right)\lambda^2 \lambda = \left(\frac{\rho_S}{\rho_M}\right)\lambda^3 \qquad (16)$$

Thus, the resistance scaling factor is the same as that for displaced mass or displaced weight (since g is constant):

$$\frac{R_S}{R_M} = \frac{\Delta_{mS}}{\Delta_{mM}} = \frac{\Delta_S}{\Delta_M} \qquad (17)$$

Equation (17) is an expression of *dynamical similitude* of the model and ship. That is, the ratio of a force on the ship to the corresponding force on the model is the same regardless of the type of force.

Model Speed Paradox

The above analysis is not yet a complete expression of dynamical similitude, however, because the influence of viscous resistance forces has not yet been considered. We must determine the consequences of equating the Reynolds numbers of the ship and its model.

$$R_{nS} = R_{nM}$$

$$\left(\frac{VL}{v}\right)_S = \left(\frac{VL}{v}\right)_M$$

$$\frac{V_S}{V_M} = \left(\frac{L_M}{L_S}\right)\left(\frac{v_S}{v_M}\right)$$

Although the water in the model tank is not usually the same as that for the ship (fresh vs. seawater, and at a different temperature), the viscosities of both will be nearly equal, so that the ratio of speeds is approximately:

$$\frac{V_S}{V_M} = \frac{L_M}{L_S} = \frac{1}{\lambda} \tag{18}$$

which requires that the model speed be greater than the ship speed:

$$V_M = V_S \lambda \tag{19}$$

This is the model test condition that would be required for dynamical similitude to apply to the viscous or frictional forces. It is immediately apparent that this model speed is greater than the ship speed, which is in conflict with the corresponding speed deduced by equating the Froude number of the model to that of the ship, as shown in equation (9). Thus *it is impossible to achieve complete dynamical similitude in a model test* because the Froude and Reynolds numbers of the model cannot be made the same as those of the ship simultaneously. To put the problem in another way, the scaling factor needed to convert the resistance force on a model to that on the ship is different for the part of the resistance caused by wave-making (Froude criterion) and the part caused by friction (Reynolds criterion). Lack of understanding of this problem caused early attempts at interpreting model test results (in order to determine the power requirement of mechanically propelled ships) to be very unreliable.

Aside from the fact that the speed required by Reynolds number equivalency is different from Froude's corresponding speed, consideration of typical model scales will show that it is not practical to test a small model at a Reynolds number equal to that of a ship. A 20-foot-long model of a 720-foot-long ship, for example, is built to a scale ratio of $\lambda = 720/20 = 36$. If this model were to be tested at the Reynolds speed corresponding to a ship speed of, say, 24 knots, the required model speed (see equation 19) would be $V_M = 24 \times 36 = 864$ knots! However, the corresponding Froude speed of this model (see equation 9) is $V_M = 24/\sqrt{36} = 4$ knots, which is practical for a model in a towing tank.

Froude's Hypothesis

Since it is impossible to make both the Froude and Reynolds numbers of the model equal to those of the ship in a model test, it cannot be said that the total resistance coefficients of the model and ship will be equal. To solve the problem of predicting ship resistance from a model test it is therefore necessary to determine the nature of the functional relationship implied by equation (6). The assumption that is made, although it is known to be not precisely correct, is that the resistance consists of only two components, one of which is a function of the Froude number and the other a function of the Reynolds number. This assumption is a modern version of what is known as Froude's hypothesis because William Froude was the first to propose it and to demonstrate that it can be the basis of a reliable model testing procedure. Froude's intuitive insight and his practical approach to solving a very difficult problem are especially remarkable because his work predated that of Reynolds, so he was unaware of the relationship of frictional resistance to the Reynolds number.

According to the Froude hypothesis, the two components of resistance are frictional (R_F) and residuary (R_R) resistance. The residuary resistance is simply defined as what is left over after the frictional resistance is subtracted from the total resistance (R_T). It therefore includes wave-making and eddy-making resistance. Froude assumed further that his "corresponding speeds" (equations 8 and 9) and the resulting resistance scaling factor (derived in equations 10 through 17) applied only to the residuary resistance. That is, he asserted that

$$\text{if} \quad \frac{V_S}{V_M} = \sqrt{\frac{L_S}{L_M}}, \quad \text{then} \quad \frac{R_{RS}}{R_{RM}} = \frac{\Delta_S}{\Delta_M} \quad (20)$$

This assumption would be exactly correct if the residuary resistance were caused entirely by wave making. The fact that it includes eddy-making resistance as well makes the assumption not exact, but since wave-making resistance usually predominates over eddy-making resistance, this simplified model of the components of resistance has been found to be satisfactory when analyzing model test measurements. Since it is not exact, an adjustment (called the model-ship correlation allowance) is added in the modern application of Froude's law of comparison, although the basic procedure used today is the same as that proposed by Froude.

The total resistance is therefore written:

$$R_T = R_F + R_R \quad (21)$$

or, in nondimensional coefficient form, each term is divided by $\frac{1}{2}\rho S V^2$ to give:

$$C_T = C_F + C_R \quad (22)$$

where $C_T = \dfrac{R_T}{\frac{\rho}{2} S V^2}$ = total resistance coefficient

$C_F = \dfrac{R_F}{\frac{\rho}{2} S V^2}$ = frictional resistance coefficient

$C_R = \dfrac{R_R}{\frac{\rho}{2} S V^2}$ = residuary resistance coefficient

The frictional and residuary resistances are assumed to be independent of one another, with C_F dependent on the Reynolds number and C_R dependent on the Froude number. Thus, the functional relationship suggested by equation (6) is

$$C_T = C_F + C_R = f_1(R_n) + f_2(F_n) \tag{23}$$

EXPANDING MODEL RESISTANCE TO THE SHIP

The only quantity in equation (23) that can be determined directly from a model test is C_T, because the towing force measured during a test run is the total resistance (R_T). But it is necessary to evaluate C_R of the model in order to expand the results to ship scale, because C_R is the quantity that is equal for the model and ship when the model has been tested at the appropriate Froude number. Neither R_F nor R_R can be measured directly, so to separate C_T into its two components, C_F must be evaluated independent of the model test and C_R deduced by subtracting C_F from C_T. Much research has been done to determine expressions for $C_F = f_1(R_n)$ that would be applicable to both models and ships. The resulting functions, known as *skin friction formulations* or *model-ship correlation lines,* are described in the section on frictional resistance.

The procedure for expanding a measured model resistance to full-scale ship resistance is as follows. It is assumed that a geometrically similar model of the ship has been built to a scale ratio of λ, and that the required ship speed is V_S.

1. Tow model at the corresponding speed,

$$V_M = \dfrac{V_S}{\sqrt{\lambda}}$$

and measure model towing force or total model resistance (R_{TM}).

2. Calculate model total resistance coefficient,

$$C_{TM} = \dfrac{R_{TM}}{\frac{1}{2} \rho_M S_M V_M^2}$$

SHIP RESISTANCE   249

3. Calculate model Reynolds number $R_{nM} = \dfrac{V_M L_M}{\nu_M}$
   and model frictional resistance coefficient $C_{FM} = f(R_{nM})$ using a skin friction formulation.

4. Calculate model residuary resistance coefficient C_{RM} by subtraction:

$$C_{RM} = C_{TM} - C_{FM}$$

5. According to the law of comparison, C_R of the ship equals that of the model.

$$C_{RS} = C_{RM} \qquad (24)$$

This is the nondimensional equivalent of equation (20).

6. Calculate ship Reynolds number (R_{nS}) and ship frictional resistance coefficient (C_{FS}) as described above for the model, using the same friction formulation as that used for the model in step 3.

7. Calculate ship total resistance coefficient by addition of its components, adding a correlation allowance (C_A) to adjust for the inexact nature of the assumption that there are only two components, and for the roughness of the ship's hull. Values of C_A are discussed later.

$$C_{TS} = C_{FS} + C_{RS} + C_A \qquad (25)$$

8. Calculate ship total resistance:

$$R_{TS} = C_{TS} \left(\tfrac{1}{2} \rho_S S_S V_S^2\right)$$

An example of the use of this procedure is included in a later section.

FRICTIONAL RESISTANCE

The ability to make an accurate prediction of ship resistance by expanding the measured model resistance rests in large part on the calculation of the frictional resistance of both the model and the ship. Research into this phenomenon has revealed a great deal about the nature of frictional resistance, but a satisfactory analytical or theoretical procedure capable of determining the forces involved without relying on experimental measurements has not yet been developed. Friction formulations currently in use are all based on measurements of the resistance of thin, flat planes or planks.

Froude's Plank Tests
Plank tests were devised by Froude, who reasoned that a thin plank was a near representation of a waveless form, so that its entire resistance would be frictional. A plank was "equivalent" to a particular model if it had the same length and wetted surface as the model, and was towed at the same speed. Recognizing that surface roughness would also influence frictional resistance, he tried various coatings, including smooth varnish, calico, and sand coatings of varying coarseness. Planks as long as 50 feet were tested so that results could be extrapolated into the range of full-scale ships.

Froude's Friction Formula. Froude's plank tests and extrapolations, augmented by further extrapolation and refinements done by his son, R. E. Froude, resulted in the following formula for frictional resistance:

$$R_F = f S V^n \qquad (26)$$

where R_F = frictional resistance, pounds
   S = wetted surface area of the plank, square feet
   V = velocity of the plank, knots

In its original form as published by William Froude, this equation contained two quantities (f and n) that were determined from the experiments and were found to be functions of plank length and surface quality or roughness. The friction coefficient (f) decreased as length increased, and generally increased with increased roughness. The exponent of the speed (n) was found to increase with surface roughness, but depended very little on length. Later refinements that were made by R. E. Froude and others established a constant exponent of speed at 1.825 and fitted an empirical equation to the coefficients of friction, resulting in the following equations:

$$R_F = f S V^{1.825} \qquad (27)$$

where, in U.S. units (pounds, feet, knots)

$$f = 0.00871 + \frac{0.0530}{(L + 8.8)} \qquad \text{[seawater]} \qquad (28)$$

$$f = 0.00849 + \frac{0.0516}{(L + 8.8)} \qquad \text{[fresh water]} \qquad (28a)$$

The formulas for f, the friction coefficient, apply to a water temperature of 15°C (59°F), which has been adopted as a standard by the International Towing Tank Conference (ITTC). For temperatures other than the standard, the friction coefficient should be decreased by 0.43 percent per +1 degree C, or by 0.24 percent per +1 degree F.

The Froude friction formulation given above was used by many ship model testing facilities to determine the frictional resistance of both models and ships through the first half of this century. At other towing tanks, new plank tests were made and the friction coefficients were modified on the basis of further analysis of the larger body of experimental results, but still using formulas similar to those of equations (27) and (28) as the only practical means of predicting the frictional resistance of ship forms.

Laminar and Turbulent Flow
A few years after the Froude plank tests were completed, an important discovery about the nature of fluid flow along a surface was made by Osborne Reynolds. In his studies of fluids flowing through glass tubes, Reynolds observed two distinct types of flow, or flow regimes, which he called laminar flow and turbulent flow. By injecting a thin stream of colored dye into water moving through a glass tube, Reynolds observed that at low velocity the dye filament moved in a straight line with little or no mixing into the water surrounding it. He called such flow *laminar flow,* because the fluid stream lines seemed to remain in layers. At higher speeds, the dye filament began to break up and mix with the surrounding water. This type of flow is called *turbulent flow,* because the fluid stream lines move in highly irregular eddies. Reynolds identified a critical velocity above which laminar flow is not possible and turbulent flow dominates after a transitional stage of partial laminar and partial turbulent flow. Furthermore, the resistance to flow is governed by different laws for the two types of flow. Frictional resistance in laminar flow is less dependent on velocity than is that in turbulent flow. The breakup of laminar flow into turbulent flow depends on the velocity of the fluid, the viscosity of the fluid (and thus its temperature), and the diameter of the tube. That is, there is a critical Reynolds number (VD/v) that marks the transition from laminar to turbulent flow.

The impact of this discovery on research into the frictional resistance of ship models and plank tests is that all experiments must be made in the turbulent flow regime because the Reynolds numbers of the ships that they represent are well beyond the critical values. If laminar flow is present during ship model tests or plank tests, the expansion of the test results to full scale will not be correct.

Although the Froude empirical formulas for determining frictional resistance were used successfully for many years for practical results on both models and ships, there are a number of theoretical objections to them (which are avoided by the use of more modern friction formulations). The principal objections are:

1. The formula is not in agreement with the results of the dimensional analysis that are summarized by equation (23), which shows that frictional resistance should be a function of the Reynolds number of the model or ship.

2. Some of the tests of the small, smooth planks involved laminar flow, so that the calculated frictional resistance of small models is likely to be too low.

3. The f coefficients are supposed to be for a smooth hull when applied to models, and for ship hulls that are typical of newly painted steel. Assessment of the effects of surface quality of ship hulls has shown that the Froude's friction values for ships are too high. Modern formulations predict the frictional resistance of a smooth ship hull and correct separately for surface roughness and type of coating.

4. The plank tests that were analyzed to determine the friction factors utilized planks that ranged in length from two feet to a maximum of fifty feet long. Coefficients for ship lengths above 50 feet up to 1,200 feet are all determined by extrapolation from these limited tests.

Modern Friction Formulations
Several friction formulations have been developed to replace the Froude plank test formula. These modern formulations overcome the objections to the Froude formula in the following ways:

1. They express C_F as a function of R_n by fitting equations to plank and pontoon measurements plotted against R_n.

2. Test measurements suspected to involve laminar flow have been eliminated. When necessary for measurements at low Reynolds numbers, turbulence is artificially stimulated in the tests. Thus the formulations are for fully turbulent flow.

3. The surface condition of all planks representative of both models and ships is smooth. No roughened surfaces are included. Thus the formulations are intended to represent smooth ship hulls. Since real hulls are not so smooth, all resistance calculations involving modern friction formulations require that an additional allowance be added to achieve proper correlation with actual ship conditions.

4. Many plank tests have been done since the time of the Froude tests, extending the lengths and the Reynolds numbers of the planks well beyond the limits of the Froude tests. In spite of this extension of the range of test values, experimental data exist at Reynolds numbers up to about 4×10^8 at the highest, while actual ship Reynolds numbers can be as high as 5×10^9.

There have been about a half-dozen smooth turbulent friction formulations proposed by researchers over the years that are consistent with the above conditions. Two that have been adopted by many towing tanks as part of their uniform procedure for frictional resistance determination will be mentioned here.

The 1947 ATTC Line. The American Towing Tank Conference, or ATTC, adopted a resolution in 1947 that recommended a friction formulation known as the Schoenherr mean line. This formulation was published in 1932 by Karl E. Schoenherr, who collected a large body of reliable plank test data, added some of his own, plotted the resulting C_F values against R_n, and fitted a curve to the data. The formula is

$$\frac{0.242}{\sqrt{C_F}} = \log_{10}(R_n C_F) \tag{29}$$

Schoenherr noted a wide scatter of points at Reynolds numbers less than 2×10^6 and deduced that they were evidence that many of the planks tested at these low Reynolds numbers experienced laminar or transitional flow. The tests involving any amount of laminar flow were not allowed to influence the mean line, which is intended to represent fully turbulent flow on smooth surfaces. The C_F value for a given R_n cannot be calculated directly from equation (29), so evaluation of this friction formulation requires either an iterative solution or the use of tabulated values. Appendix B contains a table of values of the 1947 ATTC line. After many years of experience using this line for ship resistance predictions, hydrodynamicists have pointed out that its slope at low Reynolds numbers is not steep enough, resulting in the need for illogically low (sometimes negative) correlation allowances.

The ITTC 1957 Line. The most widely used formulation today was adopted by the International Towing Tank Conference (ITTC) in 1957. The equation of this line is

$$C_F = \frac{0.075}{(\log_{10} R_n - 2)^2} \tag{30}$$

This formulation is generally agreed to be superior to the Schoenherr mean line, especially at the lower Reynolds numbers. At the high ship scale Reynolds numbers the two formulations are virtually identical. The ITTC 1957 line has the computational advantage that it can be solved directly for C_F at any given R_n. To emphasize that the purpose of this formulation is to accurately expand the results of model test measurements to the corresponding ship, and not to represent the frictional resistance of specific plane surfaces, the conference officially calls it the *ITTC 1957 model-ship correlation line.*

WAVE-MAKING RESISTANCE

The system of waves produced on the surface of a body of water by a moving object is a familiar sight to everyone, whether or not they are students of ship hydrodynamics. A duck swimming on a pond, a canoe or rowboat, even a toy boat or a person's hand or finger moving along the surface of the water in a basin or bathtub will leave behind it a series of waves that is strikingly similar in form to the wave train caused by a ship moving through calm water at sea. The causes of such waves and the effect they have on ship resistance are examined in this section.

Pressure Distribution

Waves are created around a ship moving through the water because of variations in the water pressure at different points on the ship's hull. Imagine a simplified hull form such as that shown in Figure 8-1 moving toward the right at a constant velocity through an ideal (nonviscous) fluid. If the surface of the fluid is imagined to be rigid and unyielding, the pressure distribution on the hull would be as shown in the figure, that is, high-pressure peaks around the bow and stern and a low-pressure region amidships. In this idealized case, the forces on the hull due to the high pressure there would have aftward-acting components that would cause resistance to the motion, while the symmetrical distribution of pressure forces at the stern would have forward-acting components that would exactly cancel out the resisting forces at the bow. The net resisting force would then be zero. Although such an idealized situation cannot be reproduced in real fluids or with real bodies, the pressure distribution just described provides a clue to the formation of waves around a real ship-shaped body moving through a real fluid.

What would happen if the rigid surface constraint in the above imaginary experiment were replaced with the real sea surface condition of constant atmospheric pressure? It will be apparent that under these conditions pressure variations will be replaced by variations in water level or elevation. In high-pressure regions a wave crest will appear, and low-pressure regions will produce wave troughs. The pressure distribution is then altered by the very wave pattern that it creates, and in a real fluid the viscous boundary layer and separated flow cause further interactions that change the pressure distribution. In a real case, therefore, the forces caused by hull pressures are not balanced fore and aft, and

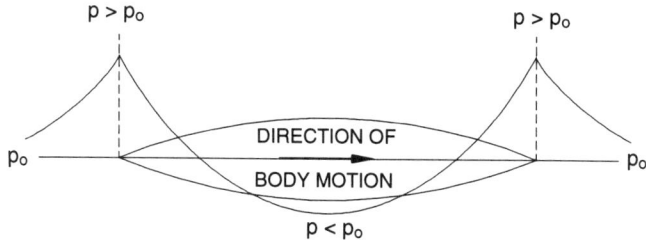

Figure 8-1. Pressure distribution, total immersion in ideal fluid.

the result is that the net aftward-acting components of the very complex pressure distribution around a ship hull cause what we know as wave-making resistance.

The simple pressure distribution model described above has another characteristic that pertains also to the real ship wave system. So long as the speed remains constant, the pressure distribution also remains unchanged, and the wave system moves with the ship. Ship-produced waves are, of course, left behind as the ship progresses, but new waves are continuously being generated by this moving pressure distribution.

The Ship Wave System

Accurate determination of the resistance caused by ship-generated waves still requires model experimentation, but our knowledge of the nature and form of a ship wave train has evolved in large part from theoretical studies. The classic investigation of the characteristics of a wave train was done by Lord Kelvin, who applied hydrodynamic theory to calculate the waves caused by a solitary pressure point moving at constant velocity at the surface of a fluid. He showed (see Figure 8-2) that the system of waves generated by the moving disturbance consists of waves of two geometrical configurations:

1. A series of *transverse waves* whose crest lines are perpendicular to the direction of motion of the point.
2. A series of *diverging waves* whose crest lines are at an angle to the direction of motion of the point.

The theory predicts that the wave length, or distance between crests of successive transverse waves, depends on the speed of the moving pressure point as follows:

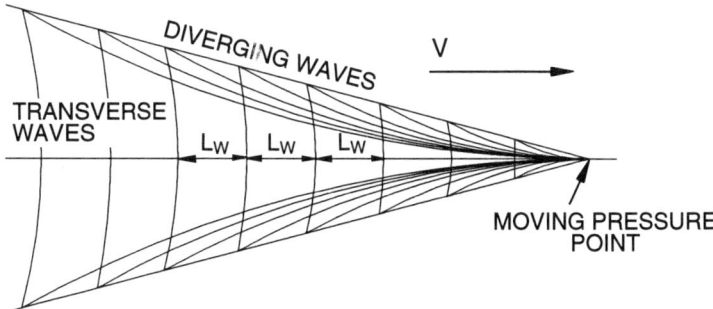

Figure 8-2. Kelvin wave train.

$$L_w = \frac{2\pi v^2}{g} \qquad (31)$$

where L_w = wave length, ft or m
  v = speed of moving pressure point, ft/sec or m/sec
  g = acceleration of gravity, ft/sec² or m/sec²

The resemblance of this equation to Froude's corresponding speed relationship for model testing is not coincidental. Equation (31) solved for the Froude number based on wave length is:

$$\frac{v}{\sqrt{gL_w}} = \frac{1}{\sqrt{2\pi}} = \text{constant} \qquad (32)$$

which shows that the speed of a disturbance is in constant proportion to the square root of the length of the waves it produces. If the moving disturbance is a ship, at a given Froude number based on ship length ($V/\sqrt{gL_S}$) there will be a constant relationship between the length of the ship and that of the waves, regardless of the absolute ship (or model) length. Thus the wave theory verifies Froude's observation that at his corresponding speeds the model wave train will be geometrically similar to the ship wave train.

The wave system produced by a moving ship is much more complex than the single pressure point system. In fact, it is complex enough that there is as yet no theoretical mathematical model to represent it with sufficient realism to calculate wave-making resistance more accurately than can be determined by expanding model test measurements. The ship hull produces many wave trains similar to the Kelvin wave train; in principle, every position along the waterline at which the flow changes direction produces another wave train. In spite of its complexity, however, an actual ship wave system bears remarkably close resemblance to a relatively simple representation—a strong Kelvinlike wave train generated by the bow and beginning with a dominant bow wave crest, plus a second and weaker wave train generated by the stern. A diagrammatic representation of a "two-train" ship wave system is shown in Figure 8-3. The lines depicting the wave trains represent wave crests of the transverse and diverging waves. Crests of the diverging waves are rather short in extent, and as each new one is created, the remains of previous ones move outward from the ship's path, leaving behind the ship an ever-widening chevronlike wave pattern that eventually degrades again into the undisturbed water surface.

Interference of Bow and Stern Wave Trains

The two-train model just described is sufficient to explain how wave-making resistance is affected by certain interactions between the bow and stern wave trains that are known as interference effects. Both wave trains advance with the ship,

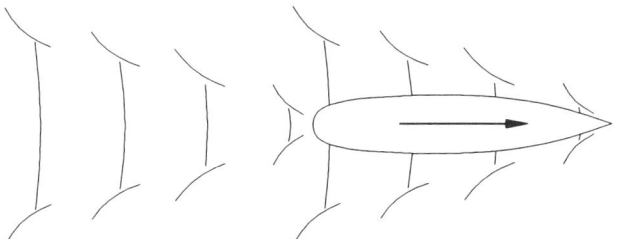

Figure 8-3. Bow and stern wave trains, showing transverse and diverging crests.

and at any given speed both have the same wavelength as determined from equation (31). The first wave crest of each of the wave trains is at an essentially fixed position relative to the ship, but the second and succeeding crests are at ever-increasing distances from the first as the speed of the ship increases. It follows, therefore, that as a ship changes speed the relative positions of the two systems will change. At speeds that cause the crests of the two systems to coincide, the waves left behind the ship will be augmented in size, and wave-making resistance will be extra high. This condition is said to produce a "hump" in the curve of residuary resistance coefficient (C_R). At speeds that cause a trough of the bow system to coincide with a crest of the stern system, the waves left behind the ship will be reduced in size, and wave-making resistance will be lower than its general trend with speed. The C_R curve will then have a "hollow." Since the distance between the initial crests of the two wave trains is determined by the ship length, while the wavelength of each depends on the square of the ship speed, it follows that the humps and hollows in the C_R curve will occur at specific values of the ratio V^2/L, or of the Froude number ($V/\sqrt{gL}$) for any given ship. The typical curve of C_R plotted against $V/\sqrt{gL}$ shown in Figure 8-4 illustrates this.

The Froude numbers at which humps and hollows occur have been calculated for a number of simplified hull forms using a more realistic model than the two-train system, but the values thus calculated cannot be applied to all hull forms. Ship designers recognize the importance of choosing a hull form such that the design ship speed occurs in a hollow of the C_R curve. Once the ship length and speed are chosen, the most effective ways for the designer to assure that this is achieved are by judicious choice of the prismatic coefficient (C_P) and of the length of parallel middle body.

EDDY-MAKING RESISTANCE

When the water flowing around a ship hull follows a path that exactly conforms to the shape of the hull, the resistance of the hull derives entirely from friction and wave making. When the streamline flow breaks away from the hull, the downstream void that is created becomes filled with eddies and the resistance is

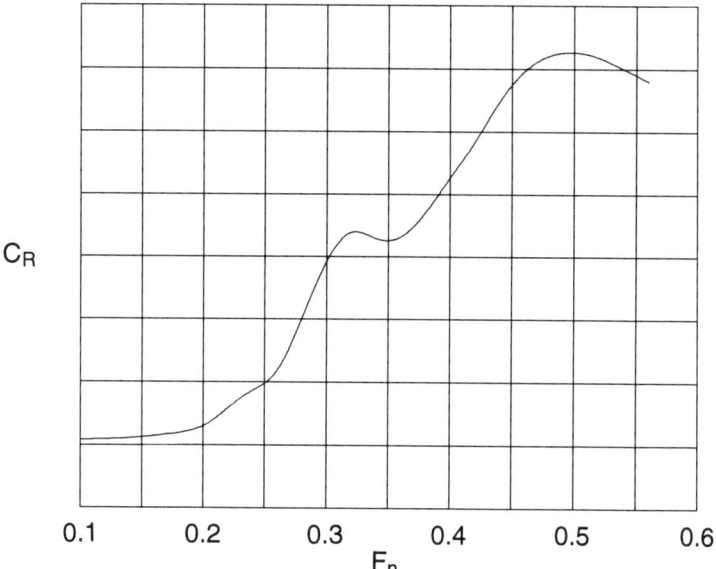

Figure 8-4. Humps and hollows in the C_R curve.

increased because of the energy drain attendant to the creation of the eddies. The amount of eddy-making resistance that a ship or model experiences at any one time is never actually calculated when model tests are expanded to full scale, since eddy-making resistance is assumed to be part of the residuary resistance. Although it is out of sight, it should never be out of mind for the ship designer, because eddy-making resistance can be substantial.

The most obvious points at which the flow will break away from the hull are where there are abrupt discontinuities in the hull form, such as at the after edge of propeller and rudder posts and behind hull appendages like the rudder, bilge keels, bossings, and struts. Since some appendages like these are necessary, a certain amount of eddy formation is unavoidable. Minimizing the additional resistance caused by appendages is accomplished by careful streamlining and fairing into the hull of any such protrusions. Well-designed bilge keels, for example, are positioned so as to follow the flow lines on the hull as much as possible. The flow lines are determined by special model tests or flow channel studies using short lengths of yarn tacked to the hull or by injecting through tiny holes in the hull streams of chemicals that discolor the paint on the model and thus mark the flow lines on the hull.

Appendage resistance is the added resistance caused by the appendages on a particular ship. It is not entirely eddy-making resistance, because each appendage has a wetted surface area and a consequent frictional resistance as well. It would seem that the resistances of the appendages would not have to be dealt with separately when determining ship resistance from a model test so long as the model is outfitted with the same appendages as the ship, but in many model tests this is not the case. The condition for correct scaling of the forces on a model appendage to full scale is that the flow patterns around the model appendage be geometrically similar to those around the ship appendage. This is impossible to achieve because the boundary layers of model and ship are not in correct scale to one another in a model test. In addition, the Reynolds numbers of the model appendages (based on appendage dimensions rather than ship dimensions) cannot be those of the corresponding ship appendages. For these reasons and others, the Froude procedure for expanding model resistance leads to inaccuracies when the model is fitted with appendages. This is particularly a problem with small models and with those that would be expected to have a large appendage resistance. In such cases, models are tested in what is called the *bare hull condition,* meaning without appendages. Approximate full-scale appendage resistances are then added to the bare hull ship resistance. Expressed as a percent of bare hull resistance, appendages may add as little as 2 percent and as much as 20 percent or more, depending on the number and type of appendages.

Eddies may also form in situations that do not involve the abrupt hull discontinuities that are associated with appendages. The viscous boundary layer that surrounds the ship hull increases steadily in thickness from the bow to the stern. As it does, the water within it moves ever more slowly with respect to the hull. As the energy of the fluid particles gets less and less, a point may be reached at which the fluid velocity becomes zero or negative, the flow separates from the surface, and chaotic flow patterns in the form of eddies fill the space aft of the separation point. Separation is more prevalent in full than in fine hulls because the more rapid change in slope of the full-formed stern aggravates the conditions that lead to separation. Separation resistance is not calculated as a distinct component of resistance. It is included in the residuary resistance when expanding model test measurements to full scale.

AIR RESISTANCE

Determining the resistance of a surface ship is a more complex problem than that of most other vehicles involving the flow of fluids because the surface ship moves through not one, but two fluids—water and air. We have already noted how that fact causes the unique phenomenon of wave making at the interface between these two fluids of vastly different densities. It remains to assess the influence on ship resistance of the air through which the upper works of a surface ship must proceed. When determining the power required to propel a ship at its designed speed, the naval architect estimates the still air resistance and adds it to the hydrodynamic resistance. Allowance for the effect of air in motion, or wind resistance, is normally not explicitly included in ship resistance calculations, nor is the secondary effect of wind on ship resistance, namely added resistance caused by ocean waves that are generated by wind. These are taken into account when powering calculations are made by adding a service power allowance, as shown in Chapter 9.

In moving through the air, the above-water body of a ship is subjected to a resistance that has viscous (frictional) and eddy-making components, so that one might expect that the calculation of still air resistance would involve the same kind of scaling problems and separate determination of its components that are required to determine the water resistance of the underwater body. As a practical matter, however, this amount of elaboration is not necessary. The phenomenon of eddy-making so dominates the still air resistance that the frictional resistance component can be considered negligible, and it may be ignored in calculations of still air resistance.

Resistance or drag forces resulting from eddy making have been shown to be proportional to the square of the velocity of the fluid, or of the body through the fluid if the fluid is at rest. Therefore still air resistance may be expressed in terms of a resistance (or drag) coefficient as follows:

$$R_{air} = C_{air}(\tfrac{1}{2} \rho_a A_{pt} V^2) \tag{33}$$

where R_{air} = still air resistance (lb$_f$ or newtons)
C_{air} = still air resistance coefficient
ρ_a = mass density of the air (lb$_f$-sec^2/ft^4 or kg/m^3)
  = 0.002377 lb$_f$-sec^2/ft^4 at 59°F = 1.226 kg/m^3 at 15°C
A_{pt} = projected transverse area of above-water portion of the ship (ft^2 or m^2). If this area is not known, it may be approximated for a ship with normal size deckhouse and superstructure by $A_{pt} = B^2/2$, where B is the ship's beam.
V = ship speed (ft/sec or m/sec)

The coefficient C_{air} must be evaluated experimentally. It depends on the shape of the hull above the waterline and on the deckhouse and above-deck equipment and gear. Numerous experiments have been done to determine values of this coefficient. They have included tests of models of the upper works of ships in wind tunnels and inverted in water, and tests of simpler two-dimensional geometries in the form of flat plates dropped through the air. Some full-scale measurements have also been made. Typical ranges of values of C_{air} published in the literature are summarized in Table 8-1.

SHIP RESISTANCE 261

Table 8-1. Still Air Resistance Coefficients

Ship Type	C_{air} Range
General cargo	0.60 to 0.85
Tanker	0.75 to 1.05
Container	0.60 to 0.75
Passenger	0.65 to 1.10
Combatant	0.40 to 0.80

There is considerable spread in the above values because of the large variation in the amount and type of above-deck cargo gear and in the extent and arrangement of deckhouses on the ships tested. For purposes of estimating still air resistance when making ship powering calculations, average values are acceptable because still air resistance is very small in comparison to hydrodynamic resistance (2 to 4 percent on average).

It should be noted that the still air resistance coefficient defined above is based on the density of air and the above-water transverse projected area of the ship. It is not directly additive to the frictional and residuary resistance coefficients C_{FS} and C_R, which are based on the density of water and the underwater wetted surface area of the ship. It is common practice in standardized procedures for ship resistance prediction to define a different still air resistance coefficient, (C_{AA}), which is consistent with and directly additive to the water resistance coefficients. It is defined as follows:

$$C_{AA} = \frac{R_{air}}{\frac{1}{2}\rho_w S V^2} \qquad (34)$$

where ρ_w = mass density of sea water
= 1.9905 lb$_f$-sec^2/ft^4 at 59°F
= 1025.9 kg/m^3 at 15°C
S = wetted surface of ship (ft^2 or m^2)
V = ship speed (ft/s or m/s)

Combining equations (33) and (34), and assuming the standard temperature for ship calculations (59°F or 15°C), the two coefficients are related as follows:

$$C_{AA} = C_{air} \left(\frac{\rho_a}{\rho_w}\right)\left(\frac{A_{pt}}{S}\right) \qquad (35)$$

$$= 0.001194 \left(\frac{A_{pt}}{S}\right) C_{air}$$

Wind Resistance

Determining the resistance caused by winds is much more difficult than the still air calculation because of several complicating effects:

1. The wind may come from any direction with respect to the ship heading, requiring that relative wind velocity and direction be determined. Wind tunnel tests must also be done at all angles, from head winds to following winds.
2. Wind velocity at sea has a large gradient near the sea surface. This means that different parts of the ship's upper works are exposed to winds of varying velocity depending on their distance above the sea, from very low velocity near the sea surface to much higher velocity aloft. The velocity profile is further complicated by the waves that the wind generates. It is difficult to model a realistic sea-surface wind velocity profile in a wind tunnel.
3. The characteristic area projected onto a plane perpendicular to the relative wind direction changes with wind direction. In beam winds, for example, the area is that of the above-water profile of the ship.
4. The ship heels and yaws under the influence of winds at oblique angles to the direction of motion of the ship, further modifying the wind forces acting on the hull.

Calculations of wind forces are not treated here, as they do not affect the determination of the installed power. Estimating wind forces accurately is necessary in the analysis of ship standardization trials. The special case of the resistance to head winds (directly on the bow) can be estimated using equation (33) by substituting for ship speed (V) the relative wind velocity,

$$V_R = V_S + V_W \qquad (36)$$

where V_R = relative wind velocity, head winds only
V_S = ship velocity
V_W = absolute wind velocity

The air resistance coefficient (C_{air}) should be reduced when head wind resistance is calculated using a relative wind velocity, because the coefficients were determined for ship motion through still air with no velocity gradient, but the wind velocity (V_W) has a velocity gradient which reduces the resultant wind force significantly. Typical reductions of the coefficient range from 25 percent for ships with relatively low above-water silhouettes such as loaded tankers to 40 percent or more for those with high silhouettes such as passenger ships and ferries.

THE CORRELATION ALLOWANCE

Since Froude's time, a great deal of research has been done, model testing techniques have been improved, and much insight has been gained into the phenomena that cause ship resistance. In spite of all of those advances, and in spite of the fact that the Froude method of separating resistance into only two components is known to be erroneous, modern methods of extrapolating model measurements to ship resistance predictions are still based on the Froude procedure because we do not have an exact analytical representation of the complex physical processes that are involved. Although the several extrapolation procedures now in use all have this common basis, they are not exact, and they do not all predict the same ship resistance when they are applied to a given set of model test data. Therefore, all extrapolation procedures require that an adjustment be made to achieve correct correlation of model measurements with ship predictions. The adjustment is made in the form of a *correlation allowance* (C_A), which is expressed like a third component of resistance, but only in the ship scale part of the expansion. Thus, the two-component equation for model resistance, expressed by equation (22):

$$C_{TM} = C_{FM} + C_{RM} \tag{22}$$

is written as a three-component equation for the ship, as shown before in equation (25),

$$C_{TS} = C_{FS} + C_{RS} + C_A \tag{25}$$

Determining appropriate values of the correlation allowance requires careful measurement of the power consumed during ship standardization trials and back calculation of C_{TS} from those full-scale measurements to compare them to the values determined from the model results. Model-testing laboratories maintain records of such analyses of ships whose powering calculations were determined using their standard methods of extrapolation, on the basis of which they derive correlation allowances appropriate to their particular extrapolation procedures. The allowances thus determined are then applied to model test analyses of new designs.

The magnitude of C_A depends not only on the extrapolation procedure used in expanding model test measurements, but also on the scaling errors inherent in using the Froude method and on the roughness of the ship hull. It is impossible to accurately scale down ship hull roughness to model scale, so the model is made with a smooth hull and the results expanded to ship scale represent a ship with a smooth hull. The actual roughness of a ship hull is influenced by structural irregularities (welds, openings in the hull for overboard discharges, etc.) and the type of paint applied to the hull, even for a newly built, freshly painted hull (that is, for a hull in trial trip condition). As a ship increases in age, the roughness of its hull increases because of corrosion and fouling; the design allowance for these service conditions is not part of the correlation allowance, but part of a service power allowance, as shown in a later section.

As might be expected, there is considerable variation in the values of the correlation allowance used by different model-testing facilities and recommended for different extrapolation procedures. Naval architects at each laboratory must determine their own best estimates and be certain that all tests are analyzed with complete uniformity of methods. In general, the collected data show that correlation allowances decrease with increasing ship size, to such an extent that for very large ships a negative allowance is often appropriate. For guidance, the following values are intended for use in conjunction with the ITTC 1957 friction coefficients.

Table 8-2. Correlation Allowance With ITTC Line*

| Ship length on waterline | | Correlation allowance |
meters	feet	C_A
50-150	160-490	$+0.40 \times 10^{-3}$
150-210	490-690	$+0.20 \times 10^{-3}$
210-260	690-850	$+0.10 \times 10^{-3}$
260-300	850-980	0
300-350	980-1,150	-0.10×10^{-3}
350-450	1,150-1,480	-0.25×10^{-3}

*From Keller, J. A., 1973. *International Shipbuilding Progress*, vol. 20

CALCULATION OF EFFECTIVE POWER

Effective power has already been defined as the power needed to tow a ship through perfectly smooth water at a given speed, or the power required to overcome the total ship resistance at that speed. It is calculated by multiplying resistance by speed:

$$P_E = R_T V \qquad (37)$$

where P_E = effective power (ft-lb/sec or watts)
　　　　R_T = total ship resistance (lb$_f$ or newtons)
　　　　V = ship speed (ft/sec or m/sec)

The units of power determined by using equation (37) are not practical for expressing ship powers. The common unit in the U.S. customary measurement system is the *horsepower,* sometimes called the English horsepower, which is defined as 550 foot-pounds per second. The effective power is then called *effective horsepower (EHP).* Thus

$$\text{EHP} = \frac{R_T V}{550} \tag{38}$$

when R is in pounds and V is in feet per second. The equivalent expression if V is in knots is

$$\text{EHP} = \frac{R_T V_K}{325.6} \tag{39}$$

In SI units, ship power is more conveniently expressed in kilowatts (kW) than in watts. The correct equation is the same as equation (37), but with R_T expressed in kilonewtons (kN).

The sample calculation that follows illustrates how a resistance force measured on a ship model in a towing basin is expanded to the corresponding full-scale ship resistance and to the effective power of the ship at its corresponding speed. The procedure, which was described previously, is the Froude method for model test expansion in modern form, using nondimensional resistance coefficients and a modern model-ship correlation line.

Example 8-1

A resistance test is conducted on a model of a containership whose characteristics are:

$$L_{WL} = 780.0 \text{ ft}$$
$$B = 101.8 \text{ ft}$$
$$T = 33.7 \text{ ft}$$
$$C_B = 0.580$$
$$S = 93,590 \text{ ft}^2$$

The model has a length on waterline of 21.97 feet. It is tested in the bare hull (no appendages) condition in 66°F fresh water. When towed at the model speed corresponding to the design ship speed of 24.5 knots, the towing force is measured at 11.54 pounds. Determine the bare hull effective horsepower of the ship in standard seawater (59°F) at design speed. Calculate also the power to be added to overcome appendage resistance (estimated to be 5 percent of bare hull resistance) and still air resistance if $A_{pt} = 5,400 \text{ ft}^2$.

Preliminary calculations are made to determine the scale ratio, characteristics, and speed of the model. The scale ratio is defined in equation (7):

$$\lambda = \frac{L_S}{L_M} = \frac{780.0}{21.97} = 35.50$$

$$B_M = \frac{B_S}{\lambda} = \frac{101.8}{35.50} = 2.87 \text{ ft}$$

$$T_M = \frac{T_S}{\lambda} = \frac{33.7}{35.50} = 0.95 \text{ ft}$$

$$S_M = \frac{S_S}{\lambda^2} = \frac{93{,}590}{(35.50)^2} = 74.55 \text{ ft}^2$$

$$C_{BM} = C_{BS} = 0.580$$

Speeds of the ship and model must be determined in feet per second for proper calculation of the dimensionless ratios when data are in U.S. customary units. The conversion from knots is

$$V_{fps} = V_K \left(\frac{6{,}080 \text{ ft/naut. mi.}}{3{,}600 \text{ sec/hr}}\right)$$

$$V = V_K \times 1.689$$

Thus the ship speed is

$$V_S = 24.5 \times 1.689 = 41.38 \text{ ft/sec}$$

As shown in equation (9), the model speed for the same Froude number as that of the ship is

$$V_M = \frac{V_S}{\sqrt{\lambda}} = \frac{41.38}{\sqrt{35.50}} = 6.95 \text{ ft/sec}$$

As an optional check, it may be verified that the ship and model Froude numbers, $F_n = V/\sqrt{gL}$, are equal:

$$F_{nS} = \frac{41.38}{\sqrt{32.2 \times 780.0}} = 0.261$$

$$F_{nM} = \frac{6.95}{\sqrt{32.2 \times 21.97}} = 0.261$$

To simplify the calculation of the resistance coefficients, the Reynolds numbers and values of $\frac{1}{2}\rho SV^2$ can be calculated for both model and ship. Tables of values of mass density and kinematic viscosity of both fresh and seawater are given in Appendix B, from which the following values are obtained:

model, fresh water at 66°F
$\rho = 1.9371 \text{ lb-sec}^2/\text{ft}^4$
$\nu = 1.1103 \times 10^{-5} \text{ ft}^2/\text{sec}$

ship, seawater at 59°F
$\rho = 1.9905$ lb-sec^2/ft^4
$\nu = 1.2791 \times 10^{-5}$ ft^2/sec

Thus we have

$$(\tfrac{1}{2}\rho SV^2)_{model} = (0.5)(1.9371)(74.55)(6.95)^2 = 3{,}488 \text{ lb}_f$$

$$(\tfrac{1}{2}\rho SV^2)_{ship} = (0.5)(1.9905)(93{,}590)(41.38)^2 = 1.526 \times 10^8 \text{ lb}_f$$

$$R_{nM} = \left(\frac{VL}{\nu}\right)_M = \frac{6.95 \times 21.97}{1.1103 \times 10^{-5}} = 1.375 \times 10^7$$

$$R_{nS} = \left(\frac{VL}{\nu}\right)_S = \frac{41.38 \times 780.0}{1.2791 \times 10^{-5}} = 2.523 \times 10^9$$

Following the steps outlined previously, the resistance coefficients are calculated.

Model total resistance coefficient:

$$C_{TM} = \frac{R_{TM}}{(\tfrac{1}{2}\rho SV^2)_M} = \frac{11.54}{3{,}488} = 3.308 \times 10^{-3}$$

Model frictional resistance coefficient, using the ITTC 1957 model-ship correlation line, equation (29):

$$C_{FM} = \frac{0.075}{(\log_{10} R_{nM} - 2)^2} = \frac{0.075}{[\log_{10}(1.375 \times 10^7) - 2]^2}$$
$$= 2.841 \times 10^{-3}$$

The residuary resistance coefficient of the model is equal to that of the ship in accordance with Froude's hypothesis:

$$C_R = C_{TM} - C_{FM} = (3.308 - 2.841) \times 10^{-3} = 0.467 \times 10^{-3}$$

Ship frictional resistance coefficient using the same friction line:

$$C_{FS} = \frac{0.075}{[\log_{10}(2.523 \times 10^9) - 2]^2} = 1.369 \times 10^{-3}$$

The ship total resistance coefficient must include a correlation allowance. Choosing $C_A = 0.10 \times 10^{-3}$ from Table 8-2 gives:

$$C_{TS} = C_{FS} + C_R + C_A = (1.369 + 0.467 + 0.10) \times 10^{-3} = 1.936 \times 10^{-3}$$

The bare hull ship resistance is

$$R_{TS} = C_{TS}(\tfrac{1}{2}\rho SV^2)_S = (1.935 \times 10^{-3})(1.526 \times 10^8)$$
$$= 295{,}400 \text{ lb}_f$$

The bare hull effective horsepower is

$$\text{EHP} = \frac{R_{TS}V_S}{550} = \frac{295{,}400 \times 41.38}{550} = 22{,}220 \text{ hp}$$

The estimated appendage resistance of 5 percent of bare hull resistance is $0.05 \times 295{,}300 = 14{,}770$ lb$_f$. For still air resistance, the estimated coefficient from Table 8-1 for a containership is $C_{air} = 0.70$. With the projected transverse area of 5,400 ft^2, the still air resistance is, from equation (33):

$$R_{air} = C_{air}(\tfrac{1}{2}\rho_a A_{pt}V^2)$$
$$= 0.70 \times 0.5 \times 0.002377 \times 5{,}400 \times (41.38)^2$$
$$= 7{,}690 \text{ lb}_f$$

Total resistance and effective horsepower including appendage and still air resistances are:

$$R_{TS} = 295{,}400 + 14{,}770 + 7{,}690 = 317{,}860 \text{ lb}_f$$
$$\text{EHP} = \frac{317{,}860 \times 41.38}{550} = 23{,}910 \text{ hp}$$

PROBLEMS

1. A 40,000-ton ship is 800 ft long. Its resistance test model is 20 ft long. Based on the principles of geometrical and dynamical similitude, determine the following, assuming 59°F seawater for the ship, and 68°F fresh water for the model.
   (a) The required weight of the model in pounds.
   (b) The model speed (ft/sec) to represent the ship at 25 knots.
   (c) The Froude numbers of the model and ship at their respective speeds.
   (d) The Reynolds numbers of the model and ship at their respective speeds.
   (e) The speed the model would have to be tested at if equality of Reynolds numbers, rather than Froude numbers, governed the model test.

SHIP RESISTANCE

2. Calculate the frictional resistances for the following ship at 20 knots and for its model at the corresponding speed, using the Froude, ATTC, and ITTC formulations, and compare the results. The ship is 600 ft long, in seawater at 59°F. The model is 15 ft long, in fresh water at 59°F. Wetted surface of the ship is 49,200 square feet. $C_A = 0.0004$.

3. A 20-ft-long model of a ship 800 ft long is tested for resistance in fresh water at 70°F, at a speed to correspond to the ship at 24 knots.

   (a) At what speed should the model be towed?
   (b) If the model resistance at the required speed is 13.5 pounds, determine how many pounds represent frictional resistance and how many represent residuary resistance.

   Base your estimates in part (b) on the ITTC model-ship correlation line. The wetted surface of the model is 65.0 square feet.

4. A model 17.0 ft long has a total resistance of 14.5 pounds when towed in 72°F fresh water at 360 ft/min. The wetted surface of the model is 46.00 ft². Calculate the effective horsepower (EHP) of the ship, which is 425 ft long, at the corresponding speed in 59°F seawater. Include still air and appendage resistances, using the following assumptions:

   ITTC model-ship correlation line
   Correlation allowance = 0.0004
   $C_{air} = 0.70$
   $A_{pt} = 1,600$ ft²
   Appendage allowance = 4 percent of bare hull resistance

5. A 20-ft-long model of the containership whose characteristics are given below is towed at the speed corresponding to the design speed of the ship. The tank water is fresh, at 70°F, and the measured model resistance is 14.75 pounds. The ship characteristics are:

   Length = 660.0 ft
   Beam = 90.7 ft
   Draft = 34.5 ft
   Wetted surface = 89,300 ft²
   Displacement = 45,140 tons
   Design speed = 20 knots

Determine, for the ship in standard seawater, the EHP of the ship, including appendage and still air resistances. Use the following assumptions:

ITTC model-ship correlation line
Correlation allowance = 0.00025
$C_{air} = 0.75$
$A_{pt} = 5{,}200 \text{ ft}^2$
Appendage allowance = 5 percent of bare hull resistance.

CHAPTER NINE

SHIP PROPULSION

In order to move a ship through the water, a propulsive force must be supplied that will oppose and overcome the ship resistance that was described and calculated in the last chapter. The propulsive force, called the *thrust,* is produced by some kind of device that we shall call here a "thruster," driven by a prime mover capable of delivering the required power. In the broadest sense the prime mover/thruster combination includes men/oars and wind/sails as well as various types of engine/propeller combinations. Since in this book the focus is on the modern merchant ship, we shall examine only the most common kind of thruster for large ship propulsion, namely the screw propeller. Other devices, such as vertical axis propellers, water jets, air screws, and paddle wheels that have been used or proposed for ship propulsion are of interest only in a historical sense or in very special applications. None can match the screw propeller in overall performance for ship propulsion.

The subject of ship propulsion as it is addressed in this chapter is independent of the type of prime mover, or engine, installed in the ship. Choosing the best type of engine for a particular ship is one of the tasks in ship design, which is not the subject of this book. We examine here the hydrodynamic aspects of ship propulsion, that is, that part of ship propulsion which is affected by the flow of water around the stern of the ship and through the propeller. The decisions as to whether the propeller shaft is to be turned by a steam turbine, diesel engine, or gas turbine, or whether heat is to be generated by burning fuel oil, coal, or by a nuclear reactor are supposed here to have been made by the design team of marine engineers and naval architects on the basis of sound engineering economics and mission requirements.

POWERS AND EFFICIENCIES

Converting the power developed in the engine, whatever its type, to the useful power output of the propeller cannot be done without some losses. There are, therefore, a number of defined powers associated with ship propulsion, depending on how they are calculated or where they are measured, as shown in the conceptual sketch, Figure 9-1. The total resistance force (R_T) acts on the ship moving at velocity V. The power needed to overcome R_T at velocity V is called the *effective power*, already defined in the last chapter. That is:

$$\text{Effective power} = P_E = R_T V \tag{1}$$

The propulsive force or thrust (T) is a measure of the useful output of the propeller, which is located in water moving at an average velocity called the *velocity of advance* (V_A). Thus the useful power output of the propeller is the *thrust power*, given by:

$$\text{Thrust power} = P_T = TV_A \tag{2}$$

As will be shown later, T must be greater than R_T, and V_A is usually less than V, therefore in general P_T is not equal to P_E. The ratio of the two is a measure of how effectively the shape of the stern has been designed to suit the propulsion arrangement. It is called the *hull efficiency* (η_H).

$$\text{Hull efficiency} = \eta_H = \frac{P_E}{P_T} = \frac{R_T V}{TV_A} \tag{3}$$

The power delivered to the propeller, called the *delivered power* (P_D), is the input to the propeller. It is given by:

$$\text{Delivered power} = P_D = 2\pi n Q_D \tag{4}$$

where n = revolutions per second of the shaft and propeller
       Q_D = torque delivered to the propeller

The efficiency of the propeller operating behind the ship is the ratio of its output (P_T) to its input (P_D). It is called the *behind efficiency* (η_B). It is important to specify that this efficiency is associated with the propeller operating behind the hull because it is not the same as the efficiency of the propeller alone. The behind efficiency is:

$$\text{Behind efficiency} = \eta_B = \frac{P_T}{P_D} = \frac{TV_A}{2\pi n Q_D} \tag{5}$$

SHIP PROPULSION

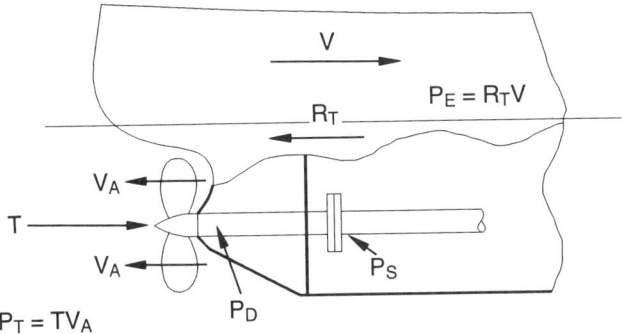

Figure 9-1. Definition sketch for powering.

If a test were made with the propeller running alone in a uniform flow of velocity V_A at n revolutions, it would produce thrust T and have the same output as that described above, but at a torque different from that required when it was behind the ship. The condition just described is called the *open-water condition,* and it is the standard condition for testing model propellers to evaluate their performance. The propeller efficiency in open water is then:

$$\text{Open water efficiency} = \eta_O = \frac{TV_A}{2\pi n Q_O} \qquad (6)$$

where Q_O is the torque in open water.

To compare the propeller performance in the two conditions, the ratio of the behind efficiency to the open water efficiency is defined. Although it is not an efficiency in the usual sense of the word, this ratio is known as the *relative rotative efficiency* (η_R).

$$\text{Relative rotative efficiency} = \eta_R = \frac{\eta_B}{\eta_O} \qquad (7)$$

It will be noted that η_R can also be expressed as the ratio of the torque in open water to the delivered torque in the behind condition, that is:

$$\eta_R = \frac{Q_O}{Q_D} \qquad (8)$$

None of the powers described above can be directly measured on a ship, but if we move into the ship to the propeller shaft inside the afterpeak bulkhead, the power can actually be measured by installing an instrument called a torsionmeter directly on the shaft. This power is called the *shaft power (P_S).* It is defined like the delivered power, except that the torque in question is the torque Q_S measured by the torsionmeter attached to the shaft. Therefore

274   APPLIED NAVAL ARCHITECTURE

$$\text{Shaft power} = P_S = 2\pi n Q_S \tag{9}$$

Since there are losses caused by friction in the stern tube bearing and in shaft bearings, the shaft power must be greater than the delivered power. The ratio of the two is the *shaft transmission efficiency* (η_S).

$$\text{Shaft transmission efficiency} = \eta_S = \frac{P_D}{P_S} \tag{10}$$

The overall efficiency of ship propulsion is known as the *propulsive efficiency* (η_P), or alternatively as the *propulsive coefficient (PC)*. It is the ratio of the effective power to the shaft power:

$$\text{Propulsive efficiency} = \eta_P = \frac{P_E}{P_S} \tag{11}$$

or, expressed in terms of the powers and efficiencies defined above,

$$\eta_P = \frac{P_E}{P_T} \times \frac{P_T}{P_D} \times \frac{P_D}{P_S}$$
$$= \eta_H \times \eta_B \times \eta_S = \eta_H \times \eta_O \times \eta_R \times \eta_S \tag{12}$$

The propulsive efficiency is thus seen to be the product of four efficiencies, namely the hull, open-water, relative rotative, and shaft efficiencies. The first three of these efficiencies depend on good hydrodynamic design of the propeller and the stern of the ship, while the fourth is a mechanical efficiency that is not associated with the propeller or how well it is matched to the ship. For evaluating the design of the propeller-stern combination it is appropriate, therefore, to separately identify an overall hydrodynamic efficiency of propulsion as the ratio of the effective power to the delivered power. This ratio is called the *quasi-propulsive efficiency* (η_D), or the *quasi-propulsive coefficient* (QPC).

$$\text{Quasi-propulsive efficiency} = \eta_D = \frac{P_E}{P_D} = \eta_H \eta_O \eta_R \tag{13}$$

In addition to the powers defined above, other names are used to specify the rated power output of the main propulsion engines used on ships. *Brake power (P_B)* is the power measured at the output shaft of a diesel engine or any type of internal combustion engine. *Indicated power (P_I)* is an internally measured power in engines that have cylinders, such as steam reciprocating engines and internal combustion engines. *Shaft power (P_S)* is the term used to rate steam turbine power plants. It is measured like the shaft power defined previously, but care must be taken to make a distinction between the shaft power at the tailshaft coupling or aftermost end of the line shafting and that at the reduction gear coupling which is at the forward end of the shaft. There are losses between the two which are accounted for by the shaft transmission efficiency.

THE SCREW PROPELLER

The screw propeller is the most common propulsive device for ships because experience has shown that a well-designed screw propeller is more efficient than any other device so far invented for ship propulsion. A screw propeller consists of several blades projecting from a hub at the end of the ship's propeller shaft. The general features of a screw propeller are illustrated in Figure 9-2. Since all blades are identical, only one blade is shown in the figure. Most ship propellers have three, four, or five blades, but in special designs blade counts from two to seven have been successfully employed.

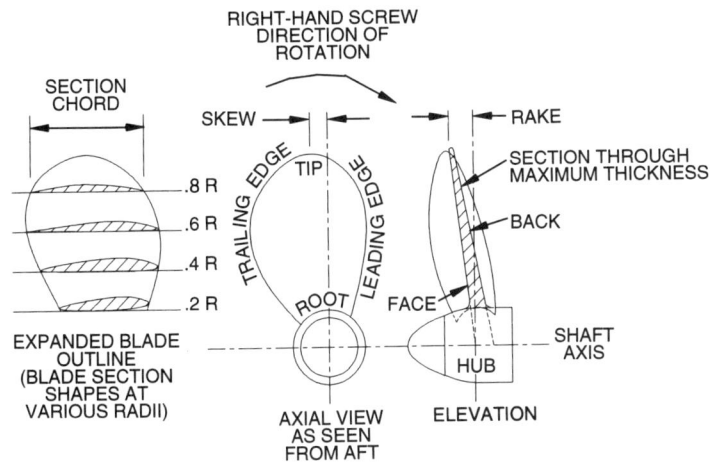

Figure 9-2. Screw propeller drawing.

Propeller Nomenclature

The following terms used to describe a screw propeller are illustrated in Figure 9-2. The blade extends from its *root,* where it is attached to the hub, to its *tip,* or outermost extremity. When it rotates, the blade edge cutting the water first is the *leading edge,* and the other edge is the *trailing edge.* The *face* of a propeller blade is the surface seen by an observer behind the propeller, that is, from aft. It is sometimes referred to as the *pressure face* or *driving face,* because when the propeller rotates, the pressure on the face increases while on the other blade surface, the *back,* the pressure decreases below that of the free stream of water surrounding the propeller. It is the pressure difference across the blade that provides the driving force of the propeller. To provide adequate clearance between the rotating blades and the ship hull, propeller blades are usually *raked* aft such that the blade tips are some distance aft of their roots. In the axial view, it can be seen that the blade tip is not aligned with a vertical line from the propeller centerline through the midpoint of the blade root, but is shifted from that line in a direction opposite to the direction of rotation. The distance of that shift is called the *skew*. Skew is often introduced to reduce unsteady or pulsating forces induced by the propeller as its blades pass in close proximity to the stern frame of the ship. A propeller that turns clockwise when viewed from aft while producing ahead ship motion is said to be a *right-handed* propeller, as shown in the figure. A *left-handed* propeller turns counterclockwise as seen from behind. On twin-screw ships, the paired propellers turn in opposite directions, the usual arrangement being *outboard-turning* propellers, in which the starboard propeller is right-handed and the port propeller is left-handed.

Propeller Blade Geometry

The characteristic twist of a propeller blade is a shape that is derived from the *helicoidal surface,* shown in Figure 9-3. A helicoidal surface of constant pitch is the surface generated by a straight-line generatrix (OA in the figure) rotating at a uniform angular velocity about an axis through one of its ends (axis OO′), while advancing at uniform rate along the same axis. While the line generates a helicoidal surface, the intersection of the line with any cylinder concentric with the axis is a spiral curve (ABC) called a *helix*. The *pitch* of a helicoidal surface is the distance its generatrix advances in one complete revolution. Since only one half revolution is shown in the figure, the height OO′ = CC′ = P/2, where P is the pitch.

A conceptual propeller blade is drawn in the figure to show its relationship to a helicoidal surface. If a propeller blade were made a constant thickness throughout, both the face and back of the blade could, in their simplest form, be true helicoidal surfaces. Actual propeller blades, however, are not so simple. Because blade sections are airfoil shaped, as described below, neither face nor back are true helicoidal surfaces. Nor is it necessary in a real propeller blade that the pitch of all parts of the blade be constant, as is true of the simple helicoi-

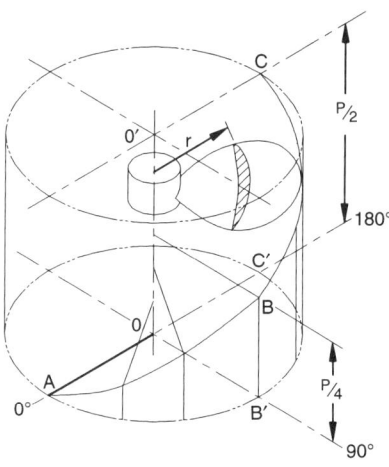

Figure 9-3. The helicoidal surface.

SHIP PROPULSION

dal surface shown. The pitch often varies radially, from the blade root to its tip, and variations in its thickness from leading edge to trailing edge produce surfaces that have circumferentially varying pitch as well.

Suppose that the face of the propeller blade shown in Figure 9-3 is a true helicoidal surface and that a cylinder of radius r concentric with the axis intersects the blade, producing the blade section shown shaded in the figure. The blade is then imagined to rotate one complete revolution, while advancing a distance equal to the pitch P as if it were a machine bolt thread being screwed into a tapped hole. If the intersecting cylinder of radius r is unrolled onto a flat surface, the path taken by a point on the face of the blade may be described by referring to the resulting diagram, Figure 9-4. The circumferential motion $2\pi r$ is laid out as the horizontal leg of the right triangle and the linear advance P is the vertical leg. The hypotenuse is the helix, which is the actual distance traveled by the point. Since it is a characteristic of the geometry of this blade section, the angle $\varphi = \tan^{-1} P/2\pi r$ is called the *geometric pitch angle* of the blade section.

Propeller Slip and the Blade Velocity Diagram

The same triangular diagram would represent the velocity diagram of a point on a machine screw thread or a propeller blade if it were advancing through an unyielding tapped hole. To convert the axial and circumferential distances moved in one revolution to their corresponding velocities, it is only necessary to multiply each distance by the revolutions per second of the propeller, designated by n. The axial velocity is then Pn feet (or meters) per second, and the circumferential velocity is $2\pi n r$. The tangent of the pitch angle thus also designates the ratio of the axial to the circumferential velocities of a point on a screw turning in a nut at a uniform velocity.

Since water is not rigid like a tapped hole, it yields to a rotating screw propeller that is placed in it, and the propeller accelerates the water passing through it. It thus advances for each revolution a distance less than the pitch, at a velocity

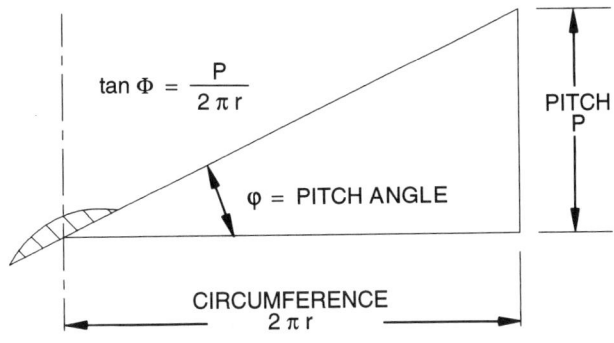

Figure 9-4. Propeller blade pitch angle.

less than Pn, as shown in Figure 9-5. This velocity is called the *velocity of advance* (V_A), because it is the axial velocity at which the propeller advances through undisturbed water. The difference in velocity between Pn and V_A is called the *slip velocity*. It is usually expressed as a fraction of the velocity Pn, the ratio being called the *real slip ratio* (s_R) or simply the *slip*. Thus

$$\text{slip velocity} = Pn - V_A$$

$$\text{real slip ratio} = s_R = \frac{Pn - V_A}{Pn} = 1 - \frac{V_A}{Pn} \quad (14)$$

Ship's officers sometimes enter a different quantity called the slip in ship's logbooks, using the ship's speed made good over a certain period of time (usually 24 hours) instead of V_A and the total number of revolutions the propeller has made during the same period of time instead of n. This "logbook slip" should be called the *apparent slip ratio* to distinguish it from the real slip ratio defined by equation (14).

Thrust and Efficiency

The phenomenon of slip leads to a kind of hydrodynamic paradox of marine propulsion devices, including screw propellers. On the one hand, the various theories of propeller action show that propeller efficiency decreases as slip increases, and it can reach 100 percent only if the slip is zero. Thus, for high efficiency, low slip is desired. But on the other hand, the same theories, and experimental measurements as well, show that the thrust produced by a propeller is proportional to the slip, so that at zero slip there would be no thrust and hence no useful output of the propeller. This comes about because the thrust, or net force on the fluid, is equal to the mass of fluid passing through the propeller times the acceleration (or change in momentum) imparted to it by the propeller, in accordance with New-

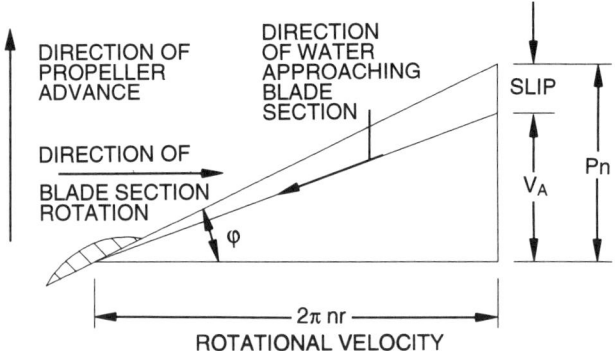

Figure 9-5. Propeller blade velocity diagram.

ton's Second Law. So to produce thrust, the fluid must be accelerated, or increased in velocity. Since the velocity increase is the slip, for good thrust high slip is desired.

How can these conflicting requirements both be met so as to produce high thrust at high efficiency? That is one of the foremost considerations in propeller design. Since we are not dealing here with the details of propeller design, suffice it to point out that, in general, it is desirable to produce a high thrust by passing a very large mass of fluid through the propeller per unit time, while increasing its velocity as little as possible. Thus in general for high efficiency a propeller should be as large as possible, turning at as low a rotational speed as is consistent with the characteristics of the main propulsion machinery.

Blade Section Shapes

The choice of appropriate cross-sectional shapes of propeller blades is critical to the design of efficient propellers. Blade section shapes refer to the sections defined by intersecting the blade with a series of cylinders concentric with the propeller shaft and at various radii from the hub outward. If the cylindrical sections thus described are "unrolled" onto a plane so that they are laid out flat, they will appear as shown in the *expanded blade view* of Figure 9-2. For each section the *blade width* is the arc length from leading edge to trailing edge, which in the expanded view is laid out as a straight line. The blade width is also called the *chord*, a term used for the similar dimension of aircraft wing sections, from which many successful propeller blade section shapes have been derived.

The blade section shapes shown in the expanded blade outline view of Figure 9-2 are typical for merchant ship propellers. Near the hub and into midblade, the sections shown are airfoil sections, characterized by the nose and tail of each section being lifted above the horizontal reference line, or *pitch face,* which is shown at each radius in the drawing. As the blade tip is approached, the airfoil sections gradually change into circular backed sections with a "flat" pitch face and generally much smaller nose and tail radii. The pitch face is not flat on the actual propeller blade because the pitch face lines shown as straight lines in the expanded view are actually the cylindrical arcs described above.

Pressures and Forces on Propeller Blades

The forces that develop on a rotating propeller blade are the hydrodynamic equivalent of the aerodynamic forces on the wings of subsonic aircraft that enable them to fly. Propeller blade geometry is more complicated than that of airplane wings, however, because the propeller blades are twisted so that, when they rotate, the flow of water across a blade section from nose to tail is oriented as shown in Figure 9-6(a). The figure is drawn as if the blade section is stationary and the water is moving. In fact, both the blade and the water are in motion, and the incident velocity V_o represents the relative velocity of blade and water.

Forces and pressures on the blade section are identical for a given incident velocity regardless of how much of that velocity is caused by blade motion and how much by water motion.

When the angle between the pitch face of the blade section and the direction of the incident flow of water is small, the flow lines around the blade are as shown in Figure 9-6(a). Over most of the back of the blade, the water velocity is greater than that of the free stream, while over the blade face, it is slowed to a velocity less than V_o. The relationship between velocity and pressure at two points along any stream line is then given by Bernoulli's Theorem:

$$p + \frac{\rho V^2}{2} = p_o + \frac{\rho V_o^2}{2} \tag{15}$$

where V and p are the velocity and pressure at a point on a streamline adjacent to the back or face of the blade.

V_o and p_o are the velocity and pressure in the free stream undisturbed by the blade.

Transposing equation (15) into a dimensionless form we can write

$$\frac{p - p_o}{\frac{1}{2}\rho V_o^2} = 1 - \left(\frac{V}{V_o}\right)^2 \tag{16}$$

In this form it is clear that if velocity V near the blade section is greater than V_o, the quantities on each side of the equation are negative. That is, the pressure at the blade is less than the free stream pressure p_o. This is what happens on the back of the blade. Conversely, on the face of the blade where V is less than V_o, the pressure difference is positive. A typical pressure distribution is shown in part (b) of the figure. The pressure differential across the blade section produces a force shown in part (c) of the figure as the resultant force dR. The blade forces are shown as differentials to emphasize the point that only a single section of the blade is depicted. The resultant force on the entire blade is the sum (that is, integral) of the dR forces on all blade elements from hub to tip.

The magnitude of the blade section forces depends on the shape and size of the section and on the angle between the pitch face of the blade section and the direction of water inflow. This angle is called the *angle of attack* or *angle of incidence*. The resultant force on a blade (or on a hydrofoil or airfoil) is not as important by itself as are its components perpendicular to and parallel to the water (or air) direction. They are, respectively, the *lift force (dL)* and *drag force (dD)*. The effectiveness of a foil as a lifting surface (such as an airplane wing or a hydrofoil) is measured by its lift-to-drag ratio (L/D), since the lift force is the useful component and the drag force represents losses. Hundreds of airfoil shapes have been tested in wind tunnels to measure these forces and their variation with the

SHIP PROPULSION 281

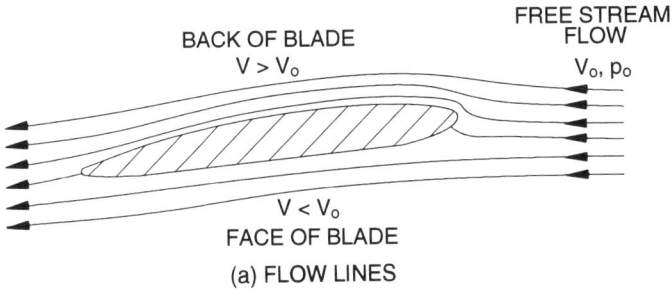

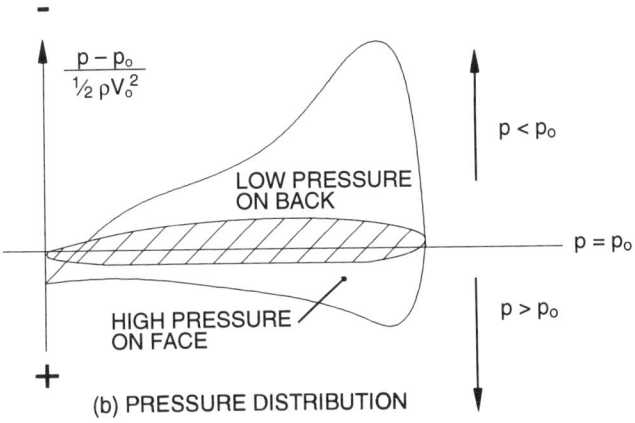

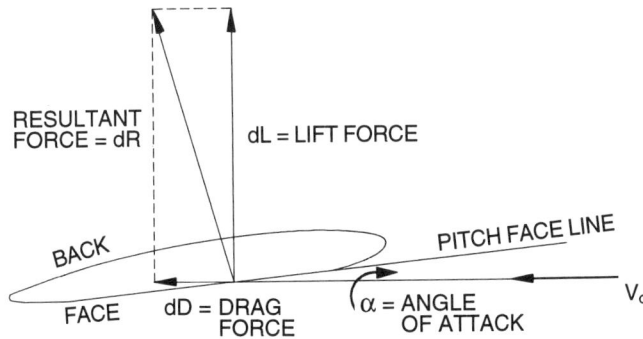

Figure 9-6. Propeller blade section mechanics.

angle of attack. In the tests, foils of constant cross section span the entire wind tunnel. A typical result is shown in Figure 9-7. Lift and drag forces are expressed as dimensionless coefficients as follows:

$$\text{Lift coefficient} = C_L = \frac{L}{\frac{1}{2} \rho A V_o^2} \qquad (17)$$

$$\text{Drag coefficient} = C_D = \frac{D}{\frac{1}{2} \rho A V_o^2} \qquad (18)$$

where L = lift force on the foil
 D = drag force on the foil
 ρ = density of the air
 A = plan view area of the foil = chord × span
 V_o = free stream velocity of the air

Optimum efficiency of the foil as a lifting device occurs at the angle of attack at which the lift-to-drag ratio (C_L/C_D), is maximum. Actual angles at which this occurs vary with the particular foil, but the angle is always a small positive angle, typically in the range of 3 to 6 degrees. Maintaining the appropriate angle of attack within close limits is important to the efficient operation of a foil. Translated into a screw propeller configuration, the efficient angle of attack for a given rotational speed of the propeller and forward speed of the ship depends on the pitch of the propeller blades and the amount of slip at which they operate.

The descriptions given above of the flow of water around screw propeller blades have been very much simplified and idealized compared with the actual flow situations around real propellers. In fact, the hydrodynamics associated with screw propellers is extremely complex. In spite of that complexity, theoretical hydrodynamicists have learned a great deal about such flow. They have developed theories of propeller action that can be used by propeller designers to assist in determining the best possible geometry for a propeller in a given application. The circulation theory of the screw propeller, based on the vortex theory of airplane wings, is the most realistic and most complex of various theories of propeller action. This theory, described in advanced treatments of the subject, continues to be developed and improved by hydrodynamicists and researchers. But the theory alone is not yet complete enough to apply to propeller design without including in the design process some information taken from experiments on airfoils or model propellers. Designers use very complex computer programs to apply circulation theory to their designs, but they have to start with a preliminary choice of the general section shapes and the main dimensions and number of blades of the propeller before they use the program to do the detailed design. The choice of the preliminary characteristics is usually based on information taken from model tests of propellers.

SHIP PROPULSION 283

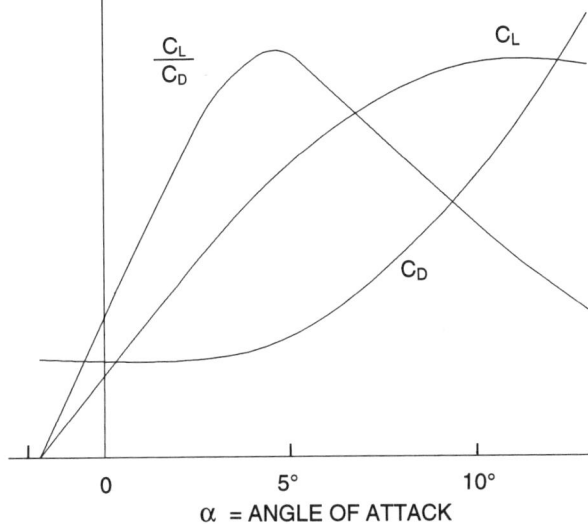

Figure 9-7. Typical blade section forces.

The Open-Water Propeller Test

To determine the performance characteristics of a propeller independent of the influence of the ship to which it may be fitted, model propellers are tested by running them through undisturbed water while measuring their torque, thrust, speed of advance, and rotational speed. This is done in a ship model basin by mounting a model propeller on the forward end of a shaft that extends forward from a special "propeller boat" attached to the towing carriage, as shown schematically in Figure 9-8. The boat carries an adjustable-speed electric motor (to turn the shaft), and propeller dynamometers that accurately measure and display or record the torque input to the propeller and the thrust, which is the useful output of the propeller. The propeller shaft extends well forward of the boat, which is moved propeller-first by the carriage so that the model propeller is well forward and thus clear of any disturbance of the water that might be made by the boat. The carriage speed or propeller boat speed is therefore equal to the velocity of advance of the propeller, a condition that could not be achieved if the propeller were located behind the ship model. Open-water propeller tests can also be done in a variable-pressure water tunnel, in which the water is moved at speed V_A past the rotating propeller.

Just as there is a law of comparison for ship model tests as shown in the last chapter, there is also a law of comparison for the proper conduct of propeller model tests. And as in the ship resistance measurements, deducing the relationships between model and full-scale propeller dimensions, velocities, and forces

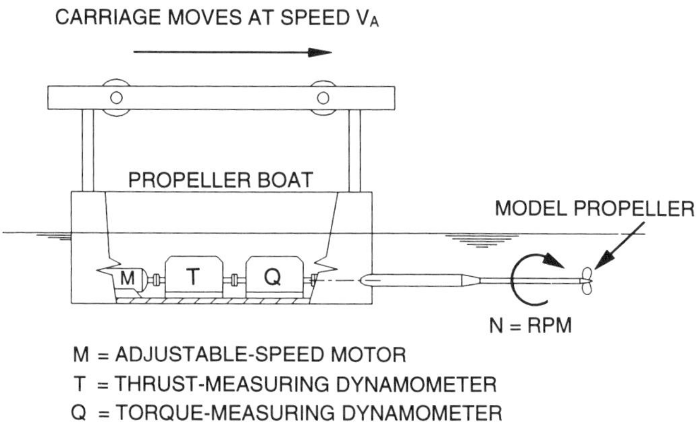

284   APPLIED NAVAL ARCHITECTURE

Figure 9-8. Open-water propeller test schematic.

requires writing a complete physical equation and applying the techniques of dimensional analysis. The results of the dimensional analyses for propeller thrust and torque expressed in terms of practical dimensionless groups are as follows:

$$\frac{T}{\rho n^2 D^4} = f_1 \left(\frac{V_A}{nD}, \frac{V_A^2}{gD}, \frac{V_A D}{\nu}, \frac{p}{\rho V_A^2} \right) \tag{19}$$

$$\frac{Q}{\rho n^2 D^5} = f_2 \left(\frac{V_A}{nD}, \frac{V_A^2}{gD}, \frac{V_A D}{\nu}, \frac{p}{\rho V_A^2} \right) \tag{20}$$

where, for either the model or the full-scale propeller,

T = thrust
Q = torque
ρ = mass density of the water
n = propeller revolutions per second
D = propeller diameter
V_A = velocity of advance
g = gravitational acceleration
ν = kinematic viscosity of the water
p = water pressure at the propeller

Since the functions f_1 and f_2 are not specified, the significance of the above equations is that if each of the dimensionless groups on the right side for a model propeller are equal in value to the corresponding groups for a full-scale propeller, then the parameter on the left containing the thrust or the torque will also be equal for the two, and model measurements may thus be used to determine full-scale performance. The fact is that it is impossible to achieve the required equality of all of the parameters at the same time, but valid model tests can still be made because the influence of some of them is small enough to be negligible. The significance of each propeller parameter is discussed below.

Thrust and torque coefficients. These are the parameters on the left-hand side of the above equations. Because a propeller that must deliver a large thrust for its size is said to be "heavily loaded," the thrust coefficient is sometimes called a thrust loading coefficient.

$$\text{Thrust coefficient} = K_T = \frac{T}{\rho n^2 D^4} \qquad (21)$$

$$\text{Torque coefficient} = K_Q = \frac{Q}{\rho n^2 D^5} \qquad (22)$$

These are nondimensional ways of expressing the propeller output (thrust) and input (torque). Other forms of thrust and torque coefficients containing the velocity of advance or the propeller disk area are used for some purposes, but K_T and K_Q are most useful for plotting the results of open-water propeller tests.

Advance coefficient. Also called the advance ratio or advance number, the advance coefficient is defined as:

$$J = \frac{V_A}{nD} \qquad (23)$$

It arises from the need to have kinematic similarity, or similarity of flow patterns, between model and propeller. V_A is the velocity at which the propeller advances, while nD is proportional to the circumferential velocity of its blades (the blade tip has a circumferential velocity of πnD, for example). Equality of the advance coefficient between model and full-scale propellers must be maintained when running open water tests. It is equivalent to maintaining equality of slip. The relationship between them can be shown by combining their definitions, equations (14) and (23). From equation (14),

$$\frac{V_A}{Pn} = 1 - s_R$$

Then equation (23) can be transformed to:

$$J = \frac{V_A}{nD} = \left(\frac{P}{D}\right)\left(\frac{V_A}{Pn}\right) = \left(\frac{P}{D}\right)(1 - s_R) \qquad (24)$$

Since the pitch ratio (P/D) is the same for geometrically similar propellers, equivalence of their advance coefficients will assure equivalence of their slip ratios as well.

Froude number. The second term on the right side of equations (19) and (20) expresses the fact that velocity of advance (V_A) of the model propeller should be determined from the ship velocity of advance according to Froude's law. Doing so will assure that the wave-making characteristics will be properly scaled. Although satisfying Froude's law in open-water propeller tests presents no problems, if the model propeller is immersed deeply enough that it produces no significant waves on the water surface, it is not necessary to be concerned with satisfying this criterion. In tests in which a ship model is fitted with a model propeller, such as certain kinds of self-propulsion tests, Froude scaling of the velocity of advance may be necessary.

Reynolds number. The Reynolds number of a propeller-based on its velocity of advance, is $V_A D/\nu$. As in the case of ship resistance model tests, this criterion cannot be satisfied at the same time as the Froude criterion. Fortunately, it can be ignored without invalidating the open-water propeller test because the viscous or frictional forces on the propeller blades are insignificant in comparison with the thrust.

Pressure ratio. The last of the dimensionless groups in equations (19) and (20) is proportional to the ratio of the static pressure (identified as p) in the fluid to the dynamic, or stagnation, pressure ($1/2\ \rho V_A^2$) caused by the velocity of advance. In an ordinary ship model basin it is not possible to adjust this ratio for the model test so that it is the same as that for the ship. The static water head can be scaled properly by scaling the immersion of the model correctly, but the static pressure includes the atmospheric pressure as well, and this cannot be adjusted unless the entire basin is enclosed in a vacuum chamber in which the air pressure can be adjusted at will. This criterion can be ignored in tests of most merchant ship propellers. It becomes important only if cavitation, a condition described later, is likely to occur. Cavitation is a problem primarily of propellers that are heavily loaded and running at high rotational speed.

The functional relationships for propeller thrust and torque in an open-water propeller test can thus be simplified from those given in equations (19) and (20) to:

$$\frac{T}{\rho n^2 D^4} = f_1\left(\frac{V_A}{nD}\right) \quad \text{or} \quad K_T = f_1(J) \quad (25)$$

$$\frac{Q}{\rho n^2 D^5} = f_2\left(\frac{V_A}{nD}\right) \quad \text{or} \quad K_Q = f_2(J) \quad (26)$$

The test is conducted by running the propeller boat at a series of speeds from zero speed ($J = 0$, $s_R = 1.0$) to the speed at which the thrust drops off to zero. Open-water performance curves are prepared as shown in Figure 9-9. Besides the thrust and torque coefficients, the open-water propeller efficiency (η_o) is plotted to indicate the advance ratio or slip ratio at which maximum efficiency is achieved in open water. Open water efficiency, defined by equation (6) can be expressed in terms of K_T, K_Q, and J as follows:

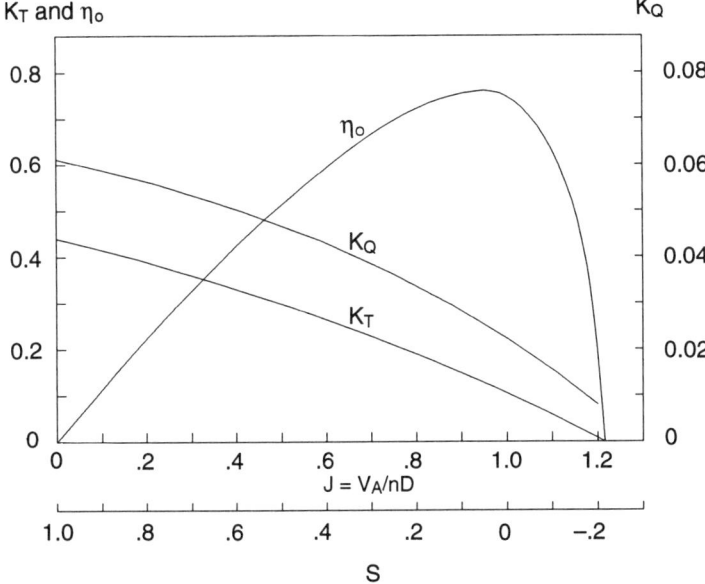

Figure 9-9. Propeller performance curves in open water.

$$\eta_o = \frac{TV_A}{2\pi Qn} = \frac{(K_T \rho n^2 D^4)(JnD)}{2\pi (K_Q \rho n^2 D^5) n}$$
$$= \frac{K_T}{K_Q} \times \frac{J}{2\pi} \qquad (27)$$

Expansion to Full Scale

Expanding the open-water test results to ship scale requires that the ratio of ship to model speeds of advance be declared. Although it is not necessary to comply with Froude speed scaling to conduct an open-water test, the equivalent full-scale results should be derived on the assumption that model and ship Froude numbers are equal for the propeller as well as for the hull. In what follows, the linear scale ratio of ship (S) to model (M) dimensions (the propeller diameters, for example) is $\lambda = D_S/D_M$.

The speed scaling is derived from the Froude number equality:

$$\left(\frac{V_A^2}{gD}\right)_S = \left(\frac{V_A^2}{gD}\right)_M$$

288 APPLIED NAVAL ARCHITECTURE

$$\frac{V_{AS}}{V_{AM}} = \sqrt{\frac{D_S}{D_M}} = \sqrt{\lambda} \tag{28}$$

Equality of the advance coefficients determines the corresponding rotational speeds:

$$J_S = J_M$$

$$\left(\frac{V_A}{nD}\right)_S = \left(\frac{V_A}{nD}\right)_M$$

$$\frac{n_S}{n_M} = \frac{D_M}{D_S} \times \frac{V_{AS}}{V_{AM}} = \frac{\sqrt{\lambda}}{\lambda} = \frac{1}{\sqrt{\lambda}} \tag{29}$$

This relationship may be used to determine the model propeller revolutions from those of the ship by solving for n_M.

$$n_M = n_S \sqrt{\lambda} \tag{30}$$

This shows that the model propeller must run at a higher rotational speed than the ship propeller. The thrust and torque of the ship propeller are derived from those of the model by equating their thrust coefficients and torque coefficients, respectively.

$$K_{TS} = K_{TM}$$

$$\left(\frac{T}{\rho n^2 D^4}\right)_S = \left(\frac{T}{\rho n^2 D^4}\right)_M$$

$$\frac{T_S}{T_M} = \frac{\rho_S}{\rho_M} \left(\frac{n_S}{n_M}\right)^2 \left(\frac{D_S}{D_M}\right)^4 = \frac{\rho_S}{\rho_M} \times \frac{1}{\lambda} \times \lambda^4$$

$$= \lambda^3 \left(\frac{\rho_S}{\rho_M}\right) \tag{31}$$

$$K_{QS} = K_{QM}$$

$$\left(\frac{Q}{\rho n^2 D^5}\right)_S = \left(\frac{Q}{\rho n^2 D^5}\right)_M$$

$$\frac{Q_S}{Q_M} = \frac{\rho_S}{\rho_M} \left(\frac{n_S}{n_M}\right)^2 \left(\frac{D_S}{D_M}\right)^5 = \frac{\rho_S}{\rho_M} \times \frac{1}{\lambda} \times \lambda^5$$

$$= \lambda^4 \left(\frac{\rho_S}{\rho_M}\right) \tag{32}$$

The useful power output of a propeller is the thrust power, given by

$$P_T = TV_A \tag{33}$$

Ship and model thrust powers are related as follows:

$$\frac{P_{TS}}{P_{TM}} = \frac{T_S}{T_M} \times \frac{V_{AS}}{V_{AM}} = \lambda^3 (\frac{\rho_S}{\rho_M}) \times \sqrt{\lambda}$$

$$= \lambda^{3.5} (\frac{\rho_S}{\rho_M}) \tag{34}$$

Methodical Series Propeller Charts
The open-water test results for a particular propeller, such as the curves plotted in Figure 9-9, can be prepared for any propeller by constructing and testing a model of the particular propeller. The curves would then be very useful for the analysis of the propeller's performance and prediction of its suitability for application to a particular ship. Individual open-water curves for propellers of varying and unrelated geometry are not very useful in propeller design, however. To help the naval architect make a preliminary choice of propeller characteristics for a ship design, several different forms of propeller design charts have been developed. They are derived by plotting, in a single chart, data from the open-water tests of a group of geometrically related propellers within which only one characteristic, usually the pitch, is varied systematically. Within the group, all other characteristics are the same for each propeller—number of blades, blade outline and section shape, blade thickness, blade area, rake, and skew for example. A simple chart is a "K_T-K_Q-J" diagram, in which open-water test curves for one group of propellers are plotted on a single sheet. Comparison of the efficiencies obtainable with different members of the group for a given set of conditions can be made from such a chart. Of course, it takes many such charts to present the complete results of testing a methodical series of propellers within which other characteristics such as number of blades and blade areas are also varied. A comprehensive series tested at the Maritime Research Institute of the Netherlands (MARIN), for example, consists of 20 groups and required testing about 120 model propellers, all of which have the same blade outline and the same blade section shape. It can be appreciated that a very large body of data exists, since many other series with different basic blade characteristics have also been tested.

Although a K_T-K_Q-J diagram is the type of methodical series chart that most closely resembles the open-water test results, it is not the most convenient to use for preliminary propeller design. For that purpose, numerous other charts using different parameters have been developed that make it easier to pick the most efficient propeller for a given set of ship characteristics. They are not described here, but they may be found in books and technical papers dealing with propeller design.

Cavitation

The procedures described above for predicting the performance of ship propellers from model propeller open-water tests are only valid if the model propeller does not cavitate. Cavitation occurs when, on the back or suction side of the propeller blade, the pressure as shown in Figure 9-6(b) becomes so low that the water vaporizes at the low pressure point and vapor-filled cavities or bubbles form in the water. In effect, there is a local "boiling" of the water, although the vaporization is not caused by high temperatures, but rather because the pressure becomes lower than the vapor pressure of the water. Cavitation can occur on propellers that are very heavily loaded or on those that have high rotational speed, and it is most likely to occur on a propeller that has both of these conditions in effect.

Cavitation bubbles and, in more severe cases, cavitation sheets forming around a propeller are very troublesome and can cause problems that go well beyond invalidating the open-water test results. Cavitation bubbles that form on a blade often collapse again as they move into higher-pressure regions. The collapse of the bubbles is accompanied by large forces on such small areas of the blade that the blade may be eroded or pitted, and the irregularities thus produced can cause even more cavitation to occur. The noise produced by collapsing cavitation bubbles is also a serious problem, especially for combatant ships and submarines. Furthermore, propeller efficiency drops off because the full amount of thrust cannot be generated by a cavitating propeller, and intermittent formation and shedding or collapsing of cavitation bubbles causes regular periodic fluctuations in the thrust, which can give rise to severe vibrations.

A propeller subject to cavitation must be tested in a special facility called a variable-pressure water tunnel in which the water pressure can be adjusted to any desired value so that a form of pressure ratio called the *cavitation number* may be made equal for model and ship propellers. Much of what is known about cavitation and how to avoid it has been learned from research done in cavitation tunnels. During the propeller design process the propeller must be checked to determine if cavitation is likely by comparing it to cavitation criteria determined by continuing research into this complex phenomenon.

THE SHIP AND PROPELLER TOGETHER

A propeller whose characteristics have been chosen for optimum performance in open water might not be the best propeller for a given ship, because the flow of water around a ship-mounted propeller is very different from that around a propeller in open water, and because the propeller itself alters the pressure distribution around the stern of the ship, thereby increasing its resistance. These phenomena are called hull-propeller interactions. Their influence on the overall efficiency of propulsion is expressed by two efficiencies mentioned earlier, namely

Wake and the Wake Fraction

In general terms, the region of disturbed fluid behind a body that is moving through a fluid is referred to as the *wake* of the body. Ship propellers are located within the wake of the ships that they propel, and since a propeller's performance is influenced by the velocity field in which it operates, the propeller designer must define the wake of a ship in specific terms in order to optimize the propeller's performance when it is propelling the ship.

The simplest representation of the wake of a ship as it affects the propeller is the average axial velocity of the water passing through the propeller disk in the absence of any influence by the propeller itself. This velocity has already been defined above—it is the speed of advance (V_A). In most cases, the speed of advance is less than the ship speed (V), because the ship hull forward of the propeller moves some water with it, especially within the viscous boundary layer surrounding the hull. This is a fortuitous circumstance because, as will be shown later, the efficiency of propulsion is increased as a result of it. This kind of wake is known as a *positive wake,* and it is defined as the difference between the ship speed and the speed of advance.

$$\text{wake} = V - V_A \tag{35}$$

The wake (or wake speed as it is sometimes called) is most conveniently expressed nondimensionally as a fraction of the ship speed. The ratio thus formed defines the *wake fraction (w)*.

$$w = \frac{V - V_A}{V} = 1 - \frac{V_A}{V} \tag{36}$$

Thus the speed of advance may be expressed in terms of the wake fraction as follows:

$$V_A = V(1 - w) \tag{37}$$

When V_A is less than V, the wake fraction is positive, consistent with the concept of a positive wake described above. The significance of a wake fraction of w = 0.30 is that, in the vicinity of the propeller, the average water velocity is 30 percent less than the ship speed. Modern wake data are invariably expressions of the quantity in equation (36), as defined by D. W. Taylor, but older sources might refer to a different definition of wake which was introduced by Froude, and which expresses the wake speed as a fraction of the speed of advance rather than of the ship speed.

$$w_F = \frac{V - V_A}{V_A} \qquad (38)$$

To be consistent with modern usage, data containing Froude wake fractions should be converted into the equivalent Taylor wake fractions as follows:

$$w = \frac{w_F}{1 + w_F} \qquad (39)$$

There are three effects of the water flow around the stern of a ship that contribute to the strength of the wake behind a ship. All three were described in Chapter 8. They give rise to three components of wake: potential wake, frictional wake, and wave wake.

Potential wake. The pressure distribution around an elongated body moving through a fluid (even an ideal, nonviscous fluid) is always high around the stern of the body where the flow lines close in after passing around the body. This is discussed in Chapter 8 and shown in Figure 8-3. Consistent with Bernoulli's principle, the velocity of the water in this region is slower than the velocity of the body or the ship. The component of wake caused by this phenomenon is called the potential wake, and it is always positive in the stern region of a ship.

Frictional wake. Within the viscous boundary layer, some water is carried along in the direction of motion of the ship. The boundary layer increases in thickness from bow to stern, so that most of the propeller (especially on a single-screw ship) of a large ship will be within the boundary layer. Frictional wake is always positive and is strongest close to the hull, getting weaker with increased distance from the hull. It is usually the strongest of the three components of wake.

Wave wake. The wave train that is generated by the ship and moves with it can cause either a crest or a trough of the transverse wave system to occur at the stern. It has been shown in Chapter 8 that, at design speed, it is preferable that a crest occur at the stern. Within each wave, the water particles move in circular orbits (see Chapter 10) such that under a crest the water moves in the direction of wave propagation and hence of the ship motion, while under a wave trough the water particles move opposite to the direction of wave propagation. If a crest of the wave system occurs at the location of the propeller, therefore, wave wake will be positive, but it will be negative if a trough occurs there.

Wake strength is influenced by the shape of the stern in ways that are not understood thoroughly enough to be able to calculate it theoretically. Experimental evidence is needed, usually from tests of large models in which an array of velocity-measuring devices is arranged at the stern in the exact position that will be occupied by the propeller disk. The test is called a wake survey, and the results of wake surveys demonstrate another characteristic of a ship wake. It is quite irregular and variable within the boundaries of the propeller, so that as a propeller rotates, its blades pass through a very irregular velocity field. Unsteady blade forces can result, and these can give rise to hull vibrations. A single wake fraction expresses only the average wake and not its variations.

Typical values of the wake fraction for single-screw merchant ships are between 0.20 and 0.40, with the great majority of cases falling between 0.25 and 0.35. It increases with hull fullness, and on very full hulls (C_B more than 0.85) the wake fraction can be even higher than 0.40. Single-screw ships have a strong positive wake because all three components are positive when the ship operates at the design speed. Such is not the case with very high speed multiple-screw ships such as destroyers and other high-speed combatant ships. Their very fine hulls (C_B around 0.50) cause the potential wake to be relatively weak, and their speed is so high ($V/\sqrt{L}$ = 2 or higher, or F_n greater than 0.6) that the wave wake is strong and negative because the stern operates in the trough of the ship's wave pattern. Furthermore, since twin screws are not nestled close behind the hull, they tend to be mostly outside the influence of the viscous boundary layer so that the frictional wake is very weak. The result is that twin-screw ships of moderate speed have wake fractions between 0.05 and 0.25, and in the high-speed types the average wake can actually be negative (a range of -0.05 to $+0.10$) because wave wake predominates.

Thrust Deduction

One of the principal differences between the conditions represented by the resistance test of a ship model and those actually present in an operating ship is that the ship is propelled by its own propeller while the model is towed by a carriage. The resistance force measured on the towed model at the required design speed is used to determine the ship resistance and then the ship effective power as described in Chapter 8. In the self-propelled condition, the thrust force produced by the propeller must be sufficient to overcome the ship resistance at the corresponding ship design speed. To determine how much thrust is necessary, one must question whether or not the resistance of the ship when it is self-propelled is the same as the resistance when it is towed.

The fact is that the resistance of a self-propelled ship is greater than the resistance of the same ship if it were towed at the same speed. Recall that there are high pressures at the stern of a ship hull, pressures which cause forces that are normal to the hull. These normal forces have components that act in the forward direction because of the shape of the hull at the stern. The high pressure surrounding the stern is therefore beneficial in reducing ship resistance. Next, recall how the propeller changes the flow of water around the stern. It accelerates the water, thereby reducing the beneficial high pressure around the stern. The resistance therefore increases when the ship is self-propelled, so that the propeller thrust must be greater than the ship resistance.

This effect is quantified by defining the *thrust deduction fraction (t)*. It is a measure of the fraction of the propeller thrust that is necessary to overcome the additional resistance imposed by the propeller itself. It is defined as follows:

$$t = \frac{T - R_T}{T} = 1 - \frac{R_T}{T} \qquad (40)$$

where t = thrust deduction fraction
T = propeller thrust
R_T = ship resistance

It can be seen that a thrust deduction fraction of 0.20 means that 20 percent of the thrust produced by the propeller is spent overcoming the additional resistance caused by propeller action, while the other 80 percent is equal to the towed ship resistance.

Equation (39) can be rewritten, solving for either R_T or T, thus:

$$R_T = T(1 - t) \qquad (41)$$

$$T = \frac{R_T}{(1 - t)} \qquad (42)$$

The quantity (1 - t) is called the *thrust deduction factor*.

Values of the thrust deduction fraction (t) are deduced from model tests of self-propelled models. Typical values are in the range of 0.15 to 0.20 for single-screw ships and 0.10 to 0.18 for twin-screw ships. As with wake fraction, the thrust deduction fraction increases with hull fullness.

Hull Efficiency

Hull efficiency has already been defined in equation (3) as the ratio of the effective power to the thrust power. That is,

$$\eta_H = \frac{P_E}{P_T} = \frac{R_T V}{T V_A} \qquad (3)$$

It can also be written in terms of the wake fraction and the thrust deduction fraction by combining equation (3) with equations (37) and (41):

$$\eta_H = \frac{R_T V}{T V_A} = \frac{1 - t}{1 - w} \qquad (43)$$

Hull efficiency depends on the shape and fullness of the stern of the ship. The normal ranges of values for conventional merchant ship forms are 1.10 to 1.25 for single-screw ships and 0.85 to 0.95 for twin-screw ships.

SHIP PROPULSION

Relative Rotative Efficiency
Relative rotative efficiency (η_R) was defined above as the ratio of the propeller efficiency behind the hull to its open-water efficiency. The two efficiencies are not necessarily the same as each other because of the differing flow patterns in the behind condition compared to the open-water condition. The principal difference in flow is attributed to the fact that the wake field behind a ship is variable, while the open-water velocity field is uniform. In spite of these differences, self-propelled model tests show that the two efficiencies are not very different from one another, so that the relative rotative efficiency is close to unity. Measurements from model tests indicate that installing a propeller on a single-screw ship with a well-designed stern actually improves its efficiency, so that η_R for single-screw merchant ships ranges from 1.0 to 1.06. Twin-screw installations generally cause the efficiency to degrade slightly from open water, typical values of η_R being 0.96 to 1.0 in this case.

PROPULSIVE EFFICIENCY AND SHAFT POWER

In the preceding sections concerning the propeller and the interactions between hull and propeller, the three hydrodynamic efficiencies defined at the beginning of this chapter and composing the quasi-propulsive efficiency have been evaluated. They are the hull efficiency, open-water propeller efficiency, and relative rotative efficiency. We come now to a description of the final stage in the process of determining a ship's powering requirements, namely the estimate of the propulsive efficiency, trial shaft power, and service shaft power.

The propulsive efficiency (η_P) has already been defined in equations (11) and (12). It is:

$$\eta_P = \frac{P_E}{P_S} = \eta_H \times \eta_O \times \eta_R \times \eta_S \qquad (12)$$

Shaft Transmission Efficiency
Only the shaft transmission efficiency (η_S) remains to be evaluated. Shaft transmission efficiency is a more familiar type of efficiency than the hydrodynamic efficiencies, in that it involves mechanical losses. If the shaft power is defined as the power at the after end of the propeller line shafting, that is, just at the coupling of the propeller shaft to the tail shaft, the losses involved are only those associated with friction in the tail shaft bearing. These losses are very small, about 1 percent, so η_S can be taken as 99 percent in this case. If the shaft power is to be determined at the forward end of the line shafting, where it is connected to the engine or reduction gears, the efficiency is often assumed to be 98 percent for machinery aft installations, and 97 percent for machinery amidships. The efficiency depends on the number of journal bearings supporting the line shafting.

Trial Shaft Power

Equation (12) may be written to show how the shaft power known as the *trial shaft power* (P_{ST}) is determined from the effective power deduced from model tests, along with a propulsive efficiency deduced from propeller open-water tests and estimates of the efficiencies described above:

$$P_{ST} = \frac{P_E}{\eta_P} \qquad (44)$$

Put in this way, it is apparent that this shaft power would be sufficient to propel the actual ship at its design speed only if the conditions of operation of the ship were similar to those of the model whose resistance was measured in the model test. Since model test conditions are ideal conditions (no wind, waves, or currents, for example), the power predicted is an ideal power. It is called the trial shaft power because the sea conditions that most nearly correspond to these ideal model test conditions are those present during the ship standardization trials, which are conducted before a new ship is delivered to the owners. Measurements taken during ship trials are, in fact, the basis on which it is judged whether or not the ship's speed and power performance satisfies the contract between the builder and owner.

Service Shaft Power

Trial trip conditions, however, do not represent the realistic service conditions that a ship encounters at sea. In order to maintain the contracted design speed on average over a ship's operational lifetime in actual sea conditions, it is necessary to install a power greater than the trial shaft power. The power actually installed is called the *service shaft horsepower* (P_{SS}). The designer of a new ship must rely on the results of trial trip measurements of similar ships and performance data gathered on ships in service to determine how much additional power must be installed beyond the trial shaft power. The relationship between trial and service shaft powers is stated in terms of a *service power allowance,* which is expressed as a fractional part of the trial shaft power and is to be added to it. The service power allowance (SA) is defined by the following expression:

$$P_{SS} = P_{ST}(1 + SA) \qquad (45)$$

The quantity $(1 + SA)$ is called the service power factor. The U.S. Maritime Administration standard service power factor for cargo ships is 1.25. The corresponding service power allowance is 0.25, more often stated as a percentage—the ship has a 25 percent service power allowance. Ship designers may use any allowance they believe to be correct for a particular design, but most ordinary ships have allowances of between 20 percent and 25 percent. Very large tankers are an exception. Experience has shown that 15 percent is more realistic for them.

SHIP PROPULSION 297

PROBLEMS

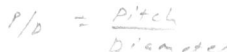
$P/D = \dfrac{Pitch}{Diameter}$

1. During an open-water test in fresh water at 68°F, a model propeller 12 inches in diameter and with P/D = 1.0 produces a thrust of 40.0 pounds when advancing at 9.80 ft/sec and turning at 690 rpm. The input torque is 7.40 ft-lb. Calculate, for these test conditions, the following:

   (a) Advance coefficient
   (b) Real slip ratio
   (c) Thrust coefficient
   (d) Thrust power, ft-lb/sec
   (e) Torque coefficient
   (f) Delivered power, ft-lb/sec
   (g) Open-water efficiency

2. Expand the open-water propeller test results of problem 1 to predict the performance of a 23.0-foot-diameter ship propeller. Determine, for the full-scale propeller in seawater at 59°F, the following:

   (a) Scale ratio
   (b) Speed of advance, ft/sec and knots
   (c) Rotational speed, rpm
   (d) Thrust, lb
   (e) Thrust power, ft-lb/sec and horsepower
   (f) Torque, ft-lb
   (g) Delivered power, ft-lb/sec and horsepower

3. A newly designed bulk carrier 852 ft long displaces 100,000 tons at design draft. Resistance tests have shown that the effective power, including air and appendage allowances, is 12,860 horsepower at the design speed of 16.5 knots. Propulsion factors measured in self-propelled model tests are:

   Wake fraction = 0.325
   Thrust deduction fraction = 0.178
   Relative rotative efficiency = 1.025

   The propeller test revealed that at its design point the propeller's open-water efficiency is 0.685. Shaft transmission efficiency is estimated at 0.990, and a service power allowance of 20 percent has been chosen.
   From the data given, calculate this ship's propulsive efficiency, trial shaft horsepower, and service shaft horsepower.

4. The effective power of a twin-screw ship 525 ft long is 14,950 horsepower including all allowances, at 21 knots. Model tests and designer estimates give the following efficiencies and propulsion factors:

   Wake fraction = 0.142
   Thrust deduction fraction = 0.170
   Relative rotative efficiency = 0.992
   Open-water propeller efficiency = 0.692
   Shaft transmission efficiency = 0.985
   Service power allowance = 25 percent

   Determine this ship's propulsive efficiency, trial shaft horsepower, and service shaft horsepower.

CHAPTER TEN

SHIP DYNAMICS

In previous chapters dealing with hydrostatics, stability, and resistance, the sea surface has been idealized to its simplest form, namely, still water. One has only to observe the ocean surface for a very brief period to conclude that such a representation is unrealistic. The sea surface is covered with waves, and while the severity of the waves can vary enormously as climatic conditions change, the smooth sea surface is the exception rather than the rule. The study of ships that are to navigate the oceans must therefore include some assessment of how ocean waves affect them. In this chapter we shall consider what ocean waves are like and how ships respond to them. A quantitative mathematical treatment of the subject is beyond the scope of this book, but the definitions and concepts included here are intended to introduce the reader to the complicated nature of the motions of ships in waves.

Ocean Waves
The observable ocean waves that affect ship motions are *wind-generated waves,* caused by winds blowing across the sea surface. They are also called *gravity waves,* so-named because the force against which the waves are raised, and which tends to restore the waves to smooth water, is the force of gravity. (Small waves, or ripples, are not gravity waves because they are restored by surface tension forces rather than gravity forces. They are too small to have any measurable effect on ship motions.)

The physical processes that take place when wind blows across the water surface and waves are created are not thoroughly understood, although much research by oceanographers and hydrodynamicists has been, and continues to be, done on the subject. Without complete knowledge of the processes by which energy is transferred from winds to waves, it is not possible to determine with certainty the wave patterns that a given storm system will produce. When a complex problem such as this one cannot be solved deterministically, analysts must resort to probabilistic techniques or models to characterize the state of the sea that ships will encounter. The probabilistic models describe the highly irregular sea surface in statistical terms; that is, they estimate the probabilities of occurrence of waves of different lengths, heights, frequencies (or periods), and directions that compose a sea of a particular severity. They are based on the notion that such a complex physical process can be represented as the sum of a very large (theoretically infinite) number of simple individual regular waves, each called a component wave, each having a particular length, height, frequency, and direction. Thus, to grasp the significance of the statistical description of the irregular sea, we must first describe the simple regular waves that are imagined to make it up.

Idealized Regular Progressive Water Waves
We seek to represent waves on the surface of a body of water with a mathematical expression which, although idealized, can be used to determine the fundamental geometric and physical properties that can be verified by measurements of actual ocean waves. Certain simplifying assumptions are necessary. The water is assumed to be nonviscous and incompressible, and the waves are called *long-crested regular waves*. "Long crestedness" implies that the wave crests are infinitely long, and "regular" implies that all waves are identical, so that all wave crests are parallel to one another and all wave heights are the same as one another. The surface might be described as a corrugated surface, such as that shown in Figure 10-1. In a long-crested sea, all wave properties are invariant along any line parallel to the crest lines, hence the problem is treated as a two-dimensional one.

The shape of the wave contours is, in this simplified theory, taken to be sinusoidal. This assumption leads to the simplest mathematical analysis while retaining acceptable realism so long as breaking waves are not involved. In fact, measure-

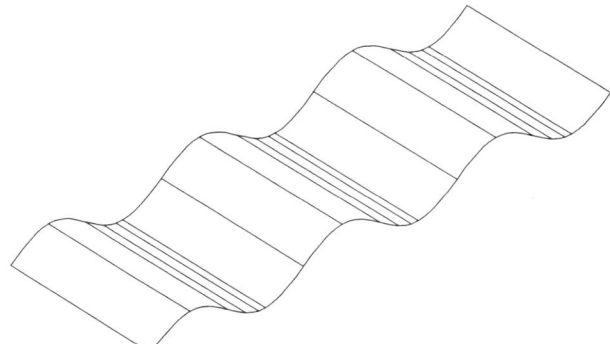

Figure 10-1. Idealized ocean wave pattern—long-crested regular waves.

ments of actual waves show that they have sharper crests and broader troughs than sine waves, but for the analyses needed by naval architects, the sinusoidal profile representation is entirely acceptable. In this regard it should be noted that, historically, the wave profile geometry assumed for naval architectural work was that of a trochoid, which has sharper crests than troughs, and which is sometimes still used in static wave bending moment calculations. For most hydrodynamic (as opposed to static) analyses, however, trochoidal wave theory cannot be used because it fails to correctly represent some of the essential dynamic properties of observable ocean waves and because it is mathematically much more cumbersome than the sinusoidal profile in complex calculations.

Surface Wave Profile. In its simplest form, the surface wave profile of an ideal sinusoidal wave train is given by the following equation and shown in Figure 10-2. It describes a simple harmonic wave of small amplitude.

$$\zeta = \zeta_a \cos k (x - V_w t) \qquad (1)$$

where
- ζ = elevation of surface of wave above still-water level
- ζ_a = wave *amplitude,* or half of the wave height
- k = *wave number* = $2\pi/L_w$
- L_w = wave length
- x = horizontal coordinate, positive in the direction of wave propagation
- V_w = wave velocity, also called phase velocity or *celerity*
- t = time

The above equation and the following equations defining the relationships among various wave properties such as length, period, and velocity, are stated here without derivation or proof. They are formally derived in textbooks on theoretical hydrodynamics and they satisfy not only the geometry of the wave profile, but the hydrodynamic requirements imposed by Bernoulli's equation for flow influenced by gravity, Laplace's equations for potential flow, and boundary condi-

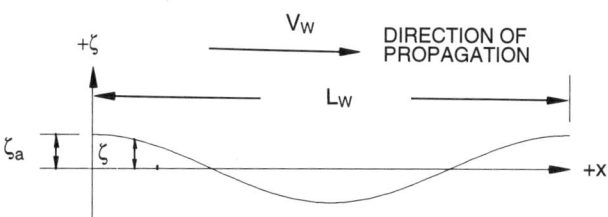

Figure 10-2. Sinusoidal surface wave at fixed time.

tions such as the uniform atmospheric pressure at the free surface and zero vertical water particle velocity at the ocean floor.

For most mathematical calculations it is preferable to express the above equation in a slightly modified form. The time for the passage of one wave past a stationary point is called the *period* (T_W) in seconds. In terms of wave length and velocity, the period is given by

$$T_w = \frac{L_w}{V_w} \qquad (2)$$

Its reciprocal is the wave frequency in cycles per second:

$$f_w = \frac{1}{T_w} = \frac{V_w}{L_w} \qquad (3)$$

For analytical calculations, the *circular frequency* (ω_w) in radians per second is preferred to f_w. Since there are 2π radians in one cycle,

$$\omega_w = 2\pi f_w = \frac{2\pi}{T_w} = \frac{2\pi V_w}{L_w} \qquad (4)$$

Thus equation (1) can be rewritten as follows:

$$\zeta = \zeta_a \cos k(x - V_w t) \qquad (1)$$
$$= \zeta_a \cos(kx - kV_w t)$$
$$= \zeta_a \cos(kx - \frac{2\pi V_w t}{L_w})$$
$$\zeta = \zeta_a \cos(kx - \omega_w t) \qquad (5)$$

Replacing the cosine function with a sine function would also yield a correct form of the equations. It would simply represent a different choice of the location of the origin shown in Figure 10-2. With the origin directly below a wave crest, the cosine is correct; with the origin at the nodal point preceding a crest, the sine would be correct. The two parts of the cosine function are of special importance in defining ocean waves. The first part, $\cos kx = \cos 2\pi x/L_w$, defines the geometrical or spatial form of the wave at a fixed time. At time $t = 0$, for example, the water elevation would be given by

$$\zeta = \zeta_a \cos \frac{2\pi x}{L_w}$$

which describes a sinusoidal wave form of amplitude ζ_a which repeats itself at $x = L_w$, or one wave length.

The second part of the cosine function, cos $kV_w t$ = cos $\omega_w t$, defines the time-varying or temporal form of the wave at a fixed point in space. At x = 0, for example, the water level rises and falls with time according to

$$\zeta = \zeta_a \cos \omega_w t = \zeta_a \cos \frac{2\pi t}{T_w}$$

which describes a sinusoidal variation in level of amplitude ζ_a which repeats itself at t = T_w, or one wave period. This term is necessary because ocean waves are *progressive* waves; that is, they travel or propagate along the water surface, advancing at velocity V_w in a direction perpendicular to their crest lines. It is the wave form that advances steadily in that direction, and not the water mass itself. There is obviously some water particle motion, and it must involve vertical motion because wave crests are higher than wave troughs. The actual water particle motion is orbital, each surface particle tracing a circle whose diameter equals the wave height. When a wave crest is passing, a surface particle moves briefly in the direction of wave propagation at a velocity much smaller than the wave velocity, and when a trough is passing, the particle moves in a direction opposite to that of the wave propagation at a similarly small particle velocity. Between crests and troughs, the particle velocities have vertical as well as horizontal components. Subsurface water particles trace similar but smaller circular orbits, whose diameters decrease rapidly (exponentially) with depth. At water depths more than about half of the wave length the orbital motion is practically negligible, so that submarines operating at depths of a few hundred feet are not subject to wave-induced motions and loads. The orbital water particle velocities can be derived from the hydrodynamic theory, but as they are of little importance as regards ship motions, the equations are not given here. It is important, however, to keep in mind the distinction between the orbital velocity of the water and the progressive straight line velocity of the wave form or pattern.

Wave Velocity or Celerity. While equations (1) or (5) describe the surface wave elevation at any point in space and at any time, they do not define the relationship of the wave velocity to its geometry. Hydrodynamic theory shows that the free-surface conditions of atmospheric pressure and vertical velocity of the surface particles result in the following relationship of wave velocity to wave length and water depth (h).

$$V_w^2 = \frac{g}{k} \tanh kh \tag{6}$$

This expression can be simplified for the case of deep water because tanh x approaches 1.0 as x increases, the difference being less than one-half of 1 percent for x greater than three (tanh 3 = 0.99503). Now if kh = $2\pi h/L_w$ > 3, then h/L_w >

3/2π, or about 1/2. Thus in water depths greater than half a wave length, waves travel at a velocity given by the following *deep-water wave velocity* formula:

$$V_w^2 = \frac{g}{k} = \frac{gL_w}{2\pi}$$

or

$$V_w = \sqrt{\frac{gL_w}{2\pi}} \qquad (7)$$

Wave velocity is thus seen to depend on wave length. Long waves travel faster than short waves. This characteristic of water waves (they are known technically as *dispersive waves* because of it) accounts for the fact that in real wave systems containing waves of many lengths and periods, the wave patterns change continuously as long waves overtake and "run through" shorter ones.

In shallow water, wave velocity is slower than the deep-water velocity, because tanh kh in equation (6) becomes less than one. In very shallow water, say at depths less than 1/20 of the wave length, tanh kh becomes nearly identical to kh, and another simplification of the equation results. The *very shallow water wave velocity* formula then becomes:

$$V_w = \sqrt{gh} \qquad (8)$$

which shows that, in very shallow water, wave velocity is independent of wavelength and depends only on water depth. This phenomenon is observable by watching waves wash up on a sloping beach, where it will be seen that the waves slow down eventually to a stop, all advancing at the same speed as they are beached.

Energy in a Wave Train. A quantity of special importance in characterizing the severity of a realistic irregular sea is the energy per unit of sea surface area. Since irregular seas are imagined to be made up of many regular wave components, the energy in a train of regular waves is an equally important concept. Imagine a surface "strip" in a regular wave train to be one unit (foot or meter) wide and one wavelength long. The energy caused by waves in the water bounded by that strip and extending to the bottom of the ocean is the sum of the potential energy associated with the raising of the wave crest and lowering of the trough, plus the kinetic energy resulting from the orbital motion of the water particles within and below the surface waves. Hydrodynamic theory shows, remarkably, that the total energy in one such wavelength is exactly half potential and half kinetic, each half being given by the following equation:

$$\text{P. E. per wavelength} = \text{K. E. per wavelength} = \tfrac{1}{4}\,\rho\,g\,\zeta_a^2 L_w$$

Thus the total energy per wavelength is $\tfrac{1}{2}\,\rho\,g\,\zeta_a^2 L_w$. To determine energy per unit area (square foot or square meter), it is only necessary to divide by L_w:

Energy per unit area $= e_w = \frac{1}{2} \rho g \zeta_a^2 = \frac{1}{8} \rho g h_w^2$ (9)

where h_w is the wave height (trough to crest). The very important conclusion to be drawn from equation (9) is that *the energy in a wave train is proportional to the square of the wave height.*

This energy can be expressed in terms of the surface elevation (ζ) instead of the amplitude (ζ_a) by computing the *mean-square surface elevation* (σ_ζ^2) which is called the *variance* in statistics. For a sine or cosine wave, the variance is given by

$$\sigma_\zeta^2 = \frac{1}{2} \zeta_a^2 \qquad (10)$$

where σ_ζ^2 denotes the mean value of the squares of many surface elevation measurements (ζ's) taken at a small time interval over the length of the record, which in this case is any number of sine or cosine cycles. For regular waves, the variance is equal to half the square of the wave amplitude. The energy per unit area can therefore be written as:

$$e_w = \frac{1}{2} \rho g \zeta_a^2 = \rho g \sigma_\zeta^2 \qquad (11)$$

Summary of Deep-water Regular Wave Properties. The properties described above for regular waves in deep water are summarized below along with additional relationships among the various properties.

Surface elevation: $\zeta = \zeta_a \cos k(x - V_w t) = \zeta_a \cos(kx - \omega_w t)$ (12)

Wave number: $k = \dfrac{2\pi}{L_w} = \dfrac{\omega_w^2}{g}$ (13)

Wave frequency: $\omega_w = \dfrac{2\pi}{T_w} = \sqrt{\dfrac{2\pi g}{L_w}} = \sqrt{kg} = kV_w$ (14)

Wave length: $L_w = \dfrac{2\pi V_w^2}{g} = \dfrac{g T_w^2}{2\pi} = V_w T_w = \dfrac{2\pi g}{\omega_w^2}$ (15)

Wave period: $T_w = \sqrt{\dfrac{2\pi L_w}{g}} = \dfrac{2\pi V_w}{g} = \dfrac{L_w}{V_w} = \dfrac{2\pi}{\omega_w}$ (16)

Wave velocity: $V_w = \dfrac{L_w}{T_w} = \sqrt{\dfrac{g L_w}{2\pi}} = \dfrac{g T_w}{2\pi} = \dfrac{g}{\omega_w}$ (17)

Wave energy per unit area of sea surface:

$$e_w = \frac{1}{2} \rho g \zeta_a^2 = \rho g \sigma_\zeta^2 \qquad (18)$$

Realistic Ocean Waves

Although the evaluation of the properties of regular sinusoidal waves described in the last section is essential as a starting point for understanding ocean waves, long-crested regular wave trains are clearly not a satisfactory model of actual ocean waves. Real waves at sea are quite irregular in their lengths, periods, and heights; hence they cannot be classed as regular waves. Nor are they long-crested, with each crest line extending indefinitely across the sea surface, and all crests dutifully marching along parallel to one another. To develop a model of ocean waves that is realistic, we first introduce the complication of the irregularity of a long-crested sea. Short-crestedness will be dealt with later. Ultimately, both phenomena must be included in a probabilistic model of realistic ocean waves.

Irregular Long-crested Wave System. Imagine that you are at the center of a large patch (one square mile, for example) of an idealized ocean surface covered with a single train of regular waves propagating across the water surface. Suppose, for example, that the waves are 200 feet long and 4 feet high (amplitude ζ_a is 2 feet). What are you observing? Wave crests are traveling past you at 32 feet per second, or about 19 knots (equation 17), and a crest passes by every 6.25 seconds (equation 16). Now imagine that another regular wave train adds to the first, the second train consisting of waves 800 feet long and 10 feet high, moving in the same direction as the first. Now what do you observe? As the two trains mix, the surface configuration will be constantly changing. This is because the longer waves travel at much higher speed than the shorter ones—in the case given, at twice the speed since the waves are four times as long. Thus the long waves, moving at 64 feet per second (about 38 knots), will overtake the shorter ones, and the relative positions of crests and troughs of the two trains will undergo continuous changes. One only has to stretch the imagination to picture ten, or sixty, or any conceivable number of such "component" wave trains, each with different lengths and heights, to understand what is meant by an irregular long-crested wave system.

The mathematical description of an irregular long-crested wave system is based on the above concept. The seaway is imagined to consist of the sum of a large number of regular progressive waves, called component waves, each having a definite frequency (or length) and amplitude (or height), and of random phase relationship to one another. In the theoretical limit, there are an infinite number of components each of infinitesimal amplitude. Each component is a harmonic (or sinusoidal) wave subject to all of the formulas relating its length, frequency, velocity, energy, and other properties to one another that have been stated in equations (12) through (18) above. In a long-crested sea, all components are progressing in the same direction. In a short-crested sea, to be discussed later, the direction of propagation of the components also varies. The component regular waves are not visible or identifiable in the seaway. The observable ever-changing irregular waves in the seaway are the result of the summation of the component waves.

Although one of the most obvious and striking characteristics of the sea surface is the randomness or irregularity of the waves, trained observers like ship's officers can estimate the relative severity of the sea at different times with remarkable consistency. The sea at a particular time and place seems to have a characteristic level of severity that remains constant even though the waves are so irregular that patterns never repeat themselves, and the characteristic appearance persists as long as the wind conditions remain constant. The mathematical model of the irregular seaway is also valid only while weather conditions remain constant. It is said to be a *short-term* model and the process described by the model is said to be a *statistically stationary* process. The mathematical model that describes such a process is the *spectrum* of the sea.

The Sea Spectrum. The sea spectrum is the mathematical model that brings some sense of statistical order to a chaotic sea surface by identifying the relative amounts of energy in the regular wave components that are imagined to compose the irregular seaway. The observer who senses that the overall appearance of the sea is holding constant is witnessing a state of constant energy per unit area of the sea. If the waves were regular, that energy (e_w) could be expressed in terms either of the wave amplitude or of the variance of the surface elevation as shown by equation (18). In an irregular sea, the total energy per unit area is the sum of the energies associated with each of the regular wave components that are present in the seaway. It is expressed as a function of the total variance of the irregular wave system, which is approximated by the sum of the variances of the component waves that compose it. The accuracy of the approximation depends on the number of component waves that are used to represent the system. The summation would theoretically be exact if the number of component waves were infinite, but experience has shown that sufficient accuracy is obtained if the irregular sea is represented by 30 to 60 components.

Given a wave record like that shown in Figure 10-3, which depicts the variation with time of the sea surface elevation at a fixed point, the severity of the sea is measured by the *variance of the record* (called E or σ^2) given by:

$$E = \sigma^2 = \frac{1}{N} \Sigma \zeta_i^2 \tag{19}$$

where the ζ_i's are measured values of *deviations from the mean,* which are the surface elevations above or below the still-water level taken at closely spaced time intervals, as shown in the closeup view of a few waves in part (b) of the figure, and N is the total number of ζ_i values in the record. Only a portion of a typical record is shown in the figure. For satisfactory analysis, a wave record must be long enough to contain all of the statistical properties of the sea surface, but not so long that the wind conditions or barometric pressure might change while the

308   APPLIED NAVAL ARCHITECTURE

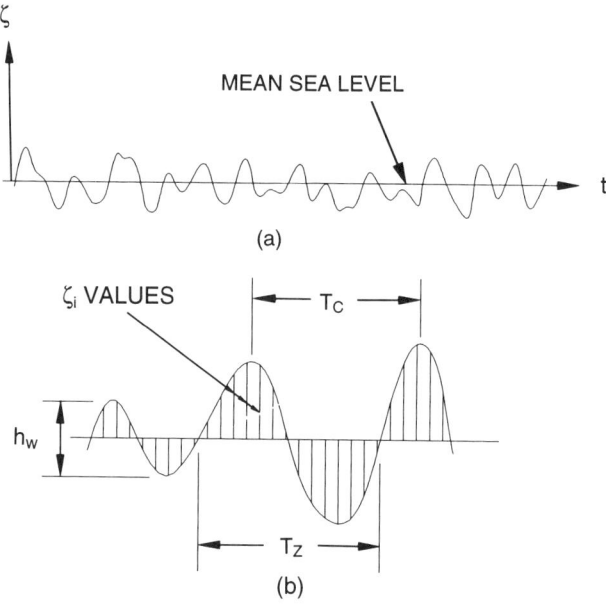

Figure 10-3. Irregular wave record.

record is being taken. This is to assure that it is a proper short-term record of a stationary process. Experience has shown that records of 20 to 30 minutes' duration are most appropriate.

Suppose a histogram of the ζ values from a wave record is made by grouping them into a number of intervals and plotting as the ordinate of each group the number of ζ's in the interval divided by the total number in the entire population. This has been done to thousands of wave records. A typical result is shown in Figure 10-4. The probability density function that has been demonstrated to fit such histograms best is the familiar *normal* or *Gaussian probability density function*. Since the measurements in question are deviations from the still-water level, their mean value is zero, and the process is known as a *zero mean Gaussian random process*.

If the same surface elevations from a wave record are subjected to an analysis procedure known as a *generalized harmonic analysis* or *spectral analysis,* it is possible to define the imaginary sinusoidal (harmonic) component waves whose sum represents the given sea. The function that results from the analysis is a unidirectional or long-crested sea spectrum, sometimes called a *point spectrum* because the information comes from a record of the sea taken at a single point. Figure 10-5 is a simplified example of a point spectrum showing only 12 compo-

SHIP DYNAMICS 309

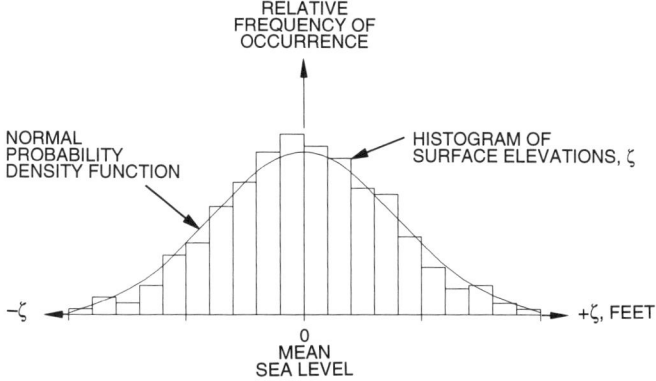

Figure 10-4. Histogram analysis of surface elevations (deviations from the mean).

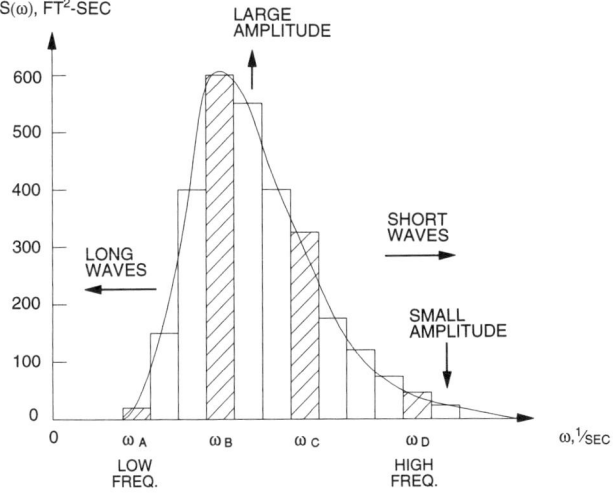

WAVE	ω rad/sec	ζ_a ft.	h_w ft.	L_w ft.
A	0.3	2	4	2240
B	0.6	11	22	560
C	0.9	8	16	250
D	1.3	3	6	120

Figure 10-5. Simplified point wave spectrum.

nent waves. Each component wave determined by the spectral analysis has a particular frequency ω and amplitude ζ_a. Each component represents all waves whose frequencies lie within a narrow frequency band of width $\delta\omega$. Long-wave components are shown toward the left, or low-frequency end, and short-wave components are at the right. The spectral ordinate $S(\omega)$ has no physical connotation in its own right, but an elemental area $S(\omega_i)\,\delta\omega$ is a measure of the variance of the regular wave of frequency ω_i. The definition of an elemental area of the spectrum, when combined with equation (10), gives:

$$S(\omega_i)\,\delta\omega = (\sigma_\zeta^2)_i = \tfrac{1}{2}(\zeta_a^2)_i \tag{20}$$

A large ordinate thus represents waves of large amplitude or height, which are those with the most energy. Since the variance of the sum of the component waves approaches the sum of the variances of the components as δw approaches zero, it follows that the total area under the spectrum plot is a measure of the variance of the system. That is,

$$E = \sigma^2 = \int_0^\infty S(\omega)\,d\omega \tag{21}$$

The area of a spectrum is thus a good measure of the severity of the seaway. In fact, since the energy per unit area contributed by each component wave is equal to ρg times the variance of the component wave as shown by equation (11), the total energy per unit area of the irregular wave system is equal to ρg times the variance of the wave system:

$$\text{Energy per unit area} = \rho g \sigma^2 = \rho g E \tag{22}$$

Because the spectral area is proportional to wave energy, the spectrum is sometimes referred to as the *energy spectrum* of the sea. Since it does not include the density ρg, however, it is more correct to refer to it as a *variance* spectrum.

In the figure, four of the defined components are described in the accompanying table to illustrate the relationships defined above. Note that the longest component wave defined by this spectrum is more than 2,000 feet long, but it is only four feet high. The most severe (highest) wave component is component "B," which is 560 feet long and 22 feet high. A study of the table of values in Figure 10-5 will give the reader some sense of what a wave spectrum portrays.

Significance of the Shape of the Spectrum. The spectral area is one of the most important properties of the sea spectrum, since it is proportional to the energy of the sea surface, and therefore to the severity of the sea, as shown in Figure 10-6(a). Certain other shape characteristics are also important. They are illustrated diagrammatically in the same figure. A spectrum is said to be *narrow* or *narrow-*

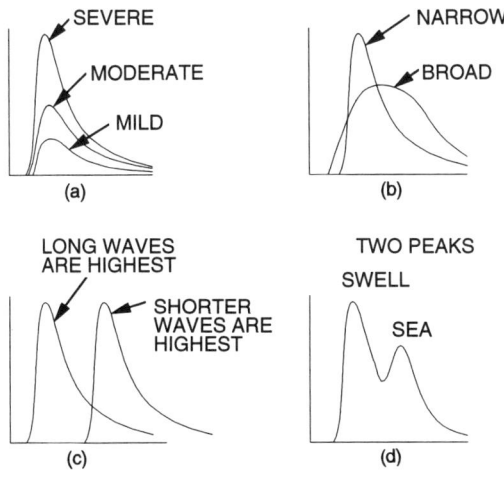

Figure 10-6. Typical spectral shapes.

band if the high-energy portions are concentrated within a small band of frequency (see part b of the figure), and *broad* if the energy is spread over a wide frequency band. Typical sea spectra tend to be relatively narrow-band spectra.

Part (c) of the figure illustrates two spectra of equal area and thus seas of equal severity, but the peaks are located at different frequencies. The peak of the spectrum is associated with the wave component having the greatest amplitude or height, so in the spectrum peaking at the lower frequency the waves of greatest amplitude are long waves, while shorter waves are the more severe ones in the spectrum peaking at the higher frequency. The location of the peak has important consequences for ships that are in a given sea. For example, longitudinal bending is most severe when ship-length waves predominate, so a spectrum with a low-frequency peak can cause severe stresses in long ships, while short ships experience their maximum bending stresses in a completely different sea with a higher-frequency peak.

A common shape found when actual wave records are processed by spectral analysis is a two-peaked spectrum such as that shown in part (d) of the figure. The lower-frequency peak is caused by the remnants of a remote storm system whose longer-wave components have traveled a great distance while its shorter components have died out. This phenomenon is a characteristic of wind-generated waves—the longer-component waves always survive longer and move farther from the storm center than do the shorter components. On the other hand, when a storm begins, it is the short (high-frequency) waves of small amplitude that develop first; and if the high winds persist for a long time, called the *dura-*

tion, the predominant components that are generated increase in both length and amplitude. A double-peaked spectrum results when the seaway consists simultaneously of a nearby storm of short duration and the remnants of a more-remote storm of longer duration. Experienced seamen are able to recognize both characteristic patterns when they take observations for their weather logs. They call the long- and the short-wave patterns *swell* and *sea,* respectively.

The Encounter Spectrum

All of the spectra so far described have been derived from the records of waves measured at a fixed location at sea. A moving ship, however, experiences the waves that it encounters as it moves through them. A record of sea surface elevations taken from a moving ship has waves whose periods or frequencies differ from those of the same waves measured at a fixed point. A spectral analysis of a wave record taken from a moving ship is called an *encounter spectrum*. The frequency of each wave component is called its *encounter frequency* or *frequency of encounter,* and its period is the *encounter period* or *period of encounter.*

The relationship between wave frequency and frequency of encounter depends on the speed of the ship and its direction of travel relative to the direction of travel of the waves. The angle of wave propagation relative to the ship's heading (μ) is defined in Figure 10-7. A ship moving in head seas, directly into the oncoming waves, is said to be at a heading of 180 degrees relative to the waves, while a 0-degree heading represents following seas. The period and frequency of encounter are derived as follows, given that

μ = ship-to-wave angle as defined in Figure 10-7
ω_e = frequency of encounter
T_e = period of encounter = $2\pi/\omega_e$
V_s = ship speed
ω = wave frequency
L_w = wave length = $2\pi g/\omega^2$ from equation (15)
V_w = wave speed = g/ω from equation (17)

From the figure it will be noted that the component of the ship's velocity in the direction of propagation of the waves is $V_s \cos \mu$. Its velocity *relative to the waves* in the direction of wave propagation is therefore $V_w - V_s \cos \mu$. In one period of encounter, T_e seconds, the component of the ship's travel in the direction of the waves is one wave length (L_w). Thus the relative velocity can also be expressed as L_w/T_e feet (or meters) per second. Equating these two expressions of the relative speed of the ship and waves, we have

$$\frac{L_w}{T_e} = V_w - V_s \cos \mu \qquad (23)$$

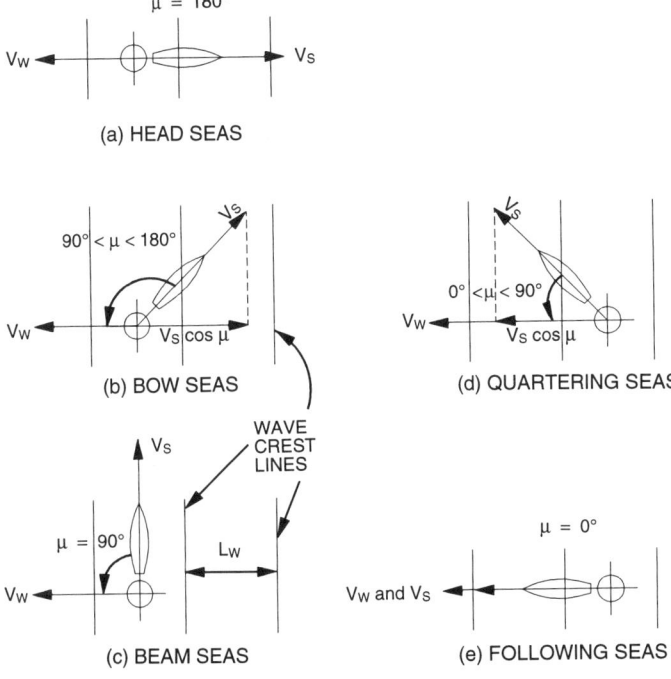

Figure 10-7. Waves encountered by a moving ship.

Substituting the expressions for L_w, T_e, and V_w given above into this equation,

$$\frac{2\pi g}{\omega^2} \cdot \frac{\omega_e}{2\pi} = \frac{g}{2\pi} - V_s \cos \mu$$

$$\omega_e = \omega - \frac{\omega^2 V_s \cos \mu}{g} = \omega \left(1 - \frac{\omega V_s}{g} \cos \mu \right) \quad (24)$$

In bow and head seas, μ is greater than 90 degrees and $\cos \mu$ is negative, so the frequency of encounter is greater than the wave frequency. In quartering and following seas, the frequency of encounter is less than the wave frequency, and in fact it can be zero in following seas if the ship speed is the same as the wave speed.

If an encounter spectrum is to be determined from the corresponding wave spectrum rather than from wave records actually measured from the moving ship, the spectral ordinates must also be transformed. This is necessary to ensure that the area of the encounter spectrum is the same as that of the wave spectrum, because the spectral area is a measure of the energy per unit area of the sea surface, which must be the same for a moving observer as for a stationary one. This can be done by equating the elemental areas of the corresponding spectra. Designating the ordinate of the encounter spectrum as $S(\omega_e)$, the requirement is that

$$S(\omega_e) \, d\omega_e = S(\omega) \, d\omega$$

or

$$S(\omega_e) = S(\omega) \cdot \frac{d\omega}{d\omega_e}$$

Differentiating equation (24) gives

$$\frac{d\omega_e}{d\omega} = 1 - \frac{2\omega V_s}{g} \cos \mu$$

Therefore

$$S(\omega_e) = \frac{S(\omega)}{1 - \frac{2\omega V_s}{g} \cos \mu} \tag{25}$$

In summary, a sea spectrum for a stationary point can be transformed into the encounter spectrum that would be experienced by a ship moving at a speed of V_s and a heading of μ degrees to the waves as follows:

Multiply wave frequencies by $(1 - \frac{\omega V_s}{g} \cos \mu)$

Divide spectral ordinates by $(1 - \frac{2\omega V_s}{g} \cos \mu)$

In Figure 10-8 encounter spectra for several ship speeds are shown compared to the wave spectrum for zero speed, assuming head seas ($\mu = 180°$). Note that the areas of all spectra are the same, but the frequency of encounter associated with the wave component of maximum energy, that is, the peak of the spectrum, depends on the ship speed. This is an important consideration for a ship that is experiencing severe motions because its natural frequency of a particular motion (roll, pitch, or heave, for example) coincides with the peak frequency of the encounter spectrum. The ship is then said to be in resonance with a wave component of high energy. Since the peak encounter frequency depends on the ship speed and heading, the severe motions can often be reduced significantly by changing either the speed or heading of the ship.

When a ship is in quartering or following seas, the transformation to the encounter spectrum is mathematically complicated by the fact that there may be three different ω values that correspond to the same ω_e value. These complications are not discussed here, because the calculations necessary to predict ship motions in irregular seas can be done without transforming sea spectra into encounter spectra, so long as the "response amplitude operators" that represent the basic ship motions are prepared in the correct way. Response amplitude operators are discussed in a later section.

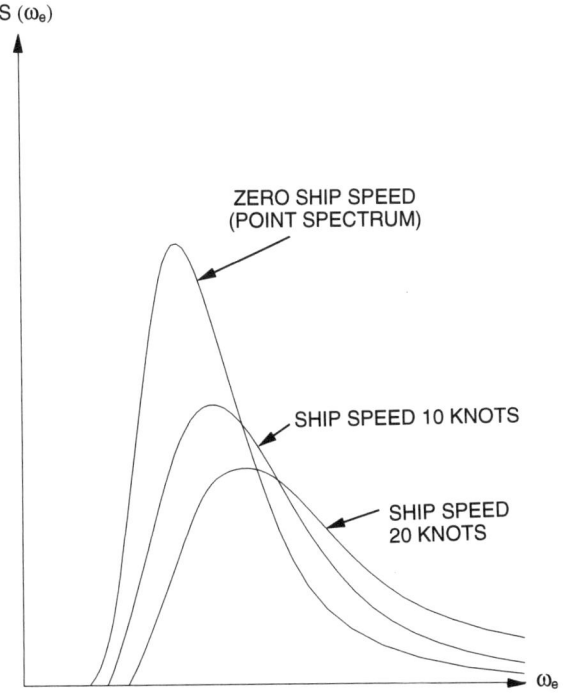

Figure 10-8. Encounter spectra, head seas.

The Directional Spectrum. A sea spectrum determined by the spectral analysis of a record of sea surface elevations at a single point at sea is called a *point spectrum* or a *one-dimensional* spectrum. A point spectrum represents the relative amounts of energy in component waves of various frequencies; thus the spectral ordinate is depicted as a function of frequency (ω) only. Measurements taken at a point with simple wave recorders cannot detect the direction of travel of the waves that make up the sea, whether one considers the apparent waves or the invisible wave components. Thus a point spectrum is not a complete representation of the ocean surface irregularity, which includes waves moving in various directions as well as those having various heights and frequencies. The simplified sea represented by a point spectrum is what has been previously called a long-crested sea, in which all waves are assumed to travel in the same direction. The spectrum of a more realistic short-crested sea, which includes information on wave direction, is a *two-dimensional spectrum* or a *directional spectrum*.

To get wave records that have directional information from actual measurements requires very special techniques and complicated analytical procedures. Several techniques have been used, including stereoscopic aerial photography of the sea surface, arrays of wave recorders operating simultaneously, and special wave rider buoys that can measure the slope and direction of the slope of the sea surface as well as its elevation. Since the complications of acquiring such measurements precludes the use of this special equipment as a general wave measurement system, the few measured directional spectra that do exist have been analyzed to determine an appropriate mathematical correction factor that can transform a point spectrum (long-crested) into an equivalent directional spectrum (short-crested). The correction factor in question is called a *spreading function*.

The spreading function approximates the percentage of wave energy that approaches a point at sea from different directions. The direction of the wind is considered the dominant wave direction, and the angle μ_w is measured from the dominant wave direction. The point spectrum $S(\omega)$ is transformed into the directional spectrum $S(\omega, \mu_w)$ by multiplying it by the spreading function $F(\mu_w)$ as follows:

$$S(\omega, \mu_w) = S(\omega) F(\mu_w) \tag{26}$$

A simple spreading function that matches directional measurements fairly well is the "cosine-squared" spreading function defined by

$$F(\mu_w) = \frac{2}{\pi} \cos^2 \mu_w \tag{27}$$

which is valid for an angular spread μ_w between –90 degrees and +90 degrees only. The function $F(\mu_w)$ is taken as zero for μ_w outside of this range. Using this spreading function preserves the amount of energy in the wave system; that is, E remains unchanged from that obtained by integrating the point spectrum.

Statistics of Irregular Wave Records

Returning to the wave record shown in Figure 10-3, we examine the statistical properties of the 20- to 30-minute record of which it is a part. To grasp the concept of the spectrum, the sea surface was imagined to be made up of many sinusoidal wave components. It must be emphasized that those harmonic components are not the waves that an observer would actually see. The crests, troughs, heights, and periods of the visible waves are shown on the wave record, from which numerical measurements can be taken. Wave heights and periods measured directly from the record are called the *apparent properties* of the waves because they are actually visible.

The apparent wave properties shown in Figure 10-3 are the apparent wave height (h_w), the peak-to-peak period (T_c), and the zero-upcrossing period (T_z). Each of these kinds of measurements occurs many times in a typical wave record, and the mean values of each can be determined from such measurements.

Theoretical studies of the probabilities associated with a stationary zero-mean Gaussian process show that these and other average apparent wave properties (wavelength and wave slope, for example) can also be determined from the spectrum by computing the moments of the spectrum about its $\omega = 0$ axis. For our purposes, we shall focus our attention only on the wave property that most influences ship motions, namely the wave height.

Apparent wave heights as shown in Figure 10-3 are the vertical measurements from a crest, or maximum point, to a trough, or minimum point on a record. It therefore follows that the probabilities associated with wave heights should be determined from a study of the statistical behavior of the *maxima and minima* of the process. Such studies have shown that the probability density function of the maxima of typical wave records is the Rayleigh probability density function, shown superimposed on a histogram of apparent wave heights in Figure 10-9. The probability density of apparent wave height [p(h$_w$)] is given by

$$p(h_w) = \frac{h_w}{4E} \exp\left(\frac{-h_w^2}{8E}\right) \qquad (28)$$

where $E = \sigma^2$ = variance of the record, or mean square value of the sea surface elevation. The probability that a wave height h_w in a wave record lies within a small interval δh_w is given by the elemental area $p(h_w)\delta h_w$.

Strictly speaking, the Rayleigh distribution is a correct fit to the distribution of maxima of a record only if the process is ideally narrow-banded. Ocean wave records tend toward narrow-bandedness, but not always to the limit. In most cases, the assumption that the Rayleigh distribution is correct for a short-term wave record is satisfactory. When it is not, a more accurate expression of the probability density function can still be determined by introducing a "spectral broadness parameter," which cannot be considered further here. Fortunately, time records of

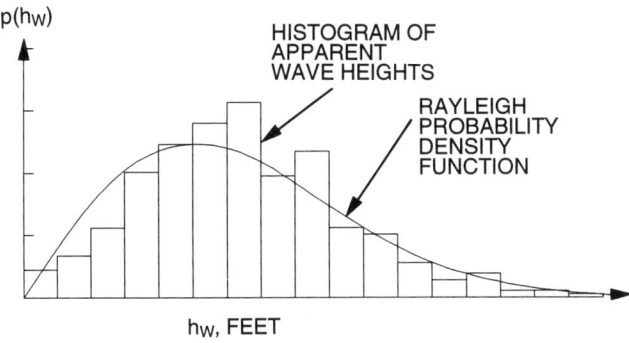

Figure 10-9. Analysis of apparent wave heights.

most ship motions in an irregular seaway are usually more narrow-banded than are the waves that produced them. Therefore the Rayleigh distribution is a very good probability model for predicting ship motions.

Many statistical properties of Rayleigh processes have been derived that apply to short-term records of waves or of ship motions. The most quoted measure of wave height is the *significant wave height* ($H_{1/3}$), which is defined as the average of the ⅓-highest apparent wave heights in a sample. The significant wave height and all of the other statistical measures listed below are functions only of the variance of the record, E or σ^2, which is also equal to the spectral area. They can therefore be determined either directly from the record by measuring apparent wave heights, or by measuring the spectral area if a spectral analysis has already been done.

Average apparent wave height $= H_{avg} = 2.5\sqrt{E}$ (29)

Significant wave height $= H_{1/3} = 4.0\sqrt{E}$ (30)

Average of ¹⁄₁₀-highest waves $= H_{1/10} = 5.1\sqrt{E}$ (31)

Predictors of extreme values of waves (or ship motions) have also been determined, based on the assumption that conditions remain statistically stationary throughout the period necessary to experience a sufficient number of waves. For example, the expected heights of the highest wave in samples of various numbers of consecutive waves are given below.

Highest expected wave in 50 waves $= 6.0\sqrt{E}$ (32)

Highest expected wave in 100 waves $= 6.5\sqrt{E}$ (33)

Highest expected wave in 500 waves $= 7.4\sqrt{E}$ (34)

Highest expected wave in 1,000 waves $= 7.7\sqrt{E}$ (35)

Highest expected wave in 5,000 waves $= 8.6\sqrt{E}$ (36)

Highest expected wave in 10,000 waves $= 8.9\sqrt{E}$ (37)

Spectra for Ship Motions Calculations

For predicting ship motions in irregular waves, it is desirable to express sea spectra by formulas that approximate measured spectra at various levels of severity of the sea. Several spectral formulations have been developed by oceanographers that can be used to represent idealized spectra in terms of some particular measures of sea severity, such as wind speed or significant wave height. Of the various formulations that have been published, one that is particularly useful to ship designers was recommended by the International Towing Tank Conference (ITTC) in 1978 for use in predicting ship motions in irregular seas. The ITTC formulation for a point spectrum is

$$S(\omega) = \frac{A}{\omega^5} \exp\left(\frac{-B}{\omega^4}\right) \tag{38}$$

The parameters A and B can be expressed in several ways, depending on how the severity of the sea is expressed. If the significant height $H_{1/3}$ and the period corresponding to the average frequency of the component waves (T_1) are known, then

$$A = 173 \cdot \frac{H_{1/3}^2}{T_1^4} \tag{39}$$

and

$$B = \frac{691}{T_1^4} \tag{40}$$

In the more usual case for practical calculations, the known information is taken from ship's logbooks, which would usually include visual estimates of wave height and period (H_V and T_V). If that is the type of data that is known, the corresponding $H_{1/3}$ and T_1 can be approximated by

$$H_{1/3} = 1.68 \, H_V^{0.75} \tag{41}$$

and

$$T_1 = T_V \tag{42}$$

The ITTC also recommends using the cosine-squared spreading function of equation (27) to convert this point spectrum to a directional spectrum.

SHIP MOTIONS

There are three sources of information on the nature of ship motions at sea:

- Hydrodynamic theory to determine equations of motion
- Model test measurements in wave basins
- Motion measurements on ships at sea

Research in all three of the above areas has produced a significant body of knowledge about the complexities of ship motions in irregular waves. The problem is an extremely complicated one, since the ship at sea is subject to motions in six degrees of freedom. That is, the six different components of motion defined in Figure 10-10 can all take place simultaneously. The three translational components along the x, y, and z axes are called surge, sway, and heave, respectively. The rotations about the axes in the same sequence are the rotational motions called roll, pitch, and yaw.

A complete theoretical solution to the ship motions problem would require writing six equations of motion, including all of the complex interactions and

320   APPLIED NAVAL ARCHITECTURE

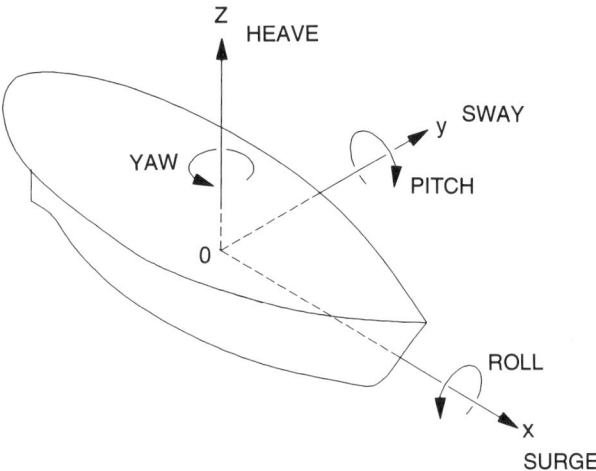

Figure 10-10. Six modes of motion (degrees of freedom).

mutual influences of each motion on the others (called cross-coupling effects), and then solving the equations simultaneously to determine the forces and moments acting on the ship. It is a monumental analytical task, requiring the tools of advanced differential equations and theoretical hydrodynamics. Theoreticians have made a lot of progress toward the ultimate goal by making certain simplifying assumptions, notably the assumption of linearity of the motions. In its simplest form, this assumption asserts that twice the wave height will produce twice the motions. Although linearity applies in practice only to small motions, the linearized theory is adequate for many purposes in predicting ship motions. Ship motions in regular waves are represented by modified versions of the equations for simple harmonic motion (the mass-spring-dashpot system and the pendulum are simple examples) in which special provisions have to be made for the effects produced by the water that is accelerated by the motions of the ship.

Supplementing the purely theoretical approach to ship motions, many tests have been made of models in special ship-testing basins equipped with wavemakers. Regular waves of all proportions and irregular long-crested wave trains matched to any given spectrum can be produced in these basins. A few major facilities also permit researchers to create short-crested irregular seas. During such model seakeeping tests, instruments installed in the model measure and record the motions that it is desired to determine in the test, while the wave pattern encountered by the model is also measured. Many special-purpose model tests have been devised to measure ship responses to waves, other than the basic ship motions. These include tests that measure accelerations, forces, or pressures at various points on the models, impact forces caused by slamming as the bow of the model emerges from and then re-enters the waves, shipping of water over the bow, racing of the propeller due to propeller emergence, internal sloshing forces in tanks, added ship resistance in waves, and wave bending moments, for a few examples. Model tests are also used to evaluate or verify some of the coefficients of the equations of motion mentioned above.

Full-scale measurements of ship responses to realistic seas have also been made to verify the theory. Measurements on ships at sea are particularly helpful in developing probability models for long-term predictions of hull stresses that result from wave-induced bending moments. Many ships of different types have been instrumented to gather stress data, some for as long as 12 years. Other types of full-scale measurements that have been made include measurements of shaft power, hull pressures at various points on the hull, accelerations to determine the forces on various parts of the ship structure or on container lashings, and shipping of water. For full-scale measurements to be most helpful in verifying theoretical concepts, the waves in the vicinity of the ship should be measured simultaneously for spectral analysis. Alternatively, it is at least as imperative that careful logbook records be kept of the observed wave heights, periods, and predominant wave direction so that the spectrum of the sea can be approximated.

The Principle of Linear Superposition
If a ship travels at a uniform speed and steady heading into a series of regular waves, the ship motions will also be regular (or sinusoidal, or harmonic). Regular motions in regular waves are the conditions simulated by the equations of motion and by tests of models in regular waves. But the ultimate goal of ship motions studies is to be able to predict how the ship will behave in realistic irregular seas such as those represented by a particular sea spectrum. The relationship between motions in regular and irregular waves must be established in order to use either the theoretical equations or the experimental studies to lead to practical predictions of ship behavior in a realistic seaway.

The needed relationship is expressed by the *principle of linear superposition,* which can be stated as follows: "The response of a ship to an irregular sea can be represented by a linear summation of its responses to the component regular waves that comprise the irregular sea."

This principle, which derives from analyses of electromagnetic wave phenomena of interest in the communications field, was first applied to the calculation of ship motions in random seas in 1953 in an outstanding paper by St. Denis and Pierson (listed at the end of this chapter), which is a compelling example of what can be accomplished by the cooperation of a naval architect (St. Denis) and an oceanographer (Pierson). Since we have seen that a random sea can be mathematically decomposed into regular wave components by spectral analysis, the power of linear superposition becomes apparent: If the ship's responses to each regular wave component of a particular sea spectrum have been determined, whether by calculation or by model testing, those response components can be added in a special way to get a prediction of the response of the ship in the sea spectrum in question. The procedure for doing this is described below.

Strictly speaking, the principle of linear superposition applies to a somewhat restrictive set of circumstances. First, the seaway must be a statistically stationary Gaussian random process. We have already noted that to a good approximation, this assumption is valid in the short term. If long-term responses are desired, predictions must be made in many spectra, each representing the sea for a particular short-term period, the whole set of spectra being chosen to represent the seaway variations over a long period of time, and each spectrum having a known frequency of occurrence. Secondly, the amplitudes of the individual responses to be evaluated (pitch, heave, and roll, for example) must vary linearly with the wave amplitude. This assumption has been shown to be correct for most responses so long as they are small to moderate in magnitude. And the technique has been shown to make acceptable, if less accurate, predictions of severe motions as well.

Response Amplitude Operators

To make use of the principle of linear superposition, a ship response to regular waves, whether it has been determined by calculation or by model measurement, is expressed in a special form often referred to as a *response amplitude operator, (RAO)*. A response amplitude operator is a measure of the response to a regular wave of unit amplitude. The response amplitude operators are defined as follows:

$$\text{RAO} = \frac{\text{amplitude of the response}}{\text{amplitude of the wave}}$$

The following examples of response amplitude operators are for heave, pitch, and roll. Of the six fundamental modes of motion or degrees of freedom, these three are the only ones that generate restoring forces or moments when wave action or the ship motion itself displaces the ship from equilibrium, hence they are most like simple harmonic motions. The RAOs are designated by the symbol Y with two subscripts, the first representing the type of response, and the second representing the wave amplitude, which is the quantity to which the response is assumed to be proportional. The symbols used for the responses and the subscripts are:

> z represents heave (feet or meters)
> θ represents pitch angle (degrees)
> φ represents roll angle (degrees)
> ζ represents the wave amplitude (feet or meters)

and the subscript "a" stands for amplitude of the wave or response.

$$\text{Heave RAO} = Y_{z\zeta} = \frac{z_a}{\zeta_a} \tag{43}$$

$$\text{Pitch RAO} = Y_{\theta\zeta} = \frac{\theta_a}{\zeta_a} \tag{44}$$

$$\text{Roll RAO} = Y_{\varphi\zeta} = \frac{\varphi_a}{\zeta_a} \tag{45}$$

It should be noted that the heave RAO is dimensionless, since heave amplitude and wave amplitude would be expressed in the same units, but the angular motion RAOs for pitch and roll are not dimensionless. They are expressed in degrees per foot or degrees per meter. (Pitch and roll angle are usually expressed in degrees, although radian measure could be used instead.) Expressing RAOs in this way is necessary for the superposition calculations described below. Experimental results of model measurements may be recorded in other forms more suitable for comparing the responses of different models to each other, but they are always converted into the "divided by wave amplitude" form before superposition calculations are performed.

An example of model test measurements converted to response amplitude operators is illustrated in Figure 10-11. A model was tested in four different regular head waves of amplitude ζ_a. The four wavelengths (L_W) varied from 0.75 times model length (L) to 1.5 times model length. Model heave amplitude (z_a) was measured in each test. Results of the tests are plotted in part (a) of the figure in a nondimensional format, z_a/ζ_a against L_W/L. The shape of the curve is quite characteristic for heave in regular waves of different lengths. In waves shorter than about half the length of the ship, heave is negligibly small. In very long waves, the ship length is small compared to the wave length and the ship motion follows that of the wave. Thus the heave amplitude approaches the wave amplitude in very long waves. The peak value of heave is reached in a condition of *resonance,* or *synchronous heaving,* which means that the frequency with which the ship encounters the regular waves matches the ship's natural frequency of heaving oscillation. This phenomenon is characteristic of forced simple harmonic motions like heave, pitch, and roll. Motions can become severe in conditions of synchronism, sometimes requiring the ship operator to change the ship's speed or heading to reduce them.

In part (b) of the figure, the heave RAO is plotted against wave frequency rather than wavelength. Since wave frequency varies inversely as wavelength, wavelength increases toward the left and the RAO curve is run in to 1.0 as it approaches $\omega = 0$. This is an expression of the fact that as the waves get extremely long (approaching infinite length as frequency approaches zero), the ship simply rises and falls with the water surface, thus the amplitude of heave becomes equal to the wave amplitude. It is in this form that response amplitude operators are

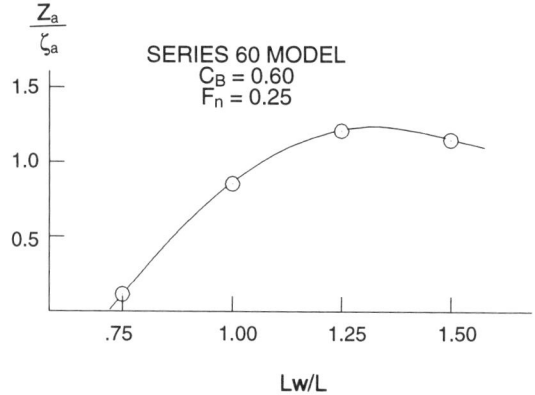

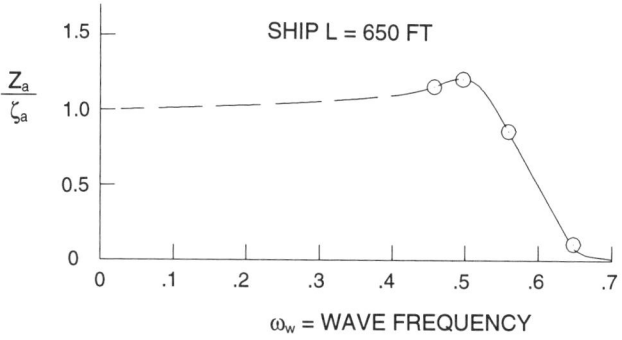

Figure 10-11. Ship motions model tests and Response Amplitude Operators.

prepared for use in superposition calculations. A sample calculation is described in a later section.

Any number of derived responses other than the three mentioned above can be expressed as response amplitude operators. Model test data of ship motions relative to the wave surface, bending moments amidships, added ship resistance, and velocities and accelerations at various points in the ship model are just a few examples of other types of measurements that have been published in the technical literature. For the purposes of this text, the above three examples of motion RAOs will suffice to illustrate the procedure for making superposition calculations.

Response Spectra

In the previous discussion of sea spectra, reference was made to the time record of the sea surface elevation at a fixed point at sea. A wave record that has been analyzed as suggested earlier can be represented numerically by its variance (E or σ^2), by its spectrum, and by the statistical predictions of its apparent properties such as its significant height. It may be recalled that the mathematical models that represent the seaway are valid if the process is a statistically stationary narrow-banded random process. Wave records in deep water tend to be such processes, and the records look like the example shown in Figure 10-3. A simplistic representation of the recording instrument for such a record would be that it represents the vertical motion of a small cork floating on the sea surface.

Imagine now that the cork is replaced by a ship that is hove-to; that is, it remains at a fixed point in the ocean. It is not underway. Suppose further that recording instruments aboard the ship are recording the ship's heave, pitch, and roll motions. If the records of these motions were not identified and no scales or units were indicated on them, it is unlikely that anyone could determine which record went with which motion. Indeed, they would also look very much like the wave record. The point to be made by this imaginary experiment is that ship motions in irregular seas can be represented by the same kinds of statistical and analytical models as the waves that caused them. Similarity of appearance is not sufficient to prove this statement, of course, but the fact is that analyses of countless records of ship motions have shown that, except for very extreme motions, the statement is correct. Therefore records of ship responses can be subjected to spectrum analysis to determine the spectrum of the response. The properties of response spectra are exactly like the properties of the sea spectra described above. For example, an elemental area of a response spectrum is a measure of the variance (or one-half the squared amplitude) of the response to the regular wave component that it represents, analogous to equation (20) for waves. Also, the integral over frequency of the response spectrum, which is its total area, is equal to the variance of the response. Furthermore, all of the predictions of extreme waves that arise out of the Rayleigh or narrow-banded assumption also apply to ship responses. Therefore the "Rayleigh multipliers" that determine significant height and extreme heights of waves from wave variance are also applicable to ship responses. Since the magnitudes of ship responses are usually expressed as *amplitudes,* while wave magnitudes are usually expressed as *heights* or *double amplitudes,* the appropriate multipliers for ship responses should be half of those quoted for wave heights in the section on the statistics of irregular wave records. This will be made clear in a later list of the multipliers.

If the ship whose motions are being measured is moving rather than being hove-to, the spectra determined from the recorded motions will be encounter spectra, and the frequencies will be encounter frequencies just as in the case of a wave record taken from a moving wave probe. Analysis of encountered response spectra involves complications that are not discussed here. Such an analysis is necessary only if full-scale or model measurements are made in irregular wave tests. The mathematical complications can be avoided if the more common model testing procedure is adopted, that is, testing models in a number of different regular wave patterns to determine response amplitude operators as illustrated in Figure 10-11, then applying the analytical techniques of linear superposition to predict responses in irregular seas.

Superposition Calculations

Based on the concepts defined above, it is now possible to formalize the procedure for predicting various statistical measures of the response of a ship to an irregular sea by calculating the variance of the response. The calculation is expressed as follows, using the response of heave as an example.

$$S_z(\omega_i) = S_\zeta(\omega_i) \times (Y_{z\zeta})_i^2 \qquad (46)$$

The procedure is illustrated graphically in Figure 10-12 for the calculation of the heave response of a ship at a given speed in long-crested irregular head seas for which the point spectrum is known. Each of the component regular waves imagined to make up the irregular sea is represented by an elemental area at its frequency (ω_i). As shown by equation (20) and on the figure, the elemental area is a measure of the variance of the regular wave component, which is equal to one-half the square of the component wave amplitude. Multiplying the wave spectrum ordinate at ω_i by the square of the RAO shown in part (b) of the figure at the same frequency yields an ordinate of the response spectrum in accordance with equation (46). Repeating the process for the entire range of frequencies determines the response spectrum shown in part (c) of the figure.

The response amplitude operators used in such a calculation could have been determined either from model tests in regular waves as described previously, or by analytical calculations. In the procedure shown here, it is important to note two characteristics of the RAO plot shown that may be different from those shown in technical papers or other texts on the subject. They are:

1. RAO defined here is the ratio of motion amplitude to wave amplitude. It must be squared during the calculations. (Some authors call the squared quantity a response amplitude operator.)

2. Values of the RAO are plotted at the frequency of the waves in which the model test was made, and not at the frequency of *encounter* of the model with the waves. Using this procedure eliminates the need to convert the wave spectrum into an encounter spectrum.

The response spectrum determined by this calculation procedure is not the spectrum that would have been determined by performing a spectral analysis of

SHIP DYNAMICS 327

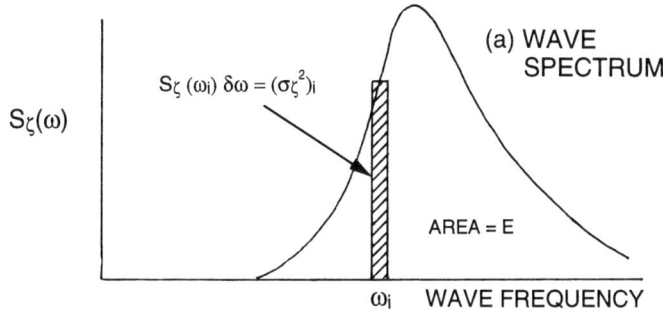

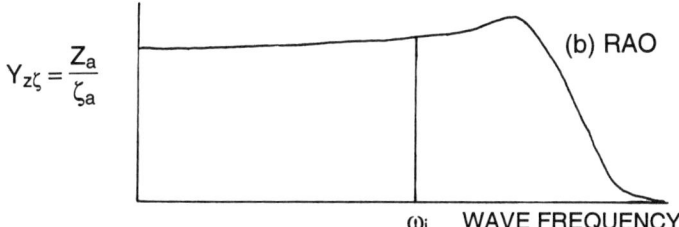

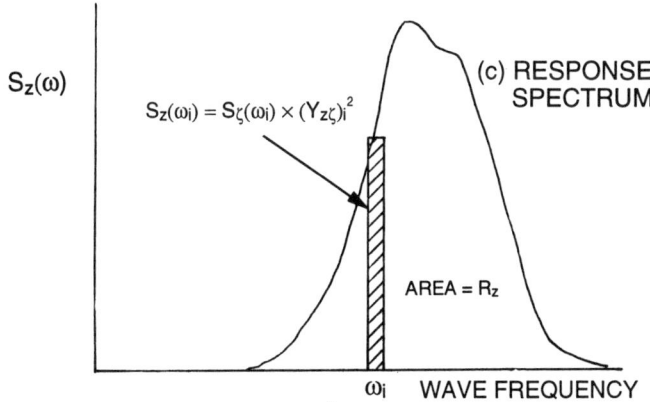

Figure 10-12. Superposition calculations.

the record of the heave motion as measured on the ship moving through the irregular waves represented by the wave spectrum. It has a different shape because the spectrum from the moving ship would be an encounter spectrum of the response. Nevertheless, the heave response spectrum of Figure 10-12 has the same area as the corresponding encounter spectrum of heave in the given seaway, and that area is a correct measure of the variance of the ship's heave response in that seaway. The heave spectral area may be calculated by integrating (by numerical integration) the response spectrum:

$$R_Z = \int_0^\infty S_z(\omega)\, d\omega \tag{47}$$

where R_Z = variance (mean square) of heave amplitude

It was noted above that the statistical distribution of the maxima of ship motions in irregular seas may be represented by the Rayleigh distribution, as for ocean waves. Therefore the statistical predictions of the amplitudes of ship motions may be determined from the variance, or spectral area, of a ship response spectrum using the relationships of equations (29) through (37), but with the multipliers halved so that motion amplitudes are calculated. The corresponding multipliers are given below. The relationships apply to any response, so the variance of the response has been generalized by using the symbol R.

$$\text{Average response amplitude} = 1.25\sqrt{R} \tag{48}$$

$$\text{Significant response amplitude} = 2.0\sqrt{R} \tag{49}$$

$$\text{Average of } 1/10 \text{ highest} = 2.55\sqrt{R} \tag{50}$$

The greatest expected amplitudes of motion in samples of N motion cycles in a given seaway are, for the following values of N:

$$N = 50,\ \text{maximum expected amplitude} = 3.00\sqrt{R} \tag{51}$$

$$N = 100,\ \text{maximum expected amplitude} = 3.25\sqrt{R} \tag{52}$$

$$N = 500,\ \text{maximum expected amplitude} = 3.70\sqrt{R} \tag{53}$$

$$N = 1{,}000,\ \text{maximum expected amplitude} = 3.85\sqrt{R} \tag{54}$$

$$N = 5{,}000,\ \text{maximum expected amplitude} = 4.30\sqrt{R} \tag{55}$$

$$N = 10{,}000,\ \text{maximum expected amplitude} = 4.45\sqrt{R} \tag{56}$$

SUGGESTIONS FOR ADVANCED STUDIES

The subject of ship motions is very complex, and the actual calculations involved in predicting the statistics of ship motions in random irregular seas requires the application of higher mathematics and complicated analytical procedures. The calculations also require the use of computers and special computer programs capable of doing voluminous numerical calculations very rapidly. Hand calculations of any significant amount of ship motions results would be very burdensome and tedious, and a complete set of calculations for all motions and for a range of sea severity would be quite impossible to do by hand. Therefore, in this elementary coverage of the subject, actual calculations have not been demonstrated.

For the reader who is interested in pursuing this subject in greater depth, the references that follow are recommended. They have been chosen not only because they contain good coverage of the subject, but because they are more readily available than the original sources of ship motions research work, which were published in technical journals. The following are all books, except for the St. Denis and Pierson paper, which is included because of its landmark status. The really zealous student might want to have a list of the original research papers. The 381 papers listed as references to Chapter 8 of the *Principles of Naval Architecture* (see below), and the more than 1,100 references cited in the Korvin-Kroukovsky book (see below) should satisfy even the most voracious appetite.

Bhattacharyya, R. 1978. *Dynamics of Marine Vehicles.* New York: Wiley Interscience.
Hogben, N., and Lumb, F. E. 1967. *Ocean Wave Statistics.* London: Her Majesty's Stationery Office.
Korvin-Kroukovsky, B. V. 1961. *Theory of Seakeeping.* New York: Society of Naval Architects and Marine Engineers.
Lamb, Sir Horace. 1945. *Hydrodynamics.* Sixth Edition. New York: Dover Publications.
Lewis, E. V., ed. 1989. *Principles of Naval Architecture.* Jersey City, N.J.: Society of Naval Architects and Marine Engineers. (Chapter VIII in Volume III, written by R. F. Beck, W. E. Cummings, J. F. Dalzell, P. Mandel, and W. C. Webster deals with ship motions in waves.)
Newman, J. N. 1977. *Marine Hydrodynamics.* Cambridge, Mass.: Massachusetts Institute of Technology Press.
Price, W. G., and Bishop, R. E. D. 1974. *Probabilistic Theory of Ship Dynamics.* London: Chapman and Hall.
St. Denis, M., and Pierson, W. J. 1953. "On the Motions of Ships in Confused Seas." Volume 61 of SNAME *Transactions.* New York: Society of Naval Architects and Marine Engineers.

APPENDIX A

EXCERPTS FROM TRIM AND STABILITY BOOKLET FOR SINGLE SCREW CARGO VESSEL, MARINER CLASS C4–S–1a

Table of principal characteristics .... *below*
Hydrostatic properties *331*
Loading table *332*
Table for free surface correction and tank capacities *333*
Example voyage summary sheet *334*

TABLE OF PRINCIPAL CHARACTERISTICS

Length, overall	563'-7¾"	Lightship VCG	31.5'
Length, B.P.	528'-0"	Lightship LCG aft F.P.	276.5'
Length, 20 stations	520'-0"	Passengers	12
Beam, molded	76'-0"	Crew	58
Depth to main deck, mld. at side	44'-6"	Grain cubic	837,305 cu. ft.
		Bale cubic	736,723 cu. ft.
Depth to 2nd deck, mld. at side	35'-6"	Reefer cubic	30,254 cu. ft.
Bulkhead deck	2nd deck	Fuel oil (D.B.'s + settlers)	2,652 tons
Machinery	turbine	Fuel oil (deep tanks)	1,156 tons
Designed sea speed	20 knots	Fuel oil, total	3,808 tons
Shaft horsepower, normal	17,500	Fresh water	257 tons
Shaft horsepower, max.	19,250	No. of holds	7
Full load draft, mld.	29'-9"	Gross tonnage	9,215
Full load displacement	21,093 tons	Net tonnage	5,367
Lightship	7,675 tons		

HYDROSTATIC PROPERTIES

MEAN DRAFT BOTTOM OF KEEL	TOTAL DISP. S.W. TONS	TRANSVERSE KM-MLD. FEET	TONS PER INCH IMMERSION	MOMENT TO TRIM 1" FT. TONS	L.C.B. AFT F.P. FEET	L.C.F. AFT F.P. FEET	MEAN DRAFT BOTTOM OF KEEL
30	21000	31.4	70	1950	269	282	30
29	20000	31.3		1900		281	29
28		31.2	69	1850		280	28
27	19000	31.1	68	1800	268	279	27
				1750		278	
26	18000		67	1700		277	26
25	17000	31.05		1650	267	276	25
24	16000	31.1	66	1600		275	24
23		31.2	65		266	274	23
22	15000	31.3		1550		273	22
		31.4					
21	14000	31.5 / 31.6	64	1500	265	272	21
20	13000	31.8 / 32.0	63			271	20
19		32.5		1450		270	19
18	12000	33.0	62		264	269	18
17	11000	33.5		1400		268	17
16	10000	34.0 / 34.5	61				16
15		35.0 / 35.5		1350		267	15
14	9000	36.0	60		263	266	14
13	8000	37.0 / 38.0	59	1300		265	13
12							12

VOYAGE NO. _EXAMPLE_ LOADING TABLE

DRY CARGO

HOLD	BALE CUBIC	TONS	KG	MOMENT	LCG F.P.	MOMENT
NO.1 MAIN DK.	16085	160	55.6	8896	59.2	9472
" 2ND "	18140	180	45.2	8136	54.8	9864
" 3RD "	12210	130	31.9	4147	56.6	7358
NO.2-2ND DK.	29255	291	43.0	12513	104.4	30380
" 3RD "	34592	369	29.1	10738	105.3	38853
" TANKTOP	25476	271	13.1	3552	106.2	28780
NO.3-2ND. DK.	42000	418	41.3	17263	161.3	67423
" 3RD. "	58150	621	28.3	17574	161.6	100354
" TANKTOP	51375	546	12.7	6934	162.7	88834
NO.4-2ND. DK.	40255	401	40.3	16160	221.5	88822
" 3RD "	60020	641	27.7	17756	221.9	142238
" TANKTOP	61140	650	12.5	8125	223.1	145015
NO.5-2ND. DK.	41775	416	40.5	16848	356.5	148304
" 26'-6" FLAT ℄	16388	175	30.8	5390	350.2	61285
" 3RD. DK. ℄	16022	171	21.4	3659	351.0	60021
" TANKTOP	38135	406	10.9	4425	353.6	143562
NO.6-2ND DK	38610	384	41.0	15744	416.5	159936
" 3RD. "	65850	703	26.9	18911	415.5	292097
" DEEP TANK P/S	11930	127	11.2	1412	402.6	51130
NO.7-2ND. DK.	25095	250	41.8	10450	469.6	117400
" 3RD. "	34220	360	28.4	10394	469.4	17800
TOTAL	736723	7676	28.5	219035	265.7	1962931

REEFER CARGO

HOLD	REEFER CUBIC	TONS	KG	MOMENT	LCG F.P.	MOMENT
NO.5-26'-6" FLAT P/S	16256	174	30.7	5342	354.4	61666
" 3RD. DK. P/S	13998	150	21.8	3270	353.4	53010
TOTAL	30254	324	26.6	8612	353.9	114676

FUEL OIL OR BALLAST

TANK	F.S.	TONS F.O.-S.W.	KG	MOMENT	LCG F.P.	MOMENT
NO.1 D.B. ℄			4.5		39.9	
NO.1A "	204	82	4.8	394	64.9	5322
" 2 " P/S			2.7		106.6	
NO.3 " "			2.5		161.6	
" " P/S			3.0		169.2	
NO.4 " ℄	943	224	2.5	560	222.0	49728
" " P/S			2.6		223.8	
NO.5 " ℄	825	196	2.5	490	278.3	54547
" " P/S ½	4096	179	2.6	465	288.3	51606
NO.6 " ℄	1021	242	2.5	605	354.4	85765
" " P/S	442	174	2.8	487	348.2	60587
NO.7 " P/S	638	189	2.7	510	412.4	77944
NO.1 D.T. ℄			16.5		40.3	
" 1A " ℄			16.8		65.1	
" 2 " P/S ⅓	40	134	19.1	2558	260.8	34947
" 3 " P/S ⅓	34	117	19.1	2235	277.0	32409
" 6 " P/S			11.4		401.2	
" 7 " P/S			11.7		430.7	
" 8 " P/S			9.6		454.0	
FORE PEAK			11.7		17.1	
AFTER PEAK			24.9		506.8	
TOTAL	8143	1537	5.4	8305	294.6	452835

(1286 TONS F.O. IN DOUBLE BOTTOM)

FRESH WATER

TANK	F.S.	TONS F.W.	KG	MOMENT	LCG F.P.	MOMENT
NO.4 D.T. P/S	5575	124	2.13	264	296.0	36704
" 5 " "	4789	108	2.09	225	312.0	33696
DIST. WATER	59	25	39.5	988	255.8	6395
TOTAL	10423	257	22.9	5886	298.8	76795

TABLE FOR FREE SURFACE CORRECTION AND TANK CAPACITIES

TANK		FRAMES	97% F.O. TONS	100% S.W. TONS	COL A i SLACK	COL B i 97%	V.C.G.	L.C.G. F.P
D.B.1	₡	14-24	48.2	52.8	106	67	4.5	39.9
D.B.1A	₡	24-36	81.9	89.8	464	204	4.8	64.9
D.B.2	P	36-57	71.2	78.1	428	158	2.7	106.6
	S	"	71.2	78.1	428	158	2.7	106.6
	₡	57-82	227.6	249.5	3777	944	2.5	161.6
D.B.3	P	"	55.6	61.0	300	120	3.0	169.2
	S	"	55.6	61.0	300	120	3.0	169.2
D.B.4	₡	82-106	224.1	245.7	3626	943	2.5	222.0
	P	"	128.1	140.5	1138	364	2.6	223.8
	S	"	128.1	140.5	1138	364	2.6	223.8
D.B.5	₡	106-127	196.2	215.1	3173	825	2.5	278.3
	P	106-134	178.0	195.2	2048	676	2.6	288.3
	S	"	180.0	197.4	2048	676	2.6	288.3
D.B.6	₡	134-160	242.3	265.7	3928	1021	2.5	354.4
	P	"	87.0	95.4	615	221	2.8	348.2
	S	"	87.0	95.4	615	221	2.8	348.2
D.B.7	P	160-184	94.6	103.7	768	269	2.7	412.4
	S	"	94.6	103.7	768	269	2.7	412.4
D.T.1	₡	14-24	125.3	137.4	134	130	16.5	40.3
D.T.1A	₡	24-36	257.6	282.5	945	680	16.8	65.1
D.T.2	P	106-113	100.7	—	20	20	19.1	260.8
	S	"	100.7	—	20	20	19.1	260.8
D.T.3	P	113-119	86.1	—	17	17	19.1	277.0
	S	"	86.1	—	17	17	19.1	277.0
D.T.6	P	160-172	201.2	220.7	1242	634	11.4	401.2
	S	"	201.2	220.7	1242	634	11.4	401.2
D.T.7	P	172-184	128.8	141.2	618	358	11.7	430.7
	S	"	128.8	141.2	618	358	11.7	430.7
D.T.8	P	184-190	50.5	55.4	68	58	9.6	454.0
	S	"	50.5	55.4	68	58	9.6	454.0

TANK	FRAMES		100% F.W. TONS	100% S.W. TONS	COL.C i SLACK	V.C.G.	L.C.G. F.P.
FORE PEAK	₡	STEM-14		110.8		11.7	17.1
AFT PEAK	₡	204-218		93.0		24.9	506.8
D.T.4	P/S	120-127	123.7	—	5575	21.3	296.0
D.T.5	P/S	127-133	108.4	—	4789	20.9	312.0
DIST. WATER	₡	106-109	24.9		59	39.5	255.8

NOTES:
FUEL OIL AT 37.23 CU. FT./TON - **97%** FULL
FRESH WATER AT 36.0 CU. FT./TON - **100%** FULL
SALT WATER AT 35.0 CU. FT./TON - **100%** FULL

FREE SURFACE CORRECTION PROCEDURE
ADD QUANTITY IN COLUMN A FOR TANKS SLACK
ADD QUANTITY IN COLUMN B FOR TANKS 97% FULL
ADD QUANTITY IN COLUMN C FOR F.W. TANKS
IF ANY TANK IS EMPTY, OR PRESSED UP WITH WATER, USE ZERO FOR THAT TANK.
DIVIDE SUM TOTAL BY THE SHIP DISPLACEMENT IN TONS TO OBTAIN FREE SURFACE CORRECTION IN FEET.

EXAMPLE VOYAGE SUMMARY SHEET

VOYAGE NO.: EXAMPLE

ITEM	TONS	KG	MOMENT	L.C.G. F.P.	MOMENT	F.S.
LIGHTSHIP	7675	31.5	241763	276.5	2122138	
CREW & STORES	50	40.7	2085	270.5	13825	
LUBE OIL	13	25.8	335	317.5	4128	
FUEL OIL & SALT WATER	1537	5.4	8305	294.6	452851	1143
FRESH WATER	257	22.9	5886	298.8	76785	1043
DRY CARGO	7676	28.5	219035	255.7	1961931	
REEFER CARGO	384	26.6	8612	369.9	114676	
DECK CARGO	200	48.4	9680	260.0	52000	
TOTAL	17792	28.0	495801	270.6	4799348	18566

MEAN S.W. DRAFT (SEE SHEET 3) 25-7' = 25'-8½"   LCG-F.P.   270.6   LCF-F.P. (SHEET 3)   277.0
KM (SEE SHEET 3)   31.05   LCB (SEE SHEET 3)   267.2   DRAFT FWD.   24'-1½"
KG   28.0   TRIM LEVER FWD, AFT   3.4   DRAFT AFT   27'-1½"
GM   3.05   MOMENT TO TRIM 1"   1700
CORR FOR F.S.   1.05   TRIM IN INCHES FWD, AFT 36
GM AVAILABLE   2.00
GM REQUIRED (SEE SHEET 6)   1.76

Legend:
- ▨ DRY OR REEFER CARGO
- ▦ FRESH WATER
- ▩ FUEL OIL
- ■ SALT WATER

APPENDIX B

PROPERTIES OF SALT WATER AND FRESH WATER, AND FRICTION FORMULATIONS

Values of mass density ρ for fresh and salt water 336
Values of kinematic viscosity ν for fresh and salt water 337
Values of C_F according to the ITTC-1957 and ATTC friction lines 339

VALUES OF MASS DENSITY ρ FOR FRESH AND SALT WATER

Values as adopted by the ITTC meeting in London in 1963.
Salinity of salt water is 3.5 percent.

U.S. Units

Density of fresh water, ρ, lb-sec^2/ft^4	Temp, deg F	Density of salt water, ρ_s, lb-sec^2/ft^4	Density of fresh water, ρ, lb-sec^2/ft^4	Temp, deg F	Density of salt water, ρ_s, lb-sec^2/ft^4
1.9399	32	1.9947	1.9384	59	1.9905
1.9399	33	1.9946	1.9383	60	1.9903
1.9400	34	1.9946	1.9381	61	1.9901
1.9400	35	1.9945	1.9379	62	1.9898
1.9401	36	1.9944	1.9377	63	1.9895
1.9401	37	1.9943	1.9375	64	1.9893
1.9401	38	1.9942	1.9373	65	1.9890
1.9401	39	1.9941	1.9371	66	1.9888
1.9401	40	1.9940	1.9369	67	1.9885
1.9401	41	1.9939	1.9367	68	1.9882
1.9401	42	1.9937	1.9365	69	1.9879
1.9401	43	1.9936	1.9362	70	1.9876
1.9400	44	1.9934	1.9360	71	1.9873
1.9400	45	1.9933	1.9358	72	1.9870
1.9399	46	1.9931	1.9355	73	1.9867
1.9398	47	1.9930	1.9352	74	1.9864
1.9398	48	1.9928	1.9350	75	1.9861
1.9397	49	1.9926	1.9347	76	1.9858
1.9396	50	1.9924	1.9344	77	1.9854
1.9395	51	1.9923	1.9342	78	1.9851
1.9394	52	1.9921	1.9339	79	1.9848
1.9393	53	1.9919	1.9336	80	1.9844
1.9392	54	1.9917	1.9333	81	1.9841
1.9390	55	1.9914	1.9330	82	1.9837
1.9389	56	1.9912	1.9327	83	1.9834
1.9387	57	1.9910	1.9324	84	1.9830
1.9386	58	1.9908	1.9321	85	1.9827
			1.9317	86	1.9823

SI Units

Temp deg C*	Density of fresh water, ρ, kg/km^3	Density of salt water, ρ_s, kg/km^3
0	999.8	1028.0
1	999.8	1027.9
2	999.9	1027.8
3	999.9	1027.8
4	999.9	1027.7
5	999.9	1027.6
6	999.9	1027.4
7	999.8	1027.3
8	999.8	1027.1
9	999.7	1027.0
10	999.6	1026.9
11	999.5	1026.7
12	999.4	1026.6
13	999.3	1026.3
14	999.1	1026.1
15	999.0	1025.9
16	998.9	1025.7
17	998.7	1025.4
18	998.5	1025.2
19	998.3	1025.0
20	998.1	1024.7
21	997.9	1024.4
22	997.7	1024.1
23	997.4	1023.8
24	997.2	1023.5
25	996.9	1023.2
26	996.7	1022.9
27	996.4	1022.6
28	996.2	1022.3
29	995.9	1022.0
30	995.6	1021.7

NOTE: For other salinities, interpolate linearly.

* deg F = 9/5 C + 32

VALUES OF KINEMATIC VISCOSITY ν FOR FRESH AND SALT WATER

Values adopted by the ITTC meeting in London in 1963.
Salinity of salt water is 3.5 percent.

U.S. Units

Kinematic viscosity of fresh water, ν, ft^2/sec × 10^5	Temp, deg F	Kinematic viscosity of salt water, ν_s, ft^2/sec × 10^5	Kinematic viscosity of fresh water, ν, ft^2/sec × 10^5	Temp, deg F	Kinematic viscosity of salt water, ν_s, ft^2/sec × 10^5
1.9231	32	1.9681	1.2260	59	1.2791
1.8871	33	1.9323	1.2083	60	1.2615
1.8520	34	1.8974	1.1910	61	1.2443
1.8180	35	1.8637	1.1741	62	1.2275
1.7849	36	1.8309	1.1576	63	1.2111
1.7527	37	1.7991	1.1415	64	1.1951
1.7215	38	1.7682	1.1257	65	1.1794
1.6911	39	1.7382	1.1103	66	1.1640
1.6616	40	1.7091	1.0952	67	1.1489
1.6329	41	1.6807	1.0804	68	1.1342
1.6049	42	1.6532	1.0660	69	1.1198
1.5777	43	1.6263	1.0519	70	1.1057
1.5512	44	1.6002	1.0381	71	1.0918
1.5254	45	1.5748	1.0245	72	1.0783
1.5003	46	1.5501	1.0113	73	1.0650
1.4759	47	1.5259	0.9984	74	1.0520
1.4520	48	1.5024	0.9857	75	1.0392
1.4288	49	1.4796	0.9733	76	1.0267
1.4062	50	1.4572	0.9611	77	1.0145
1.3841	51	1.4354	0.9492	78	1.0025
1.3626	52	1.4142	0.9375	79	0.9907
1.3416	53	1.3935	0.9261	80	0.9791
1.3212	54	1.3732	0.9149	81	0.9678
1.3012	55	1.3535	0.9039	82	0.9567
1.2817	56	1.3343	0.8931	83	0.9457
1.2627	57	1.3154	0.8826	84	0.9350
1.2441	58	1.2970	0.8722	85	0.9245
			0.8621	86	0.9142

NOTE: For other salinities, interpolate linearly.

VALUES OF KINEMATIC VISCOSITY ν FOR FRESH AND SALT WATER

Adopted by the ITTC in 1963;
Salinity of salt water is 3.5 percent.

SI Units

Temp deg C*	Kinematic viscosity of fresh water, ν, $m^2/sec \times 10^6$	Kinematic viscosity of salt water, ν_s, $m^2/sec \times 10^6$
0	1.78667	1.82844
1	1.72701	1.76915
2	1.67040	1.71306
3	1.61655	1.65988
4	1.56557	1.60940
5	1.51698	1.56142
6	1.47070	1.51584
7	1.42667	1.47242
8	1.38471	1.43102
9	1.34463	1.39152
10	1.30641	1.35383
11	1.26988	1.31773
12	1.23495	1.28324
13	1.20159	1.25028
14	1.16964	1.21862
15	1.13902	1.18831
16	1.10966	1.15916
17	1.08155	1.13125
18	1.05456	1.10438
19	1.02865	1.07854
20	1.00374	1.05372
21	0.97984	1.02981
22	0.95682	1.00678
23	0.93471	0.98457
24	0.91340	0.96315
25	0.89292	0.94252
26	0.87313	0.92255
27	0.85409	0.90331
28	0.83572	0.88470
29	0.81798	0.86671
30	0.80091	0.84931

* deg F = 9/5 C + 32

NOTE: For other salinities, interpolate linearly.

VALUES OF C_F ACCORDING TO THE ITTC-1957 AND ATTC FRICTION LINES

Rn	$C_F \times 10^3$ (ITTC)	$C_F \times 10^3$ (ATTC)	Rn	$C_F \times 10^3$ (ITTC)	$C_F \times 10^3$ (ATTC)
1×10^5	8.333	7.179	6×10^7	2.246	2.229
2	6.882	6.137	7	2.195	2.180
3	6.203	5.623	8	2.162	2.138
4	5.780	5.294	9	2.115	2.103
5	5.482	5.057	1×10^8	2.083	2.072
6	5.254	4.875	2	1.889	1.884
7	5.073	4.727	3	1.788	1.784
8	4.923	4.605	4	1.721	1.719
9	4.797	4.500	5	1.671	1.670
1×10^6	4.688	4.409	6	1.632	1.632
2	4.054	3.872	7	1.601	1.600
3	3.741	3.600	8	1.574	1.574
4	3.541	3.423	9	1.551	1.551
5	3.397	3.294	1×10^9	1.531	1.531
6	3.285	3.193	2	1.407	1.408
7	3.195	3.112	3	1.342	1.342
8	3.120	3.044	4	1.298	1.299
9	3.056	2.985	5	1.265	1.266
1×10^7	3.000	2.934	6	1.240	1.240
2	2.669	2.628	7	1.219	1.219
3	2.500	2.470	8	1.201	1.201
4	2.390	2.365	9	1.185	1.186
5	2.309	2.289	1×10^{10}	1.172	1.172

$$C_F = \frac{0.075}{(\log_{10} Rn - 2)^2} \quad \text{(ITTC-1957)}$$

$$\frac{0.242}{\sqrt{C_F}} = \log_{10}(Rn \times C_F) \quad \text{(ATTC)}$$

$$Rn = \frac{VL}{\nu}$$

ANSWERS TO SELECTED PROBLEMS

CHAPTER 3
1. 0.688
   0.749
   1,897,000 ft^3
   3.718 × 10^7 ft^4
3. 29.67 ft
   58,656 ft^2
   3,039 ft^2
   0.712
   6.285 × 10^7 ft^4
   139.7
5. 9.35 ft
7. 5.66 ft
   59 tons
11. 22'2"
   26.98 ft
   4.38 ft
13. 22.19 ft
15. 0.79 ft
   4.13 ft
17. 9.75 ft
   9.46 ft
19. 3.6 degrees
21. 3.0 degrees
23. 3.53 m
25. (a) 2.37 ft
   3.15°
   (b) 2.37 ft
   0°
   (c) 2.50 ft
   0°
27. 25'10½"
   1.82 ft
   6.0°
29. 2.94 m

1,813 MT
1.71 m
31. 4.42 ft
33. 5.45 ft
35. 16.04 ft
   6.5°

CHAPTER 4
1. (a) 0 to 76°
   (b) 45°
   (c) 3.9 ft
3. curve, 14.5°
   formula, 17°
5. curve, 9°
   formula, 12°
7. curve, 5°
   formula, 5.6°
9. (a) 0 to 64°
   (b) 11 to 65°
   11° to port

CHAPTER 5
1. 27'3.9" F
   28'8.6" A
3. (a) 4.39 ft
   160.8 ft
   (b) 11.88 ft F
   10.78 ft A
   7.6°
5. 0.256 m
7. 13'2.0" F
   12'6.5"A
9. 152.5 tons
   Sufficient
   19'9.1"

11. 613.6 tons
13. 170.2m
   8.93m F
   9.34m A
15. 15,200 tons
   28.10 ft
   272.06 ft
19. 3.01 ft
   24'1.6" F
   25'3.7"A

CHAPTER 6
1. (a) 12.50 ft
   2.42 ft
   (b) 11.76 ft
   2.31 ft
   (c) 11.11 ft
   2.25 ft
3. 13.06 ft F
   10.57 ft A
   2.12 ft
5. 21.55 ft F
   11.96 ft A

CHAPTER 7
1. 8,120 psi Tens. Dk.
   8,667 psi Comp. Bot.
3. $p_{max} = 2\Delta/L$
   $V_{max} = \Delta/4$
   $M_{max} = \Delta L/12$
5. $V_{max} = -40$ tons
   $M_{max} = 1,097$ ft-tons
7. $V_{max} = 410$ tons

$M_{max} = 25,400$ ft-tons
9. $M_{max} = 36,000$ ft-tons
   $\sigma_{BOT} = 8,359$ psi
   $\sigma_{DECK} = 10,080$ psi
11. 1,903 in^2 ft
   2,129 in^2 ft
   2.41

CHAPTER 8
1. (a) 1,362 lb
   (b) 6.68 ft/sec
   (c) 0.263
   (d) 1.2366 × 10^7 M
   2.6409 × 10^9 S
   (e) 1,427 ft/sec
3. (a) 6.41 ft/sec
   (b) $R_F = 7.50$ lb
   $R_R = 6.00$ lb
5. 27,205 hp

CHAPTER 9
1. (a) 0.8522
   (b) 0.1478
   (c) 0.1562
   (d) 392 ft-lb/sec
   (e) 0.02889
   (f) 534.7 ft-lb/sec
   (g) 73.3%
3. 0.8466
   15,190 hp
   18,228 hp

BIBLIOGRAPHY

For those who want to learn more about some of the topics introduced in this book, the publications listed below are suggested. All are readily available, with the exception of *The Speed and Power of Ships*, which is included here because of its historical significance. An additional bibliography of ship dynamics references is provided at the end of chapter 10. Up-to-date information and the results of recent research can be found in *The Society of Naval Architects and Marine Engineers Transactions* and other material published by SNAME, 601 Pavonia Avenue, Jersey City, N.J. 07306. The Society publishes a comprehensive index by subject and author of all of their publications, including *Transactions* (annual), *Marine Technology* (bimonthly), *Journal of Ship Research* (quarterly), and papers presented at special meetings and symposia.

Abbott, I.H., and A.E. von Doenhoff. *Theory of Wing Sections*. New York: Dover, 1959.
American Bureau of Shipping. *Rules for Building and Classing Steel Vessels*. Paramus, N.J.: American Bureau of Shipping, annual.
Bhattacharyya, R. *Dynamics of Marine Vehicles*. New York: Wiley, 1978.
D'Arcangelo, A.M. *A Guide to Sound Ship Structures*. Centreville, Md.: Cornell Maritime Press, 1969.
George, William E., ed. *Stability and Trim for the Ship's Officer*. 3d ed. Centreville, Md.: Cornell Maritime Press, 1983.
Gertler, M. *David Taylor Model Basin Report 806*. A reanalysis of the original test data for the Taylor Standard Series. Washington, D.C.: Government Printing Office, 1954. (David Taylor Model Basin is now called the David Taylor Research Center.)
Gillmer, T.C. and B. Johnson. *Introduction to Naval Architecture*. Annapolis, Md.: Naval Institute Press, 1982.

Harvald, Sv. Aa. *Resistance and Propulsion of Ships.* New York: Wiley, 1983.

Hughes, O.F. *Ship Structural Design.* Jersey City, N.J.: The Society of Naval Architects and Marine Engineers, 1988.

Institution of Naval Architects. *The Papers of William Froude, 1810-1879.* London: The Institution of Naval Architects (now the Royal Institution of Naval Architects), 1955.

Lewis, E.V., ed. *Principles of Naval Architecture.* Jersey City, N.J.: Society of Naval Architects and Marine Engineers, 1989.

Muckle, W. *Naval Architecture for Marine Engineers.* London: Butterworths, 1975.

Newman, J.N. *Marine Hydrodynamics.* Cambridge, Mass.: The MIT Press, 1977.

Rawson, J., and E.C. Tupper. *Basic Ship Theory.* 4th ed. London: Longman, 1993.

Saunders, H.E. *Hydrodynamics in Ship Design.* Jersey City, N.J.: The Society of Naval Architects and Marine Engineers, 1957.

Taggart, R., ed. *Ship Design and Construction.* Jersey City, N.J.: The Society of Naval Architects and Marine Engineers, 1980.

Taylor, D.W. *The Speed and Power of Ships.* 2nd rev. Washington, D.C.: Government Printing Office, 1943.

Todd, F.H. "Series 60 - Methodical experiments with models of single-screw merchant ships." *David Taylor Model Basin Report 1712.* Washington, D.C.: Government Printing Office, 1963. (David Taylor Model Basin is now called the David Taylor Research Center.)

INDEX

A

aluminum in ship structures, 211
amidships
    definition of, 23
angle of vanishing stability, 115
Archimedes' principle, 48, 49. *See also* flotation, law of
area curve, sectional, 33
axis, neutral, 224

B

barge(s), cargo, 21
barge carriers, 9-10
    advantages of, 10
    LASH, 9
    loading, 9–10
    Seabee, 10
    sizes of barges on, 10
beam (of a ship), 28
    maximum, 28
beam theory, 212–19
bending moment, 212–14, 217–18
    curve, 222–23
    diagrams, 214–16
    relationship to load and shear, 216–18
    still-water, 231
    total maximum, 231
    wave-induced, 231
bending stress(es), longitudinal
    causes, 224–26
    nominal, 232–33
    permissible, 230–33
bilge
    radius, 29
    turn of the, 29
block coefficient, *34*–35
body plan, *24*, 25
breadth. *See* beam (of a ship)
bulk carrier(s), dry, 17–20
    cargo-handling gear on, 19
    characteristics of, 17–18
    combination, 20
    ore, 18–20: configuration of, 19; Great Lakes, 19–20; size of, 20
bulkhead(s)
    deck, 170: definition, 188
    subdivision, 170, 175–76
    transverse, 4
buoyancy, 48–49
    active, 170
    lost, 170, *171*
    regained, 170–71
    reserve, 170, *171*
    transference of, *65*–66
buoyancy, center of, 33, 58, 64
    location of, 43–44, 64
    longitudinal, 33, 43, 45
    vertical, 33, 44, 47
buoyant force, 50, 52, 58, 59
buttock(s), 5, 26. *See also* plane, buttock
    height offsets, 30

C

camber, 28
capacity plan, 72
capsize, capsizing, 115–16, 119, 123
cargo(es), liquid, 5
    loading and unloading, 13

cargo ship, 5
　history of, 3–4
　size of, 4
cavitation, 286, 290
　number, 290
centerline, 63
center of ——. *See* individual properties, such as buoyancy, gravity
collision, 169
comparison, laws of, between model and ship, 241–42
compartment, one- or two-, ship, 190
container(s)
　on deck, 8
　handling, 8
　size of, 7–8
　stowage of, 8–9
containership(s), 6–7
　advantages of, 6–7
　history of, 7
cornering. *See* free surface, effect of cornering on
correlation allowance, 263–64
couple. *See* moment
crane(s), 5
cross curves of stability, 108–09

D

deadrise, 29
deadweight, total, 72–73
density
　salt water and fresh water, table, 51, 336
　water, effect of, on draft, 54–57
depth, molded or at side, 28
derrick(s), cargo, 5
dimension(s), of a ship, 26, *27*, 28
displacement, 45, 48–49, 50–51
　calculation of: change, when trimmed, 143–46; change, when loading and discharging small weights, 149–58; from drafts, 160–62
　change in per inch of trim. *See* trim by stern
　curves of. *See* hydrostatic curves
　total, 33. *See also* volume of displacement

double bottom(s), 4, 13, 18
double side skin(s), 13
downflooding, 116, 169–70. *See also* flooding
draft(s), 29, 49. *See also* displacement
　changes in, when trimming, 140–42
　difference in, 134
　design, 29
　determining, 55–57: final, from lost buoyancy method, 183–86; final, from added weight method, 186–87; from displacement and longitudinal center of gravity, 163–65
　effect of water density on, 54–57

E

English system of units. *See* U.S. system
equilibrium
　disruptions to, 74–76
　floating, 48–51, 57–60
　neutral, *62*
　static, 50, 57–59: conditions of, 60
　unstable, 60, 117

F

fiberglass. *See* plastics, glass-reinforced
flare, 29
flexure formula, 224–25
floodable length
　calculating, 187–90
　curves, 189–90
　definition, 188
flooded spaces, flooding
　calculations, computer method, 187
　causes of, 169–70
　effect(s) of, 170–74: on draft and freeboard, 172; on heel, 173–74; on stability, 172–73; on trim, 172–73
　limiting by subdivision, 174–76
　permeability of, 177–78
flotation, center of, 32, 39, 40
　longitudinal, 32, 46
flotation, law of or principle of, 48–49, 55
flow
　laminar, 251–52
　Reynolds number as transition, 251
　turbulent, 251–52

INDEX 345

foil(s), effectiveness of as a lifting device, 280–81
forklift truck(s), 4–5
form. *See also* hull form
   coefficients of, 34–37, 47
   curves of. *See* hydrostatic curves
   molded, 26–27
framing systems, 199–204
   combination, 203–4
   comparison of, 201–3
   longitudinal, 201, *202*
   transverse, 199–*200*
freeboard, 29
   statutory, 29
free surface
   correction to GM, 84–91
   effect of cornering on, 88–*89*
   liquids with, 84–91
   moment of, 84–85
friction coefficient. *See also* resistance, frictional 250–51
Froude number, 243–44, 246, 247–48, 257, 285–86
   based on wave length, 255–56
Froude's hypothesis, 247

G

grain ships, 20
gravity, center of, 33, 58, 61, 119
   calculation of, 69–70, 71, 90
   height of, 79, 80–81
   longitudinal, determining, 162: from drafts, 160–62
   movement of, 76–77, 125
   rise in, caused by free liquids, 86–88
grounding, 169

H

half-breadth(s)
   definition of, 23
   plan, *24*, 25, 30
   of the waterplane, 37, 64
half-siding, 29
hatch(es), 4
heel, heeling, 64, *65*
   angle(s), 125, 130
   arm(s), 118–*19*, 124–30: wind, 125–28

   effect of on righting arms, 113-116
   moment(s), 61, 119–30
   during a turn, 129–30
height(s) above baseline, 23, 30
hogging, 219, 224
horsepower
   effective, 265
   service shaft, 296
hull efficiency, 272, 290–91. *See also* wake fraction; thrust deduction fraction
hull form
   characteristics, represented as integrals, 35–44
   coefficients, 34–35
hydrostatic curves, 45–47
hydrostatics, 30–47

I

icing, 125
immersion
   deck edge, 114–15
   tons per centimeter, 46, 52, 53
   tons per inch, 46, 52, 53
inclining experiment, 79–82
inertial forces, 196
intermodal transport, 6
International Convention on Load Lines, 55
International System of Units, 50–51

K

keel, draft(s), 29
kingpost(s), 5

L

LASH barge carrier, 9–10
length (of a ship), 28
   overall, 28
   between perpendiculars, 28
   on waterline, 28
light ship
   characteristics, 81–82
   conditions, 70–71
lines
   displacement, 26
   drawing, 23, 24

lines *(continued)*
    plan, 37
liquefied gas carriers, 15–17. *See also* LPG ships; LNG ships
list, angle of, 91, 121–22
load line, subdivision, definition, 188
load(s), loading, 125
    categories of, 72
    diagram, 216–17
    distributed, 215–16
    dynamic, 196–97
    effects of on ship structure, 195–97
    occasional and unusual, 197
    relationships to shear and bending moment, 216–18
    static, 196
    waterline. *See* waterline, design
loll, angle of, 118, *119*
longitudinal strength, 212–24
LNG ships, 16–17
LPG ships, 16

M

margin line, definition, 188
Maritime Research Institute of the Netherlands (MARIN), 289
mass, 50
    light ship, calculation of, 71–72
mast(s), 5
metacenter. *See also* metacentric height
    longitudinal, 136, 139: height of, above keel, 47
    transverse, 139: definition of, 63; height of, above keel, 46
metacentric height, 63–69, 113–14
    "as-inclined," 79, 80
    calculation of, 66–69, 73–74, 90–91
    negative, 117–119
metacentric radius, 64–66
midship section coefficient, 35, 36
model, ship('s)
    dynamical similitude, 245, 246
    -ship correlation: allowance, 247; lines, 248. *See also* resistance, correlation allowance
    scaling between, and ship, 241
    testing in bare hull condition, 259
modes of ship motion, 319–21

mold(s), molded form, 26–29
moment. *See* heeling moment, righting moment
moment to change trim, 146–48: one inch/centimeter, 46
    of free surface, 84–85
moment of inertia
    longitudinal, waterplane, 32, 40–41
    structural, 226
    tank surface, 86
    transverse, waterplane, 33, 41

O

OBO ships. *See* bulk carrier(s), dry
offset(s)
    table of, 29–30, *31*
ore/bulk/oil ships. *See* bulk carrier(s), dry

P

permeability. *See also* flooded spaces, flooding, permeability of
    surface, 178
    volume, 177
perpendicular(s)
    aft, after, 23, 28
    forward, 23, 28
    length between, 28
plane(s)
    baseline, 23
    buttock, 25, 26
    centerline, 23
    diagonal, 26
    midship section, 23, 24
    station, 25
plank tests, Froude's, 250–51
plastics, glass-reinforced, 211–12
plating, stiffened, as structure, 197–99
plunge, 173
pollution (anti) regulations, 13
power
    allowance, service, 296
    brake, 274
    delivered, 272
    effective, calculation of, 264–68, 272
    indicated, 274
    shaft, 273, 274–75: ratio to delivered power, 274
    thrust, 272

pressure distribution, as cause of waves, 254–55
    on propeller blade, 280–81
prismatic coefficient, 35, 36
profile plan. *See* sheer plan
propeller, 275–290. *See also* thrust; power; screw propeller blade
    behind efficiency of, 272
    dynamic pressure, 286
    open water efficiency of, 273, 286–87; test, 283–87; expansion of test, 287–89
    performance: characteristics of, 283–85; curves, *287*
    pressure ratio, 286–87
    relative rotative efficiency of, 273, 290–91, 295
    stagnation pressure. *See* dynamic pressure
    static pressure, 286
propulsion, ship
    coefficient or efficiency, 274: quasi-propulsive, 274

R

range of stability, 115–16
resistance, frictional, 240, 247–48, 249–51
    formulations for, 252–53: ATTC Schoenherr mean line, 253; ITTC 1957 mean line, 253
    Froude's formula for, 250–52: theoretical objections to, 251–52
resistance(s), ship
    air, 240, 259–62
    appendage, 258–59
    coefficient(s), 242–43, 245, 247–48
    components of, 239–41
    correlation allowance, 263–64; with ITTC line, table, 264
    definition, 239
    eddy-making, 240, 257–59, 260
    factors affecting, 241–42
    frictional. *See above*
    residuary, 247–48
    wave-making, 240, 253–57
    wind, 262
response amplitude operator, 322–24, 326–28
revolution, body of, stability of, 101–3

Reynolds number(s), 243, 245, 246, 247–48, 253, 286–87
righting arm, 63, 100–1, 118–*19*, 120, 121. *See also* moment, righting and statical stability
    determining, for ship forms, 103–7: methods of, 105–7
    maximum, 115
righting moment, 61, 63, 100–1. *See also* righting arm
    curves, 122–24
roll-on/roll-off ship(s), 10–11
    configuration of, 11–12
RO/RO. *See* roll-on/roll-off ship(s)
round of beam or round down. *See* camber

S

sagging, 219, 224
sail area, 125, *126*, 127, 128
samson post(s), 5
scantlings, 229
screw propeller blade. *See also* propeller
    angle of attack of, 280, *281*
    angle of incidence of. *See* angle of attack of
    cross-sectional shapes, 279
    drag force of, 280–82
    efficiency of, 278–79
    geometry of, 275–77
    hydrodynamics associated with, 282
    lift force of, 280–82
    nomenclature of parts, 275
    pressures and forces on, 279–82
    slip, 277–78
    thrust, 278–79
    velocity: of advance, 277–78; diagram, 277–78
Seabee barge carrier, 10
section modulus, 226: calculation of, 226–30; required, 233
shaft
    power: service, 296; trial, 296
    transmission efficiency, 274, 295
shear, 212–14
    diagrams, 214–16, 221–22
    force, 217
    relationship to bending moment and load, 216–18

sheer, 28
  plan, 26
SI system. *See* International System of Units
sinkage, parallel, 52, 149: determining after flooding, 179–80
skin friction formulations, 248
slack tank(s), 84
slip ratio
  apparent, 278
  real. *See* screw propeller blade, slip
spectrum, spectra
  broad, 311
  directional, 315–16
  encounter, 312–15: frequency, 312; period, 312; transformed from sea, 314
  narrow or narrow–band, 310–11
  one-dimensional. *See* spectrum, point
  point, 315, 318–19
  response, 325–26
  sea, 307–12: shape, significance of, 310–12; transforming into encounter, 314
  two-dimensional. *See* spectrum, directional
  two-peaked, 311–12
speed(s), corresponding, between ship and model, 244
speed-length ratio, 244
stability. *See also* statical stability
  change in, when taking on liquids, 89–90
  cross curves of, 108: based on assuming watertight integrity of weather deck, 116
  damaged, calculation of after flooding, 183–86
  equation of, 107
  after flooding, determining: by lost buoyancy method, 178–83; by added weight method, 186–87
  indifferent. *See* equilibrium, neutral
  initial transverse, 62–63
  longitudinal, 133, 136–42
  maximum, angle of, 115
  negative initial, 118
  large-angle, 114–15
  range of, 115–16
  test for, *61*
  vanishing, angle of, 115–16, 118
statical stability
  curve, 102, 117–22: constructing, 108–12; anatomy of, 112–16; two–sided, 118–*19*
  definition of, 60
station(s), 25
  areas, immersed, 33: integration, *41*–42,
steel
  mechanical properties of, 206–10
  ship, approved, 210–11
  shapes, standard, 228
stem, 28
structure, ship's. *See also* plating, stiffened; framing systems
  materials, 204–12: requirements for, 204–6;
subdivision
  load line, 188
  regulations: integer compartmentation, 191; factor of, 191–92; probability of survival, 192–93
  standards, 190–91
Suezmax tankers, 13–14
superposition, linear: calculations, 326–28; principle of, 321–22
swell, 312

T

tank(s), three possible conditions of, 88–89
tanker(s)
  chemical, 14–15
  configurations of, 13
  crude oil, size of, 12–13
  product carriers, 14
  ULCC and VLCC, 13
thrust, 271
  coefficient or thrust loading coefficient, 285
  deduction, 293: factor, 294; fraction, 294
tipping center. *See* trim, axis of
trim, 133. *See also* moment to change trim
  angle of, 135–36, 140
  axis of, 137–39
  calculations: small weights method, 149–58; large weights method, 158–65

trim *(continued)*
    change of, 136: determining after damage, 181–83
    definitions, 133–35
    diagram. *See* trim table
    and stability booklet, 72, 88
    by stern, 47
    geometry of, 133–35
    mechanics of, 136–42
    table, 157–58, *159*
'tween deck(s) space(s), 4

V

viscosity
    absolute or dynamic, 241
    kinematic, 241
volume
    center of. *See* gravity, center of
    of displacement, 33, 42–*43*, 44, 49,
    immersed, of the hull. *See* volume of displacement
    molded, 33
    moment of, longitudinal, 42–43: about keel, 44

W

wake, 291
    fraction, 291–93: and hull efficiency, 294
    frictional, 292
    potential, 292
    survey, 292–93
    wave, 292–93
wall-sided(ness) of a ship, 52–53
waterline(s), 25, 134. *See also* waterplane(s)
    design, 28
    determining after damage, 178–86
waterplane(s), 25, 134. *See also* waterline(s)
    area (of), 32, *37*, 38–*39*: centroid of. *See* flotation, center of; loss of due to flooding, 177–78
    coefficient, 35, 37
    half-breadths of, 37
waves, ocean, 299-300
    circular frequency, 302
    dispersive, 304
    energy in, 304–5;
    heights: apparent, *317*; significant, 318
    patterns of long and short, 311–12
    period, 302
    profile, surface, 301
    progressive, 300–1, 303
    properties, 305: apparent, 316–17
    records, irregular, statistics of, 316–18
    sinusoidal, 301
    system, irregular long-crested, 306–7
    velocity, 303–4: very shallow water, 304
waves, ship-generated, 255-56
    caused by pressure distribution, 254–55
    diverging, 255, *256*
    train(s), 255, *256*–*57*
    transverse, 255, *256*
wave(s). *See also* resistance, wave-making; spectrum
weather deck, assumption of perfect watertight integrity of, 116
weight(s). *See also* load(s)
    of a body. See displacement
    classification of, 70–73
    distribution, 223–24
    force, 58
    freely suspended, 82–*83*
    loading or discharging, 75–76, 158–65
    off-center, effect of, 76–78, 124–25
    shifting, 74–75, 85
wetted surface, 14
wind(s)
    beam, 125–28. *See also* heel, wind

Z

"zero–moment" condition, 40

ABOUT THE AUTHOR

Robert B. Zubaly, Professor of Engineering at the State University of New York Maritime College, received his Bachelor of Science in Naval Architecture and Marine Engineering from the Webb Institute of Naval Architecture. While pursuing advanced studies in mechanical engineering at Columbia University, he began a career devoted to teaching and research in naval architecture that has lasted for forty years.

He has lectured or conducted research at all of the maritime-related institutions of higher education in the New York metropolitan area, at several east coast shipyards, at the American Bureau of Shipping, and at the Seamen's Church Institute. He was awarded the SUNY Chancellor's Award for Excellence in Teaching in 1973. Many of his students have become prominent merchant ship officers, maritime business executives, flag-rank officers in the United States Navy and Coast Guard, and admiralty attorneys.

Professor Zubaly is a fellow of the Society of Naval Architects and Marine Engineers, which awarded him the Centennial Medallion in 1993 and the William H. Webb Medal for outstanding contributions to education in naval architecture in 2001.